Materials Forming, Machining and Tribology

Series Editor

J. Paulo Davim, Department of Mechanical Engineering, University of Aveiro, Aveiro, Portugal

This series fosters information exchange and discussion on all aspects of materials forming, machining and tribology. This series focuses on materials forming and machining processes, namely, metal casting, rolling, forging, extrusion, drawing, sheet metal forming, microforming, hydroforming, thermoforming, incremental forming, joining, powder metallurgy and ceramics processing, shaping processes for plastics/composites, traditional machining (turning, drilling, miling, broaching, etc.), non-traditional machining (EDM, ECM, USM, LAM, etc.), grinding and others abrasive processes, hard part machining, high speed machining, high efficiency machining, micro and nanomachining, among others. The formability and machinability of all materials will be considered, including metals, polymers, ceramics, composites, biomaterials, nanomaterials, special materials, etc. The series covers the full range of tribological aspects such as surface integrity, friction and wear, lubrication and multiscale tribology including biomedical systems and manufacturing processes. It also covers modelling and optimization techniques applied in materials forming, machining and tribology. Contributions to this book series are welcome on all subjects of "green" materials forming, machining and tribology. To submit a proposal or request further information, please contact Dr. Mayra Castro, Publishing Editor Applied Sciences, via mayra.castro@springer.com or Professor J. Paulo Davim, Book Series Editor, via pdavim@ua.pt.

More information about this series at http://www.springer.com/series/11181

Sunil Pathak · Gobinda C. Saha
Editors

Cold Spray in the Realm of Additive Manufacturing

Springer

Editors
Sunil Pathak
Faculty of Manufacturing
and Mechatronics Engineering Technology
Universiti Malaysia Pahang
Kuantan, Pahang, Malaysia

Gobinda C. Saha
Nanocomposites and Mechanics Laboratory,
Department of Mechanical Engineering
University of New Brunswick
Fredericton, NB, Canada

ISSN 2195-0911 ISSN 2195-092X (electronic)
Materials Forming, Machining and Tribology
ISBN 978-3-030-42755-9 ISBN 978-3-030-42756-6 (eBook)
https://doi.org/10.1007/978-3-030-42756-6

This Springer imprint is published by the registered company Springer Nature Switzerland AG
The registered company address is: Gewerbestrasse 11, 6330 Cham, Switzerland

Preface

Additive manufacturing of engineering parts is essential to fulfil the requirements of improved operating performance and longer service life. Manufacturing to the desired size, shape, surface quality and mechanical properties is possible with the help of precision additive manufacturing process such as "Cold Spray Additive Manufacturing" (CSAM). This book sheds light on the development of CSAM in the field of additive manufacturing and development of functional material properties. This book covers the bonding mechanism, powder–substrate interface and salient features of CSAM to manufacture near net-shaped parts. The latest research in this area and possible future research avenues are also highlighted to encourage further research and development in this field.

This book consists of eight chapters on Cold Spray Additive Manufacturing (CSAM). Chapter "Cold Spray: Its Prominence as an Additive Manufacturing Technology" describes the fundamentals of cold spray and displays its benefits over other competitive processes. Chapter "Commercial Cold Spray Equipment" sheds details on the development of commercial cold spray equipment. Chapter "The Influence of Feedstock Powder" provides a comprehensive review on the influence of feedstock powder in cold spray. Chapter "Cold Spray Deposition on Polymeric and Composite Substrates" sheds light on capabilities of cold spray in deposition on polymeric and composite substrates. Advanced modelling and simulation of the cold spray process to resolve the additive manufacturing issues are discussed in Chapter "Advanced Modeling and Simulation Tools to Address Build-Up Issues in Additive Manufacturing by Cold Spray". Chapter "Laser Assisted Cold Spray Deposition" is focused on laser-assisted cold spraying. Hybridization of cold spray with laser remelting for development of shape memory alloys is discussed in Chapter "Development of CuAlNi Shape Memory Alloy Structures Using Cold Spray Deposition Technique with Laser Remelting". Metrological aspects and dimensional aspects of cold spray additive manufacturing are presented in Chapter "Dimensional Analysis and Laser-Ultrasonic Inspection of Cold Spray Additive Manufacturing Components".

We sincerely acknowledge Springer for this opportunity and their professional support. Finally, we would like to thank all the chapter contributors for their availability and valuable contributions.

Kuantan, Malaysia
Fredericton, NB, Canada
December 2019

Sunil Pathak
Gobinda C. Saha

Contents

About the Editors

Dr. Sunil Pathak is specialized in gear engineering and advanced/hybrid manufacturing processes. He has beenworking in the field of gear engineering since last 9 years and advanced manufacturing for over 6 years. He has conducted extensive research on advanced finishing of gears. He possesses specialized skills in gear finishing, gear metrology (micro, and macro-geometry) and measurement of gear accuracy. Presently he has been working on machining of difficult-to-machine materials using advanced machining processes such as EDM and WEDM. He is also working in developing cold spray coatings as sustainable process for manufacturing of 3D additive manufacturing components and repair/remanufacturing engineering, where he has specially gained experience in materials and remanufacturing engineering. He has developed analytical models of manufacturing processes andoptimizing parameters to achieve overall sustainability in manufacturing.

Dr. Gobinda C. Saha is an Associate Professor and Director of Nanocomposites and Mechanics Laboratory at the University of New Brunswick, Canada. In recent past, Dr. Saha has led the R&D division at two successive nanocomposites manufacturing companies in conjunction with academic appointment at the University of Calgary, Canada. His focused research and innovation work in bio and nanostructured materials and additive manufacturing is dedicated to serve as bridge building between industry and academia, on the divulsion of theoretical knowledge into applied solutions. Pertaining work in nanostructured materials has contributed to design, synthesis, testing, and commercialization of two surface protective coating systems for oil and gas industry, developed on the high velocity oxy-fuel and cold spraying methodology. Simultaneously, Dr. Saha's research in design, fabrication, processing, and characterization/testing of smart optical sensor reinforced FRP composites and modelling has contributed to the dissemination of knowledge through peer-review publication in over 50 research papers, two patents, as well as a strong supervision of graduate thesis research track record. Dr. Saha has given numerous invited talks on the topic of 'Nanostructured Cermet Composites and Biocomposites'. He has served as the Chair of ASME Composites & Heterogeneous Materials Technical Committee, as well the founding member of the Joint Canada-USA Regulatory Cooperation Council on developing strategies for nanomaterials including risk assessment and management and uses of industrial nanomaterials.

Cold Spray: Its Prominence as an Additive Manufacturing Technology

Sunil Pathak and Gobinda C. Saha

Abstract Cold gas dynamic spray (or the 'cold spray') technology has brought the attention to effectively realize additive manufacturing concept by better aligning innovative material design with precision manufacturing. The evolutionary technology with the revolutionary concept is permitting the creation of substrate surface unprecedented to human history. The objective of this chapter is to highlight the state-of-the-art of cold spray additive manufacturing (CSAM), current research challenges in both feedstock and spray infrastructure, and more importantly, the potential. A comparative chart on contemporary process technologies involving CSAM and other nanoscale additive manufacturing technique-based material deposition processes is presented at the end.

Keywords Innovation · Nanostructured materials · Cold spray · Additive manufacturing · Applications

1 Introduction

1.1 Cold Spray Additive Manufacturing (CSAM)

In contrast to conventional manufacturing processes, i.e. material removal process to manufacture an engineering component, additive manufacturing (AM) depends on a novel material incremental manufacturing method. AM strategy has been widely investigated for quite a long time to create functionalized segments, to deliver complex structures or shapes that are in fact impractical using the conventional subtractive

---

S. Pathak
Faculty of Manufacturing and Mechatronics Engineering Technology, University Malaysia Pahang, 26300 Pahang, Malaysia
e-mail: sunilpathak87@gmail.com

G. C. Saha (✉)
Nanocomposites and Mechanics Laboratory, Department of Mechanical Engineering, University of New Brunswick, 15 Dineen Drive, Head Hall E-25, Fredericton, NB E3B 5A3, Canada
e-mail: gsaha@unb.ca

S. Pathak and G. C. Saha (eds.), *Cold Spray in the Realm of Additive Manufacturing*, Materials Forming, Machining and Tribology,
https://doi.org/10.1007/978-3-030-42756-6_1

machining processes. This inventive strategy of manufacturing brings few advantages, including decrease in both assembling cost and material utilization, liberty to use different sets of materials and their combinations. However, the intricate benefit of the AM method lies in its ability to create layer-by-layer deposits with excellent material and geometric controls of the underlying shapes.

There are now a handful of AM-based techniques available to support industrial scale manufacturing. They can be categorized along their operational task, type of materials handled, or complexity involved in the deposition process. Material plays an important role in deciding what AM technology to be utilized. For example, with regards to the AM for metals, there are primarily two methods: (1) non-powder-based; and (2) powder-based. In non-powder feed method, such as wire arc AM (WAAM) or laser melt deposition wire (LMD-w), a wire is fed through a nozzle that is melted by corresponding plasma arc or laser. The method incorporates inert gas shielding in either an open environment or in a sealed gas chamber. Compared with powder-based method, the system provides a higher deposition rate; however, lacks in control of variable parameters. All the process parameters lead to produce bead geometries, and it is the manipulation of beads that results in desired component shape. Unfortunately, and unlike other AM processes, bead geometry is affected by more than just the process parameters. The residual heat as the part is built results in a rapidly-varying thermal field that must be accounted for if a deposit layer is to be accurate and free from defect. Whereas, in the powder-based deposition method, the powder as feedstock is sprayed onto a substrate in a supersonic atmosphere to develop coatings and bulk parts through self-consolidation. The method is thus used as a direct AM process in high productivity requirement. It is estimated that thermal spraying of powder-based deposition method constitutes an annual growth of 10%, reaching US$28 billion by 2024. This demand is largely attributed to the industrial acceptance of thermally-sprayed coated parts required to protect from corrosion, erosion, and highly-elevated temperatures and pressures. The rise of acceptance of thermal spraying has given birth to the process technologies, namely, high velocity oxy-fuel (HVOF), flame spray, detonation spray, laser spray, high-pressure gas spray, and recently the cold spray. All the processes in thermal spray technology family, except the cold spray, uses a combination of thermal and kinetic energies for the formation of deposits. On the other hand, cold spray works solely on the particle kinetic energy making the deposit far-superior cleaner, as well as widening the material design composition stunningly innovative. In other words, this revolutionary distinction puts the cold spray technology being considered as a solid-state additive method. The conducive benefit of CSAM is that the feedstock powder retains its solid-state during deposition and self-consolidation.

Cold spray uses low deposition temperature against industry-standard thermal spraying techniques involving elevated temperature gas spray. This enables the powder to retain their physical characteristics during deposition. Powder granulometry's suitability is less restrictive compared to standard thermal spraying. The powder particles cover a wide range of size distribution, from micron to submicron. Further,

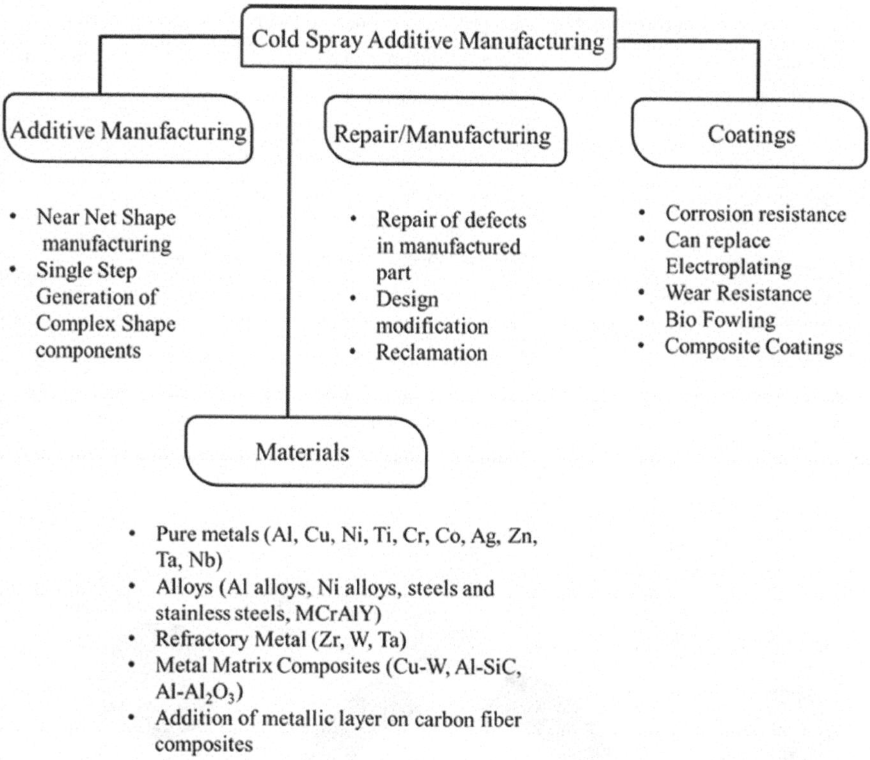

Fig. 1 Key features of cold spray [1]

the technology allows for thermally-sensitive materials as well as dissimilar materials (e.g. ceramic-metallic or cermets) to be manufactured. The key features and capabilities of cold stray is presented in Fig. 1.

As compared with technologically-advanced comparable AM technologies, cold spray neither includes high temperature (as in selective laser melting (SLM) and direct metal deposition (DMD)) nor does it engage into complex chemical processes (such as in electroplating). Hence, cold spray can be suggested as the best fit for making different geometries of complex shapes and simultaneously accomplish deposition without thickness limitation. Table 1 presents the details of AM processes as per ASTM F2793-12A standard which suggests cold spray as a promising AM technology.

Figure 2 shows the dissemination of procedures as indicated by their different added substance producing abilities; parts external to the parabola indicate AM in a wider sense, whereas parts within the parabola demonstrate AM in a narrow sense.

Table 1 Highlights of different AM based processes, as per ASTM F2793-12A standard [2]

AM based process	Process fundamental as per ASTM F2793-12A standard
Material extrusion	Material is delivered selectively through a nozzle or aperture
Material jetting	Building material droplets are deposited selectively
Binder jetting	To join powder materials, a liquid bonding agent is selectively placed
Sheet lamination	Material sheets are attached to create an item
Vat photopolymerization	Light-activated polymerization selectively cures liquid photopolymer in a vat
Powder bed fusion	Thermal energy fuses areas of a powder bed selectively
Directed energy deposition (DED)	When the material is deposited, focused thermal energy is used to fuse components
Cold spraying	Powdered material is driven at a substrate with sufficient speed to adhere and build up material

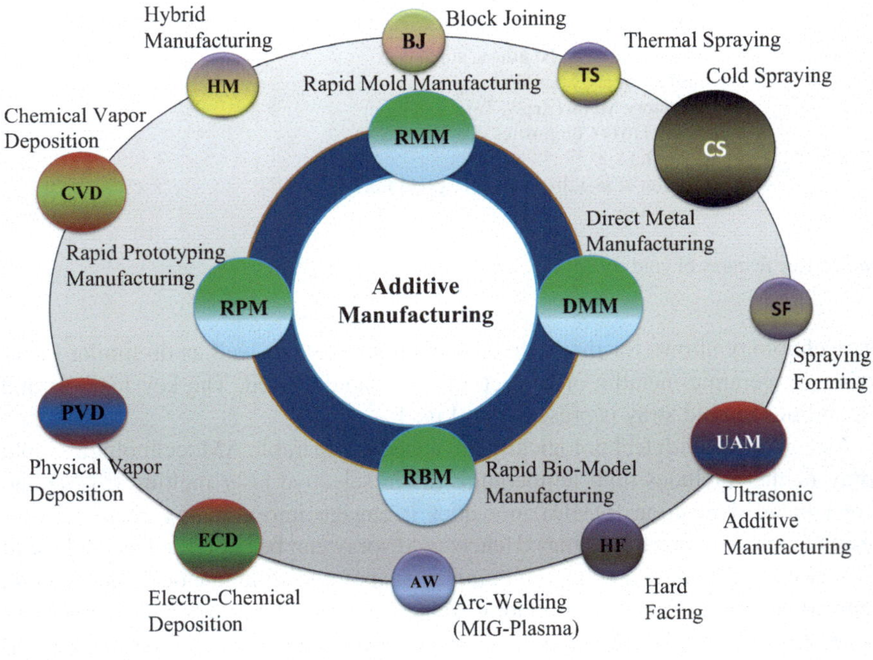

Fig. 2 Division of procedures as indicated by their different AM abilities

1.2 Overview of Cold Spray Process

In cold spray, solid-state particles are delivered to substrate via a carrier gas at a supersonic speed to build a layered deposit or a freeform object. Figure 3 portrays the schematic of a simple cold spray setup.

High velocity is essential for ideal deposition of particle and achieve coating density. Parameters like gas condition, particle characteristics, and nozzle geometry influence heavily to reach to the required velocity [3]. In this chain, if contrasted with other spraying methods, deposition of coating happens at moderately low temperature, enabling sprayed particles to stay in solid state [4]. In contrast to thermal spray that includes either total or partial powder particle melting, cold spray takes out thermal imperfections, for example, high residual stress in the coating because of solidification shrinkage, potential harm for the substrate caused by molten metal effect, and high-temperature oxidization [5]. Thus, the process is especially suitable for coating of thermally-sensitive materials, for example, amorphous and nanocrystalline materials, and for oxygen-sensitive materials, for example, magnesium, aluminum, and titanium composites [5]. The essential utilization of cold spray coatings is for the surface upgrade of metals to enhance properties, for example, resistance from wear and corrosion, conductivity of electrical/thermal, and so on [6]. This process is also appropriate for coatings on light metal substrates, for example, magnesium alloys because of its low temperature tolerance [6]. As a result, particle plastic deformation during flight disturbs the thin oxide film at substrate arrival, helps promote close conformal contact under high local pressure, and allows for occurrence of bonding, which is thought to equivalent to that in explosive welding or shock wave powder compaction [5].

The cold spray system can be designed in either automated or manual operation mode, making it flexible for fixed or portable operation. In many respects, a generic cold spray setup looks very similar to some of the traditional thermal spray systems. The gases having aerodynamic properties are generally used to propel the feedstock particles, including helium, nitrogen, combination of helium and nitrogen, and dry air (79% nitrogen + 21% oxygen). The major components making up the system include:

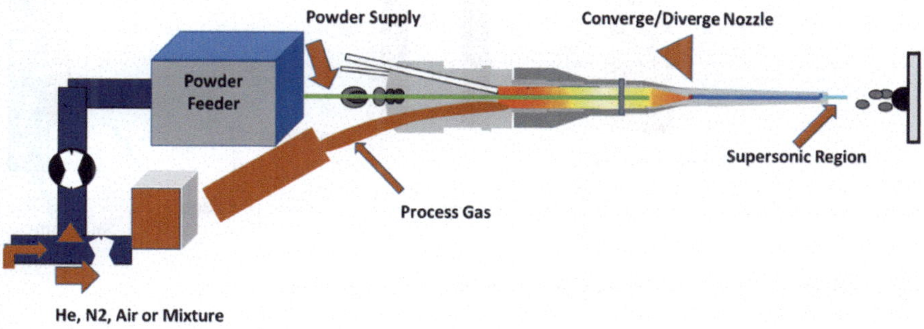

Fig. 3 Schematic diagram of a cold spray gun [1]

- Spray gun and powder feeder;
- Compressed gas source;
- Gas heater pre-nozzle entry, to compensate for cooling owing to fast nozzle expansion;
- Supersonic deLaval type nozzle;
- Spray chamber with motion mechanism;
- Monitoring and control system for spray parameters (measuring and controlling gas temperature and pressure).

Figure 4 presents a connection among various process components. Compressed gas such as N_2, O_2, or He passes through a diverse system consisting of a gas heater and a powder feeder at pressures in the range 1.4–3.4 MPa and maintained at minimum 1.7 MPa. The pressurized gas is warmed electrically to between 100 and 600 °C and then goes through a deLaval type converging/diverging nozzle until the gas reaches a supersonic speed. The powder particles are brought to the gas stream just before the converging region of the nozzle and extending gas quickens this process. The metering instrument passes on the powder feedstock on the high-pressure side of the nozzle, which is heated and kept up at the raised pressure of the manifold. A diminishing temperature takes place in the midst of the supersonic expansion through the deLaval nozzle. Therefore, the gas stream temperature is often underneath the melting point of the particulate material, giving the development of the coatings essentially from particles in the solid state with virtually no oxidation [7].

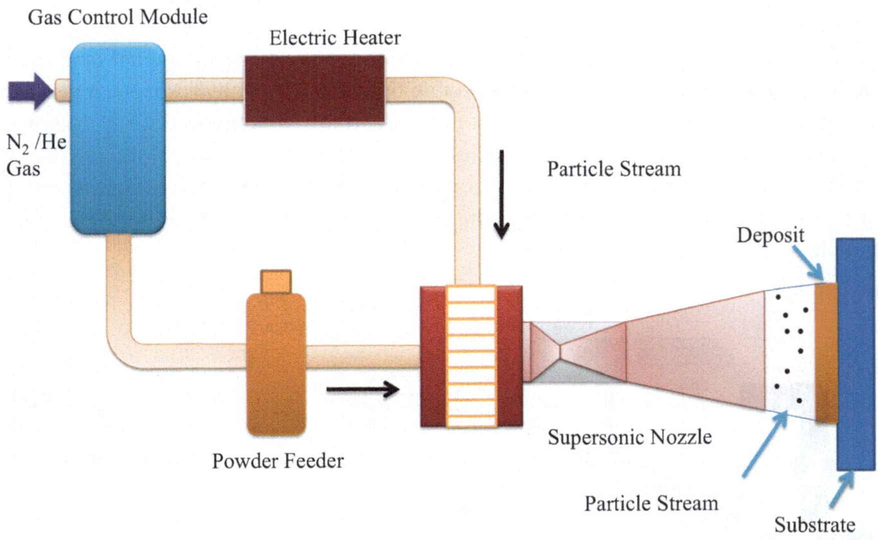

Fig. 4 Schematic of the connectivity among cold spray components

2 Mechanism of Cold Spray Process

Cold spray uses the principle of energy conversion. It utilizes particle kinetic energy, and converts into plastic deformation, strain, and residual heat [8]. The process mechanism finds some similarity with explosive cladding and shock wave powder compaction because of its high-rate-strain phenomenon. Figure 5 shows a schematic of the process principle.

The main concern with any AM based technologies is the state of bonding between feedstock and substrate or feedstock-feedstock interface. In cold spray, the bonding of particles is attributed to adiabatic shear instability (ASI), occurring at the particle-substrate or particle-particle interface due to the presence of high kinetic energy. ASI also takes place at a loss of strength when thermal energy due to supersonic collision rises past the effect of strain and strain-rate hardening from the deformation [5]. This presence of high-strain and high-strain-rate deformation is particularly advantageous for cold spray as ASI covers the complete surface area by impacting particles with various velocities (yet in supersonic state) direct to the target. The mechanical bonding of the particles is a result of the tangled structure occurred on the outermost layer due to ASI. The harsh plastic deformation further implies the microstructure evolution from the criterion of microbands to the processing of sub-grain area [5].

Plastic deformation of the particles makes deposition process possible, edging cold spray as a superior technology as compared with other thermal spray family members. The involvement of lesser heat helps in reducing the negative effect on the deposit, such as residual stress, surface distortion, oxidation, void, phase transformation. Under high pressure, metallurgical bond between interacting surfaces most likely is strengthened as a result of removal and destroy of oxide layers at the contact interfaces. A comprehensive analysis of the bonding mechanism and impact phenomena occurring during the cold spray deposition is summarized by the following four events, also shown schematically by Fig. 6.

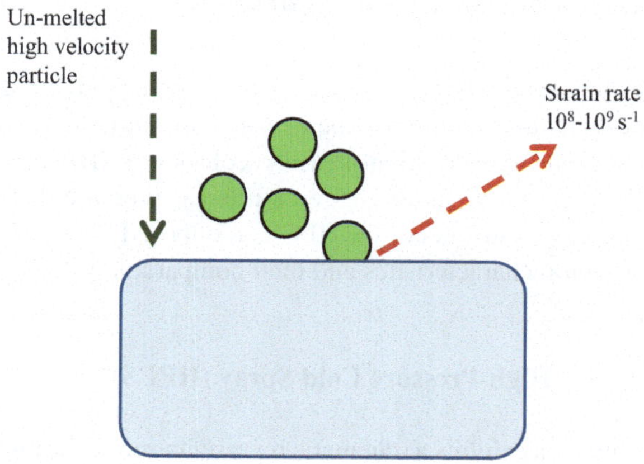

Fig. 5 A schematic showing cold spray process principle

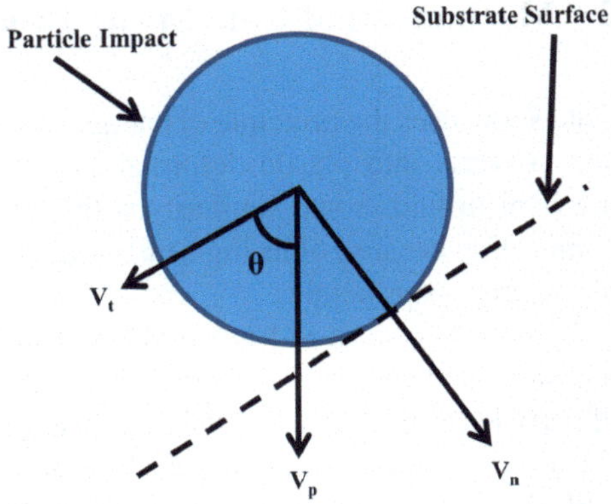

Fig. 6 Schematic of the bonding phenomena in cold spray

(1) Collision plastic strain energy is kept in the system as the heat raises the substrate's temperature and softens substrate.
(2) Flow stress (instant stress needed to deform a body plastically) reaches a maximum at a certain stress value and then starts to decline with added stress.
(3) Strain softening (shear or heat) is localized under actual circumstances, i.e. fluctuating stress, pressure, temperature, microstructure. In this case of localization, the flow stress drops rapidly to zero (i.e. plastic deformation occurs very easily in a narrow area surrounding the particle-substrate interface). This causes the injection of an interfacial jet, like a splash of an object hitting water, made of the same material as the particle, which is now heavily deformed.
(4) The jet removes the particulate oxide film and allows direct contact between the material surfaces (particle-substrate or particle-particle), supporting bonding in-between with comparatively elevated contact pressure (shown by schematic in Fig. 6) and adiabatic softening of the interfacial region [9].

2.1 Cold Spray Configuration

When dealt with the practicality of the cold spray process, two distinct system technologies are patented in line with how the particulate feedstock is carried out through the spray nozzle: high-pressure cold spray (HPCS) and low-pressure cold spray (LPCS). As the name suggests, it is gas pressure that is required to propel injected particle axially, hence the HPCS, as opposed to radially to synonym with LPCS. The common characteristics and their comparable values are given in Table 2.

2.1.1 High-Pressure Cold Spray (HPCS)

Figure 7a exhibits a schematic representation of the high-pressure cold spray (HPCS) system, where the main gas stream and the powder stream are both introduced into

Table 2 Comparative HPCS and LPCS system parameters

System parameters	HPCS	LPCS
Process gas	Nitrogen, Helium	Air
Pressure (bar)	7–40 (50)	6–10
Feedstock preheating temperature (°C)	20–550–800–1000–1100	20–650
Gas flow (m^3/min)	0.85–2.5 (N_2), maximum 4.2 (He)	0.3–0.4
Powder feed rate (kg/h)	4.5–13.5	0.3–3.0
Spray distance (mm)	10–50	5–15
Electric power (kW)	17–47	3.3

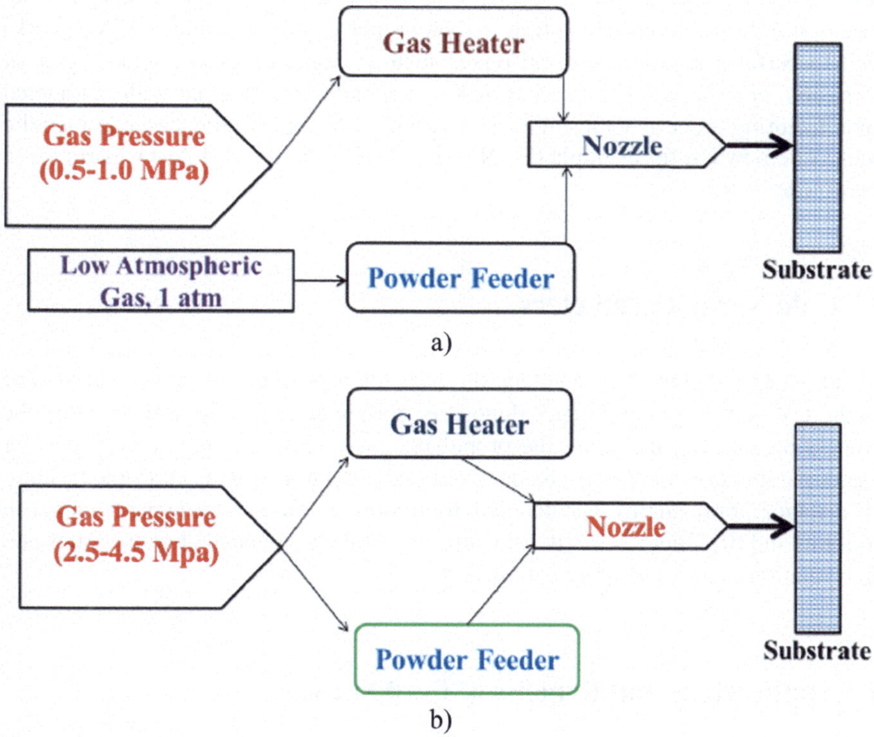

Fig. 7 Schematic of **a** high-pressure cold spray; and **b** low-pressure cold spray process

the inlet chamber of the nozzle. High-pressure systems utilize higher pressure gases (2.5–4.5 MPa) and often have a dedicated gas compressor. A low-molecular-weight gas, such as helium, is sometimes used as the accelerating gas when particles must be brought to very high velocity. The use of a wide spectrum substrates such as stainless steel, titanium, Inconel, gold and silver is unique to the HPCS method, even though it's potential could extend to most metal, glass, and ceramic substrates. Because it is feasible to process high regions, high-pressure offers a high demand from the aerospace, marine, automotive, and electronic sectors, with so far tried feedstock design based on Cu, Al, Ta, Ni, Cu(Sn), Ni(Cr), and Ni(Al).

2.1.2 Low-Pressure Cold Spray (LPCS)

Figure 7b shows a schematic of LPCS system in which powder stream is injected into the nozzle at a point where the gas has expanded to low pressure. Atmospheric pressure air, drawn by the lower pressure nozzle injection point, is used to transport powder from the feeder. Since the system does not require a pressurized feeder, it is often used in portable cold spray mode. LPCS generally utilizes readily-available compressed air but can run on nitrogen as well featuring prime advantage being ease of use. In situations where time and part replacement cost money, LPCS works well to perform repair jobs of all types, such as dents, dings, scratches, spots of corrosion, cracks, etc. The method makes it possible to integrate with automated manufacturing process, with somewhat advantageous for spraying ceramic-metallic composite powder, for example Cu, Ni, Zn, Al with additions of Al_2O_3 particles in the powder mix.

3 Cold Spray Parameters

By adjusting the control parameters, the cold spray process can be optimized. The method of deposition, and hence deposit performance is greatly influenced by the process parameters, including the propulsive gas (pressure, temperature), powder feeder parameters (dual/single feeder, feed rate), and nozzle (trajectory path, stand-off distance, spray angle). The detailed discussion on effect of these parameters can be found in [10]. Table 3 briefly sums up some standard deposit characteristics and their relation to the cold spray parameters.

4 Applications and Benefits of Cold Spray

It is anticipated that cold spray technology will complement and expand the variety of thermal spray applications. Its application includes both manufacturing and repair in medical, aviation, electronics, automotive, and petrochemical industries. This technology can readily remove any flaw to save the quality of manufacturing. The repair of casting defects is one of the significant application. Both casting defects and removing machining flaws can save small batch or distinctive manufacturing. This technology also makes casting molds simple to restore or modify. Table 4 presents the materials with prominent benefits from cold spraying, and corresponding application targets.

Table 3 Cold spray deposit characteristics as a function of process parameters

Parameter	Value	Deposit strength	Adhesion	Deposit efficiency	Porosity	Residual stress
Gas pressure	High	High	High	High	Low	High
Gas temperature	High	High	High	High	Low	High
Molecular weight of the gas	High	Low	Low	Low	High	Low
Travel velocity	High	High	High	High	High	Low
Powder feed rate	High	Low	Low	Low	High	High
Stand-off distance	High	No common view	No common view	No common view	No common view	No common view
Spray angle	High	High	High	High	Low	High

Supersonic cold Spray provides many technical advantages compared to other thermal spray processes, a detailed comparison is presented in Table 5.

There are other distinctive characteristics of the cold spray technology that makes it superior to competitive additive manufacturing-based technologies, including:

- **High end quality product**: the cold spray process runs at below metal melting point, thus avoiding crystallographic transformation which ultimately can result in high porosity in the deposit.
- **Manufacturing of large-scale components**: large parts and structures can be fabricated using the cold spray technology that are hard for the present powder-bed AM techniques owing to size restriction. Cold spray has the prospect of scaling up, and the current restriction is the choice of feedstock powders that could be applied.
- **Reduced manufacturing time**: cold spray is a much faster process than powder-bed or other powder-fed process technologies, making deposition rate up to 1000 times faster than the current state of direct metal laser (DML) sintering technology.
- **Green technology**: There is no need for extensive power source; or with limited gas variability the cold spray process can accomplish the industry-scale part fabrication. Research to-date has revealed that up to 98% of feedstock particles is useable/reusable, creating a less-material-intensive process. This has the cumulative edge of meeting stricter environmental regulations by contributing reduced greenhouse gas emissions.

Table 4 Applications and properties of cold spray process technology [1, 11]

Coating materials	Primary property benefits	Applications
Ceramics (glass, cement linings)	Enhanced protection for corrosion and wear	High temperature environment like furnace parts and heat treatment apparatus, nozzles of rocket and other parts of rockets and jet, etching equipment, chemical processing equipment, several components of nuclear power plants, etc.
Hard chromium coatings	Better protection against heat abrasion, wear and corrosion	Engine components such as pistons, piston rings and cylinders. Different polyethylene made components, etc.
Zinc-aluminum alloys	Sacrificial galvanic protection for corrosion resistance, oxidation and sulfidation resistance	High-temperature and high-load carrying applications with exposure to chemical and different environmental aspects, i.e. construction sites and thermal power plants etc.
Cobalt based alloys	Better strength and wear protection	Shipping and offshore industries, engine components of automobile, hydraulic power plant, and various chemical processes, etc.
Nickel based alloys	Helps in restricting high temperature corrosion and provide resistance to wear and oxidation	Aviation, automobile, biomedical devices, farming equipment, rolling process, bushing of pumps, mold and glass industries, etc.
Iron based alloys	Better wear characteristics	Gas turbines, steam and hydro power plants, rolling process, chemical and textile applications, cement and other construction equipment manufacturing industries etc.
Copper, silver, aluminum, and titanium alloys	Exceptional resistance to corrosion and good electrical conductivity	All electrical and electronics applications
Organic coatings (paints and polymeric or elastomeric coatings and linings)	Enhanced resistance to corrosion and wear, excellent aesthetic appearance and improved impact resistance	Aviation, construction, farming, oil and offshore industries, repair and remanufacturing, etc.

Table 5 Comparison of cold spray with other thermal spray processes

Cold spray process	Other thermal spray process
It is a solid state process	It is a liquid or molten state process
No heat affected zone is formed	Heat affected zone is formed
No oxidation of cold sprayed materials	Oxidation of thermal spray materials happens
Process done at room temperature	Requires preheating to reach semi-molten state when they reach substrate
Very low heat input is required	High heat input is required
It can be carried out in atmospheric pressure	It may require lesser pressure such as vacuum to achieve similar quality
Stand-off distance is less	Stand-off distance is comparably high
Porosity commonly below 1%	1–25% porosity is normal but can be further manipulated by changes in process and materials
Hand operable	Machine or robot operable
No toxic fume	Toxic fume is generated
Higher strength coatings for most metal alloys i.e. coating strengths >40 ksi	Lesser strength of metal alloys compared to cold spray
Compressive residual stresses in coating, rather than tensile	Usually tensile stress is developed
Intricate shape is possible	Intricate shape complexity 3D
No limit on deposition thickness	No limit on deposition thickness
Minimal masking requirement due to focused particle spray path	Minimal masking requirement due to focused particle spray path

5 Cold Spray Additive Manufacturing (CSAM) and Its Prominence

The idea that cold spray can be used as a new component development technique or repair/restore an otherwise obsolete (e.g. dimensionally-incorrect) one has born from the observation that it is capable to produce dense deposit. Consequently, in recent years, more and more companies in aerospace and defence industry has adopted the technology as an AM tool to produce freestanding metal parts and the restoration of damaged metal parts [12]. CSAM has a distinct advantage over other well-researched fusion-based AM technologies like laser beam melting (LBM), electron beam melting (EBM), and laser metal deposition (LMD). Table 6 equates CSAM and other additive manufacturing technologies with their strengths and weaknesses. CSAM's most important advantages over other AM processes are less production time, unrestricted product size, high adaptability, and adequacy for repairing dam-

Table 6 Comparison of CSAM with SLM and EBM

Comparative measure	Selective laser melting (SLM)	Electron beam melting (EBM)	Cold spray additive manufacturing (CSAM)
Mode for powder feeding	Powder bed feeding	Powder bed feeding	Direct deposition
Feedstock restrictions	Processing difficulty with elevated reflectivity and low flow metals	Inadequate for non-conductive and low fusion metals	High hardness and resistance metal handling difficulty
Melting of powder	Yes	Yes	No
Manufacturing size capability	Large	Small	Small
Geometrical accuracy	Moderate	High	High
Mechanical characteristics: (a) as-fabricated; and (b) after post-processing	(a) Moderate (b) High	(a) High (b) High	(a) High (b) High
Processing time	Low	High	High
System flexibility	High	Low	Low
Adaptability for repair/remanufacturing	Yes	No	No

aged components. In addition, CSAM is appropriate for high-reflectivity metals like copper and aluminum, that are quite difficult to be produced using laser-based additive techniques. Though, CSAM's challenges are also apparent. CSAM typically manufactures a rough surface or semi-finished product that needs post-processing operations. Moreover, owing to the intrinsic failures, CSAM deposits have lower mechanical characteristics in their as-fabricated condition. Consequently, heat treatments are frequently used to enhance mechanical characteristics. Table 6 presents a brief comparison of CSAM with commonly used fusion-based additive manufacturing processes such as selective laser melting (SLM) and electron beam melting (EBM). Whereas, Table 7 presents a detailed comparison of cold spray deposition with other commonly used deposition processes.

Table 7 Comparison of cold spray deposition with other deposition processes

Comparative measure	Gaseous state processes				
	PVD	PAPVD	CVD	PACVD	Ion implantation
Deposition rate (kg/h)	Up to 0.5 per source	Up to 0.2	Up to 1	Up to 0.5	N/A
Component size	Limited by chamber size				
Substrate material	Wide choice	Wide choice	Limited by deposition temperature	Some restrictions	Some restrictions
Pre-treatment	Mechanical/chemical plus ion bombardment	Mechanical/chemical plus ion bombardment	Mechanical/chemical	Mechanical/chemical plus ion bombardment	Chemical plus ion bombardment
Post treatment	None	None	Substrate stress relief/mechanical properties	None	None
Control of deposit thickness	Good	Good	Fair/good	Fair/good	Good
Uniformity of deposit	Good	Good	Very good	Good	Line of sight
Bonding mechanism	Atomic	Atomic plus diffusion	Atomic	Atomic plus diffusion	Integral
Distortion of substrate	Low	Low	Can be high	Low/moderate	Low

Comparative measure	Solution processes		Molten or semi-molten processes			Cold spray deposition
	Sol-gel	Electroplating	Laser	Thermal spraying	Welding	
Deposition rate (kg/h)	0.1–0.5	0.1–0.5	0.1–1	0.1–10	3.0–50	Up to 6
Component size	Limited by solution bath		May be limited by chamber size			Larger parts can be made
Substrate material	Wide choice	Only for electrically conductive material	Wide choice	Wide choice	Mostly steels	Wide choice

(continued)

Table 7 (continued)

Comparative measure	Solution processes		Molten or semi-molten processes			Cold spray deposition
	Sol-gel	Electroplating	Laser	Thermal spraying	Welding	
Pre-treatment	Grit blast and/or chemical clean	Chemical bleaching and etching	Mechanical and chemical cleaning			Mechanical/chemical
Post treatment	High temperature calcine	None/thermal Treatment	None/substrate Stress Relief		None	None
Control of deposit thickness	Fair/good	Fair/good	Fair/good	Manual variable automated-good	Poor	Excellent
Uniformity of deposit	Fair/good	Fair/good	Fair	Variable	Variable	Excellent
Bonding mechanism	Surface forces		Mechanical/chemical		Metallurgical	Mechanical
Distortion of substrate	Low	Low	Low/moderate	Low/moderate	Can be high	None

Acknowledgements This research was funded by a Natural Sciences and Engineering Research Council of Canada (NSERC) grant #RGPIN-2018-04440.

References

1. Pathak, S., & Saha, G. C. (2017). Sustainable development of cold spray coatings and 3D additive manufacturing components for repair/manufacturing applications: A critical review. *Coatings, 7*(8), 122–149.
2. ASTM F2792-12A. (2012). *Standard terminology for additive manufacturing technologies.* West Conshohocken: ASTM International.
3. Li, W., Yang, K., Yin, S., Yang, X., Xu, Y., & Lupoi, R. (2018). Solid-state additive manufacturing and repairing by cold spraying: A review. *Journal of Materials Science and Technology, 34*(3), 440–457.
4. Wang, Q., Birbilis, N., & Zhang, M. X. (2011). Interfacial structure between particles in an aluminum deposit produced by cold spray. *Materials Letters, 65*(11), 1576–1578.
5. Luzin, V., Spencer, K., & Zhang, M. X. (2011). Residual stress and thermo-mechanical properties of cold spray metal coatings. *Acta Materialia, 59*(3), 1259–1270.
6. Rolland, G., Sallamand, P., Guipont, V., Jeandin, M., Boller, E., & Bourda, C. (2011). Laser-induced damage in cold-sprayed composite coatings. *Surface & Coatings Technology, 205*(21–22), 4915–4927.
7. Zhou, X. L., Chen, A. F., Liu, J. C., Wu, X. K., & Zhang, J. S. (2011). Preparation of metallic coatings on polymer matrix composites by cold spray. *Surface & Coatings Technology, 206,* 132–136.
8. Chen, X., Wang, H. T., Ji, G. C., Bai, X. B., & Dong, Z. X. (2016). Deposition behavior of nanostructured WC—23Co particles in cold spraying process. *Materials and Manufacturing Processes, 31*(16), 1507–1513.
9. Champagne, V. K., Helfritch, D. J., & Trexler, M. D. (2007). Some material characteristics of cold-sprayed structures. *Research Letter Materials Science, 2007*(3), 1–3.
10. Grigoriev, S., Okunkova, A., Sova, A., Bertrand, P., & Smurov, I. (2015). Cold spraying: from process fundamentals towards advanced applications. *Surface & Coatings Technology, 268,* 77–84.
11. Tang, J., Saha, G. C., Richter, P., Kondás, J., Colella, A., & Matteazzi, P. (2018). Effects of post-spray heat treatment on hardness and wear properties of Ti-WC high-pressure cold spray coatings. *Journal of Thermal Spray Technology, 27*(7), 1153–1164.
12. Raoelison, R. N., Verdy, C., & Liao, H. (2017). Cold gas dynamic spray additive manufacturing today: Deposit possibilities, technological solutions and viable applications. *Materials and Design, 133,* 266–287.

Commercial Cold Spray Equipment

Julio Villafuerte

Abstract Two variants of cold spray powder injection point (downstream and upstream) came about from two distinct development threads in Russia; These two groups produced two distinct families of commercial equipment, each one with its own unique merits and attributes. Based on these concepts, commercial cold spray equipment became available in the late 1990s with several machine suppliers, some of them already gone, whereas others became established. This chapter is an update to the latest commercially deployed equipment, keeping in mind that the rapid evolution of cold spray technology may likely result in availability of additional upgrades by the time of publication. This Chapter discusses machine design features of each one of the families which determines capacity, reliability, portability, and cost. As every cold spray application requires a unique set of conditions, for it to be commercially viable, each type of equipment offers advantages and disadvantages in relation to an application; therefore, proper equipment selection is always the best practice.

Keywords Cold spray · Nozzle design · Automation · Controller

1 The Heart of the Method: Cold Spray Nozzle Design

The nozzle in a cold spray gun is a key element. It plays a primary role in converting high-enthalpy, high-pressure, low-speed gas into low-enthalpy, low-pressure, high-speed gas jet essential for particle acceleration. A cold spray nozzle's ultimate aim is to create optimal gas jet conditions that boost the ability of particles to combine effectively with the substrate when impacting. The physics of the process has been well explained elsewhere. A converging-diverging type (DeLaval) nozzle is typically used at the exit of the nozzle to generate such favorable supersonic gas flow [1]. The supersonic flow characteristics, including its kinetic and thermal energy content, are a function of (i) geometry of nozzle and; (ii) parameters of the gas (i.e. gas size, pressure and temperature). The amount of energy (thermal and kinetic) eventually

J. Villafuerte (✉)
CenterLine (Windsor) Ltd., 6290 Morton Industrial Drive, Windsor, ON N9J 3W3, Canada
e-mail: julio.villafuerte@cntrline.com

S. Pathak and G. C. Saha (eds.), *Cold Spray in the Realm of Additive Manufacturing*, Materials Forming, Machining and Tribology,
https://doi.org/10.1007/978-3-030-42756-6_2

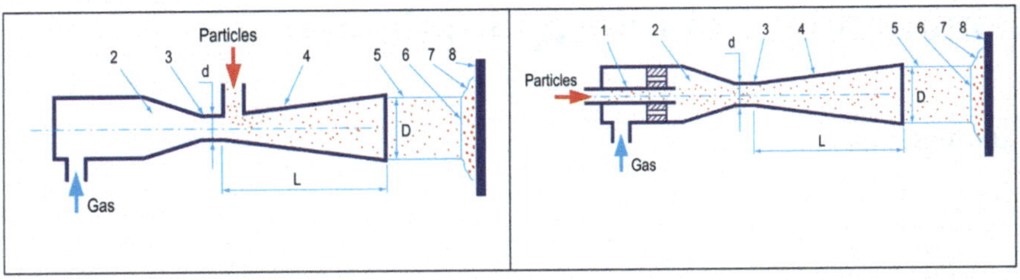

Fig. 1 Downstream and upstream injection nozzle parameters. 1—Tube for injection of particles, 2—pre-chamber, 3—nozzle throat (d = diameter), 4—diverging supersonic section, 5—free jet, 6—bow shock wave, 7—compressed layer, 8—substrate [3, 4]

transmitted to the particles depends heavily on the ratio of the powder to the mass of the gas, the physical characteristics of the powder (i.e. density, thickness, form, size distribution, etc.) and where and how these particles are introduced into the gas stream.

Some key geometrical parameters of the nozzle (Fig. 1) are the divergence ratio (D/d), diverging shape, and length of the diverging section. Such geometric features decide to a large extent the characteristics of the gas jet inside and at the exit of the nozzle. Other conditions prevail as the gas jet leaves the nozzle, including bow-shock waves and compressed layers that may affect the substrate's particle impact conditions. In practice the diameters of the round nozzle-exit vary between 2 and 12 mm, while the diameters of the throat range from 1 to 3 mm.

On the one hand, large deposition areas are often achieved by using a raster approach of subsequent passes with a step-over pattern, which is typically 25% of the pass's total width, due to the relatively small nozzle diameter. The width of any over-spray is relatively small due to the sharp adhesion threshold versus re-bounce conditions. A spray pass then generally displays well-defined sharp edges with a width near the exit diameter of the nozzle.

On the other hand, to attain the fine detail required by some metal additive manufacturing, computational work has suggested that there may be, in fact, physical constrains when attempting to down scale the nozzle size, in order to improve spray pattern resolution [2]. This has been brought about as cold spraying is one of several methods proposed for metal additive manufacturing applications. The attractiveness of cold spray is the fact that solid-state metal deposition can be quite rapid in comparison to traditional methods for metal additive manufacturing.

The repeated use of a nozzle eventually causes clogging or erosion of the inside surfaces. Nozzle clogging with a few pure metals like silver, aluminum, nickel and indium is more prevalent. Nozzle clogging mechanisms are not fully understood, but it is accepted that nozzles play a role in surface finish, surface temperature and surface chemistry. Therefore, manufacturers are using one of two approaches to reduce or prevent nozzles clogging. (a) nozzles made of high performance polymers, such as polybenzimidazole (PBI), exhibiting non-clogging properties, or (b) metal nozzles with ultra-high internal surface finish along with water-cooling.

With the former approach (a), fabricators must take note that, even high-end polymers typically show service temperatures at a maximum of around 400 °C. Therefore, dual-material designs may be necessary, where the hot section (converging section of the nozzle) must be made of metal while the cold section (diverging tube) can be made of a polymeric material. Per the second approach (b), surface finish along with water cooling has proven effective in suppressing clogging; however, the added complexity of the design, increased maintenance, increased cost, and other factors make this method somewhat less practical.

2 Upstream Injection

In upstream injection cold spray, helium or nitrogen at high pressures (up to 70 bar) are preheated (up to 1100 °C). Both, the high enthalpy gas along with the injected powder are forced through the nozzle to convert into supersonic expanding gas that (depending on gas type and powder characteristics) accelerate particles to supersonic velocities (typically around 1000 m/s). Different to downstream injection, the powder feedstock is axially pumped into the gas stream upstream of the nozzle throat using a pressurized powder feeder (Fig. 2).

One of the main advantages of this approach is that spray particles can achieve higher velocities compared to downstream injection due to the higher gas pressure and the upstream injection point, simply because of the longer interaction between particles and gas jet. Another advantage is the powder preheating effect, where the feedstock particles are likely to be preheated in the nozzle's high-pressure side, resulting in higher particle temperature compared to downstream injection, thus reducing the critical velocity of the powder.

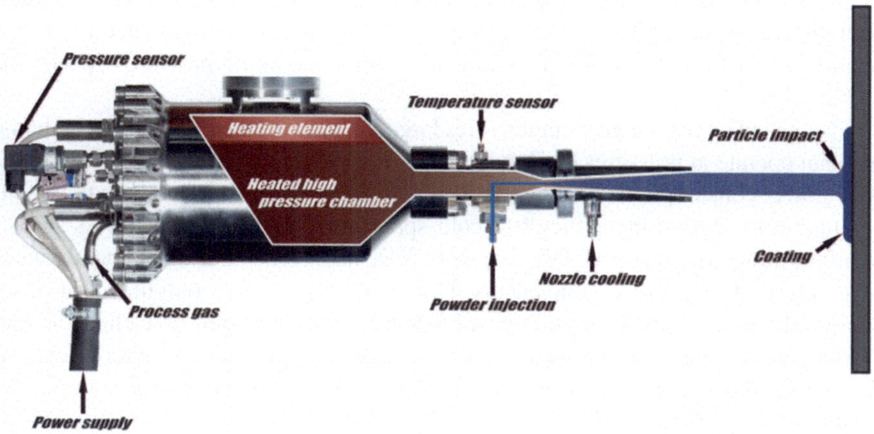

Fig. 2 Schematic showing the operating principle of upstream injection equipment. Courtesy of Impact Innovations

The high cost and limited availability of helium, on the other hand, adversely affects the economics and long-term viability of any cold spray process that requires this gas. While these drawbacks have been somewhat mitigated by helium recovery systems, the fact is that for most cold spray applications, including upstream and downstream injection equipment, nitrogen has become the gas of choice, naturally requiring applicators to compensate by working at higher temperatures and pressures.

On the process consumable side, upstream injection implies having particles go through the small nozzle throat which causes wear and, with some materials, clogging of this orifice. Additionally, higher operating gas pressures and temperatures result in nozzle service temperatures high, making nozzles prone to clogging, therefore the need of water-cooling.

Heating up pressurized gases at relatively high gas flow volumes requires high power capacity on the gas heater, which results in large heaters that must be located externally is the gun weight is to be minimized for practical reasons. One challenge with remote external heaters is the reliability and durability of flexible conduits capable of delivering the high temperature/high pressure gas to the spray gun with minimum heat losses and acceptable mechanical performance.

Upstream injection equipment is typically less mobile than downstream injection hardware, simply because of size of the required gun gas heater necessary to supply such temperatures at such high pressures. Consequently, upstream injection has traditionally been used as a mechanically driven tool that sprays well specialty materials that display high critical velocities, which require high impact velocities for bonding. For better portability of the spray gun, manufacturers of upstream injection equipment have developed low pressure/low temperature (e.g. 20–35 bar/400 °C) upstream injection systems. Alternatively, others have focused in lightening the upstream injection gun by placing the heater remote to the gun.

Impact Innovations [5] Germany, produces upstream injection systems 5/8 (Fig. 3) and 5/11 which use proprietary powder feeding technology. Impact 5/8 system features a maximum operating temperature of 800 °C (34KW) along with a maximum operating pressure of 50 bar. Whereas Impact 5/11 features maximum operating temperature of 1100 °C (40 KW) in conjunction with a maximum pressure capacity of 50 bar.

Plasma Giken [6], Japan, manufactures large capacity ***upstream*** injection systems that can operate at pressures of 50 bar providing gas temperatures of up to 1000 °C with power consumptions from 35 to 70 KW (Fig. 4). This commercial equipment is suitable for depositing difficult to cold spray materials such as stainless steels, titanium alloys (e.g. Ti-6Al-4V), Inconels (e.g. 625, 718), and other metals with relatively high deposition efficiencies. Due to its high power output, it can also deposit thick at relatively high deposition rates, using nitrogen or Helium as the carrier gas. Nozzles must be water-cooled to avoid clogging and/or other effects at the nozzle. Also, due to the size of the heater at the gun, its operation can only be robotic or mechanized.

VRC Metal Systems, USA, manufactures the VRC Gen III™ upstream injection cold spray system (Fig. 5). The system was specifically developed to address repair and maintenance activities in both fixed shop installations and in-field application

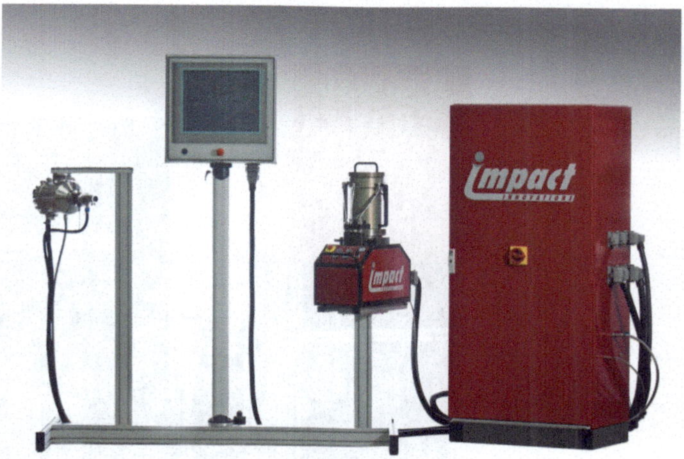

Fig. 3 Upstream injection cold spray system 5/8 (34 KW/50 bar/800 °C) showing control unit, and powder feeder, with a 34 KW gun-mounted gas heater. Photo courtesy of Impact Innovations [5]

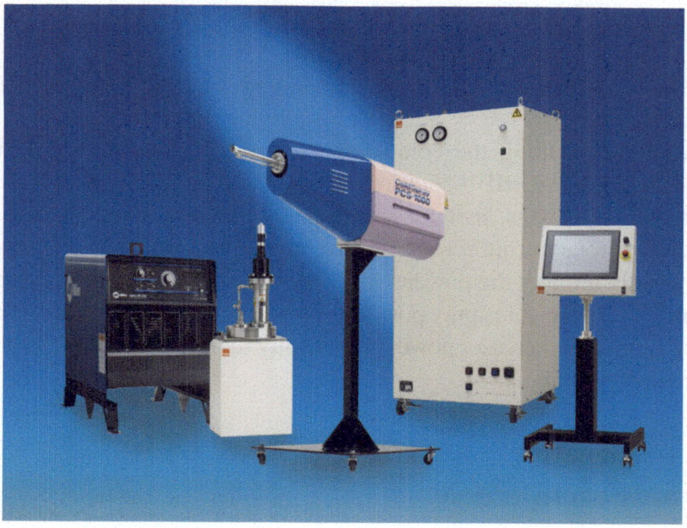

Fig. 4 Plasma Giken PCS 1000 upstream injection system (70 KW/50 bar/1000 °C) showing water-cooled spray gun with 70KW combined (on-gun plus remote) gas heating. Photo courtesy of Plasma Giken Co. [6]

that required increased flexibility and mobility. The VRC Gen III™ is a hand-held capable high-pressure (70 bar/650 °C, 21 KW) upstream injection system, based on a design that lightened the gun by removing the heater from the gun and supplying a high temperature/pressure flexible connection between the heater and powder feeder to the nozzle assembly. The benefit of having a wide pressure range, however, is so

Fig. 5 VRC Gen III upstream injection system (15 KW/70 bar/900 °C). Photo courtesy of VRC Metal Systems [7]

that softer materials, which do not need high pressure can also be deposited with the same system.

Beyond light-weighting the cold spray gun, two other patented features are included in the VRC Gen III™ cold spray system; (a) an internally insulated heater which reduces heat losses, increases the heater efficiency, and improves its field portability; (b) a rotary drum powder feeder, which is angled at approximately 30°, and continuously tumbles the powder while feeding so that particle segregation due can be eliminated from cold spray powders, particularly in the case of blended powders. This feature also increases powder flowability for more consistent powder flows during the spray operation.

3 Downstream Injection

In downstream injection, air, nitrogen, or Helium at low to medium pressures (4–35 bar) can be preheated to about 600 °C. As in upstream injection, when high enthalpy gas is passed through the DeLaval nozzle it accelerates to supersonic velocities (typically around 300–1000 m/s) as its temperature drops to below 200 °C. The unique characteristic of this family of equipment is that the powder feedstock is introduced downstream (after the nozzle throat) into the diverging section of the nozzle (Fig. 6).

Downstream injection guns use *venturi effect*-based powder feeding, which can operate consistently without the need of a pressurized powder feeder, but only at gas

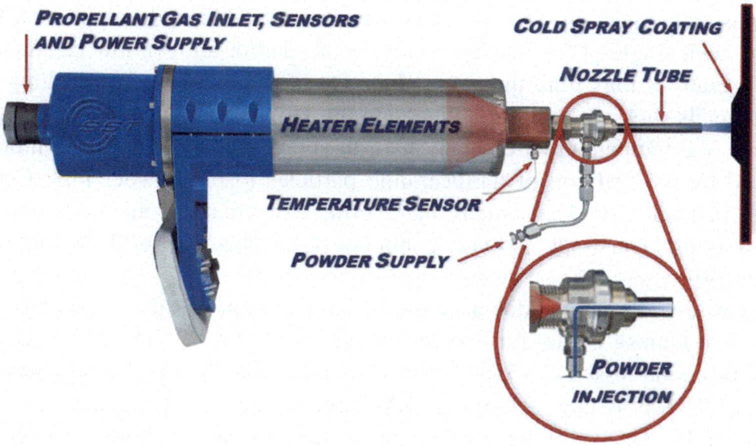

Fig. 6 A schematic drawing showing the operating principle of downstream injection equipment [8]

pressures below 10 bar, using either air or nitrogen. Above this pressure, the *venturi*-based powder feeding becomes unstable and unpredictable, making the process only suitable for manual operation unless a pressurized external feeder is used to bypass the venture feeder.

Today, most engineering metals and blends require more than 10 bar for acceptable properties; on top of that, there is an increasing use of automation for many cold spray applications. Therefore, *venturi*-based feeding became limited to a small number of manual applications, while external low-pressure pressurized powder feeders, such as the ones used in traditional thermal spray, became the industry standard for downstream injection cold spray systems.

Over the last few years, there has been a significant increase in newer cold spray production applications where factors such as cost, reliability, repeatability, maintainability, and productivity become paramount. One challenge to meet these requirements was the availability of economical, repeatable, consistent, and measurable powder feeding systems. One manufacturer of downstream injection systems developed a novel non-pressurized powder feeding method, which can provide extraordinary feeding stability over the full range of operating pressures (6 through 35 bar), more precise control over feed rate, high reliability, and easy maintenance. This was done in conjunction with the development of a new enhanced vacuum design (EV) nozzle, which extended the range of pressures over which the venture-effect can be effective.

Only the diverging section of the nozzle is subject to erosion due to the direct impact of the spray powder in downstream injection. The nozzle design can therefore be divided into the nozzle holder containing the converging section plus the nozzle throat or "orifice" and the nozzle tube, the diverging section of the nozzle. The life of the orifice is virtually unlimited, as the nozzle holder only passes gas.

Downstream injection systems can be compact, portable, and economical. When spraying with single-phase powders (pure metals and/or alloys), low gas pressures and temperatures may limit the range of spray-able materials to low-melting point ductile metals such as aluminum, zinc, and tin. However, the spray-ability (at low pressures and temperatures) of many hard-to-spray materials can be significantly enhanced by the addition of hard/ceramic particles to the powder mix. Ceramic particles in the mix produce a micro-hammering effect that not only helps compacting underlying layers, but also may continuously activate the substrate, improving adhesion [9].

The earliest commercial downstream injection system was developed by Dymet [9] Russia [Obninsk Center for Powder Spraying Ltd. (OCPS)] (Fig. 7). Today, this system uses compressed air (5–8 bar) @ 400 l/min heated in the gun to a maximum of 600 °C. The spray gun includes 3.5 KW light air heater and a replaceable nozzle expansion tube. The life of the nozzle for a powder feed rate of about 0.5 g/s is about 1 spray-hour. This downstream injection system utilizes a standard non-pressurized venturi powder feeder that works up to about 8 bar because of the low pressure required. This system has been widely used in Russia and many other countries, primarily for field portable corrosion damage repair and dimensional reconstruction using commercially pure aluminum, copper, zinc, nickel, tin, and lead mixed with sufficient quantities of ceramic particles to optimize deposition performance. Typical efficiency of deposition for these materials is 20–30% with a deposition rate of 3–10 g/min.

Today, downstream injection equipment offers extended operating pressure capacity up to 35 bar along with improved powder feeding technology, which translates into higher deposition efficiencies and extended range of spray-able materials, including

Fig. 7 Dymet 423 portable downstream injection equipment (3.3 KW/8 bar/600 °C). Courtesy of DYMET Corporation [10]

stainless steels, titanium, and nickel-based alloys and blends, as well as better process consistency and repeatability.

CenterLine SST™ Canada, produces SST™ PX and EPX cold spray industrial machines, which are production-ready downstream injection machines. Based on the same principle as Dymet machines [9], these machines operate at extended ranges of gas pressures (4–35 bar) using air, nitrogen, or helium at up to 550 °C gas heating with a power consumption of 3.8, 4.2, and 15 KW, and without the need of a *pressurized* feeder. The new PX and EPX systems use a re-designed *venturi*-effect powder feed (Figs. 8 and 9) and are engineered to satisfy the needs of volume production. Since automation is the desirable modus operandi, these systems exceed minimum requirements for productivity, cost, reliability, repeatability, and maintainability. The new non-pressurized powder feed method was implemented along with a novel nozzle holder design that permits non-pressurized powder feeding up to 35 bar gas pressure. The powder feed is integral part of the machine, requires no pressure, is free from clogging, can be easily removed for service, can be swapped in minutes, comes with multi-hopper capability, and feeds up to 120 g/min. Table 1 summarizes the state of commercial cold spray systems, with system capacities.

Fig. 8 SST PX downstream injection manual system (3.8 KW/17 bar/550 °C) integrated with the new dual-non-pressurized powder feeder. Photo courtesy of CenterLine (Windsor) Limited

Fig. 9 SST PX downstream injection robotic system (4.2 KW/17 bar/550 °C) integrated with the new dual-non-pressurized powder feeder. Photo courtesy of CenterLine (Windsor) Limited

Table 1 Summary of commercial cold spray system capacities

Manufacturer	Model	Power (KW)	Temperature (°C)	Pressure (bar)	Notes
Dymet (Russia)	423	3.3	600	8	Downstream-integrated dual hopper non-pressurized feeder
CenterLine SST (Canada)	PX	3.8–4.2	550	17	Downstream-integrated dual hopper feeder-non-pressurized
CenterLine SST (Canada)	EPX	15	550	35	Downstream-integrated dual hopper feeder-non-pressurized
VRC (USA)	Gen III	21	650	70	Upstream-gas heater remote/pressurized single-hopper feeder
Impact Innovations (Germany)	5/8	34	800	50	Upstream-Standalone pressurized single-hopper feeder
Impact Innovations (Germany)	5/11	40	1100	50	Upstream-nozzle water-cooling/pressurized stand-alone single hopper feeder
Plasma Giken (Japan)	PCS 800	35	800	50	Upstream-split gas heating/standalone single hopper pressurized heater
Plasma Giken (Japan)	PCS 1000	70	1000	50	Upstream-split gas heating/standalone single hopper pressurized heater

4 Automation of the Cold Spray Process

4.1 Industrial Automation

Industrial automation refers to the use of mechanical systems for performing routine manufacturing activities without or with the supervision of a human operator. The fact that mechanical systems powered by Programmable Logic Controllers (PLC) are able to perform repeated tasks faster and more effectively than human operators leads to increased efficiency and repeatability.

Automation can be applied in various ways, from a simple single axis linear drive to complicated Multi axis computer numerical control (CNC) machining centers. In this section we will also make reference to dedicated automation and flexible automation. Dedicated automation is generally designed to perform single or multi tasks aimed at performing a specific process following a specific pattern. On the other hand, flexible automation has the capability of being re-programmed to do many other functions completely different from its original tasks.

The main advantages of automation are:

- Increased productivity by eliminating the dependency of production on how skilled and/or the emotional state of the operator.
- Improved quality by removing the element of human error.
- Increased process and/or product consistency by allowing the automation to doing repetitive motions or process more consistently.
- Reduce direct labor expenses.

The main disadvantages of automation are:

- An automated system may have a limited level of intelligence, and is therefore more susceptible to committing errors outside of its immediate scope of knowledge.
- Unpredictable development costs.
- Automation cost for a new product or plant typically requires a very large initial investment in comparison with the unit cost of the product, although the cost of automation may be spread among many products and over time.
- Increased indirect labor to maintain more automation.

In manufacturing, the purpose of automation has shifted to issues broader than productivity, cost, and time. It has shifted to focus on quality with consistency and repeatability of the process. Therefore, users have exerted pressure on automation suppliers to build automation components that are more accurate and consistent. This trend has also been reflected in the case of Cold Spray, as users of this technology are increasingly demanded to produce more consistent and higher quality cold spray deposits. This can only be achieved by having more control over the cold spray process parameters as well as tighter control of the characteristics of the feedstock material, as further described in the following sections.

4.2 Automation of Cold Spray Process Controls

Control of an automated cold spray process can be attained by monitoring and controlling process parameters such as gas pressure, gas temperature, feedstock feed rate, and gun travel speed. Commercial cold spray systems use a fixed diameter of the DeLaval nozzle throat (or "orifice"), which is contained within the spray gun itself. This orifice operates in the choked condition for the various gases.

The actual mass flow rate of the carrier gas is determined by the gas density, gas pressure, and gas temperature. Therefore, to control gas mass flow for a given gas, pressure and temperature must be controlled within certain tolerances. This is generally achieved with the implementation of closed loop control using pressure transducers and thermocouples in the gas circuit. Other alternatives include the implementation of closed loop control using a mass flow controller and thermocouples.

Feedstock rate control can be attained by volumetric feeding or weight loss delivery of the spray powder. Volumetric powder feeding is the most common technique as it is more economical. However, volumetric feeding is typically not consistent and monitored flow rates may vary up to 10% depending on powder characteristics. A better method for controlling feed rate includes the weight loss technique, which uses load cells to monitor feed rate and then controls the feedstock delivery rate to maintain a more constant feed rate. Other indirect but consistent methods include Auger feed of powder in combination with negative pressure into the nozzle.

Monitoring and control of gun travel speed and raster step are desirable in order to produce smooth finish and consistent thickness of the deposit. The gun travel speed, together with the powder feed rate, determine the thickness of the deposit during one pass. It is generally desirable that the deposited thickness be within 0.1–0.5 mm per pass. Gun motion need to be generated from an automated drive, which may be capable of closed-loop feedback control.

The raster steps over is important to define the thickness variation per gun raster. Depending on the surface smoothness required the raster step over is typically set from 6 to 50% of the nozzle exit diameter (Fig. 10).

4.3 Gun Manipulators

There are many possible combinations of automating the cold spray process. The following are some of the most common combinations:

- Single linear drive for the gun with linear movement of the substrate or part.
- Single linear drive for the gun with rotational movement of the substrate or part.
- Multiple axis robotic manipulation of the gun or the substrate with stationary substrate or part.
- Multiple axis robotic manipulation of the gun with substrate manipulation by auxiliary axis.

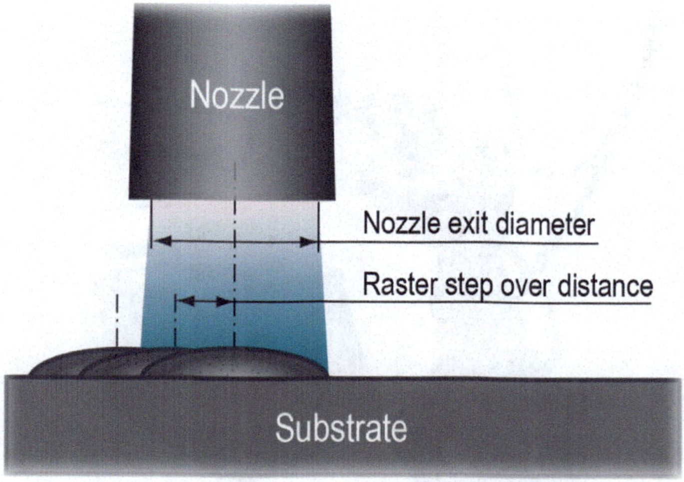

Fig. 10 Illustration of the raster step over method to attain smooth surface finish with cold spray

For a smooth and consistent coating on cylindrical substrates the use of a variable speed linear drive for the gun and a variable speed rotational drive for the substrate is the most economical method for best results. This method is only useful when the coating thickness permits a large variability tolerance. However, when thickness variability has a tight window of tolerance then a two-axis servo drive system for both the gun and substrate would be required along with a weight-loss powder feeding system.

For substrates that have surface contour, multiple axis robotic manipulation of the gun would be desirable for the tilt, travel and raster movements. If the coating thickness variability is critical, then a weight loss powder feeding would be required.

A more sophisticated level of automation consists of a gun mounted on a multi axis robotic arm with the substrate or part mounted on a single or two-axis manipulator with coordinated axis control from the robot controller. Generally, the substrate manipulator has a rotational axis and a tilt axis. The rotational axis may either have operated as a continuous rotation or rotational position. The tilt axis generally has a 90 degree of freedom from vertical to horizontal positioning as a coordinated axis with the robot controller (Fig. 11). This scenario would also require the use of a volumetric or a weight-loss feedback powder feeding system.

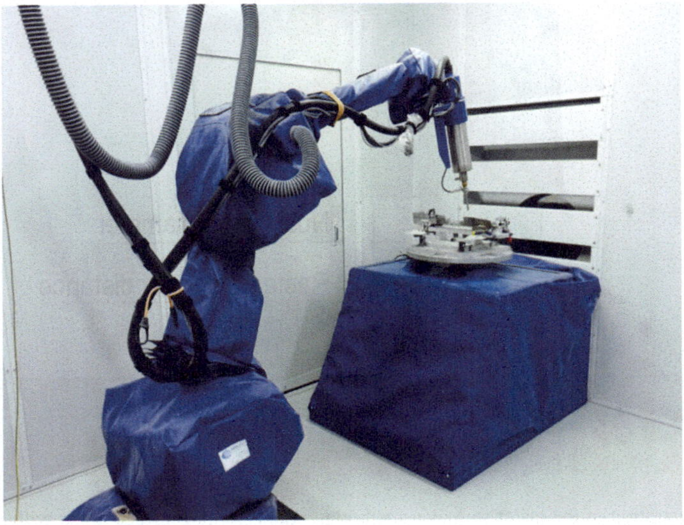

Fig. 11 Cold spray gun mounted on a multi-axis manipulator with the substrate or part mounted on a two-axis manipulator. Courtesy of Able Engineering

References

1. Pathak, S., & Saha, G. C. (2017). Sustainable development of cold spray coatings and 3D additive manufacturing components for repair/manufacturing applications: A critical review. *Coatings, 7*(8), 122–149.
2. Nasif, G., Barron, R. M., Balachandar, R. J., & Villafuerte, J. (2019). Numerical assessment of miniaturized cold spray nozzle for additive manufacturing. *International Journal of Numerical Methods for Heat & Fluid Flow, 29*(7), 2277–2296.
3. Papyrin, A., Kosarev, V., Klinkov, S., Alkimov, A., & Fomin, V. (2007). *Cold spray technology*. Oxford: Elsevier.
4. Papyrin, A. (2001). Cold spray technology. *Advanced Materials Process*, p. 49.
5. Impact Innovations. (2019). https://www.impact-innovations.com/en/index_en.html.
6. Plasma Giken. (2019). http://www.plasma.co.jp/en/products/coldspray.html.
7. VRC. (2019). https://vrcmetalsystems.com/products/.
8. Villafuerte, J. (2010). Current and future applications of cold spray technology. In *Metal finishing*.
9. Maev, R. G., & Leshchynsky, V. *Introduction to low pressure gas dynamic spray: Physics & technology*. Weinheim, Germany: Wiley VCH Veriag Gmbh & Co KGaA.
10. Dymet Corporation. (2019). http://dymet.info/indexe.html.

The Influence of Feedstock Powder

Aleksandra Nastic, Daniel MacDonald and Bertrand Jodoin

Abstract Additive manufacturing is based on the concept of freeform structures built up using a consecutive layer-by-layer material deposition approach, enabling the production of complex and functional components in a single manufacturing step. It allows the creation of high complexity components with minimal time and cost, as opposed to traditional subtractive manufacturing techniques. Current metallic AM technologies include selective laser melting, directed energy deposition, laser engineered net shaping, and plasma spraying. Although used commercially, these processes all suffer from the detrimental effects of high temperature processing, generally resulting in component distortion, uncontrolled phase transformations, undesirable residual stresses, and non-uniform mechanical properties. The cold spray process has recently gained attention in the additive manufacturing field as it may mitigate the undesirable thermal effects of current freeform manufacturing techniques, as well as drastically increase the available deposition rates. In cold spray, feedstock particles are injected in a supersonic gas flow and accelerated to velocities as high as 1200 m/s prior to impact. This high impact velocity is responsible for the material consolidation in the cold spray process. The particle impact velocity is dictated by the particle/gas flow interaction, which can be altered through the modification of the gas stagnation properties and spray nozzle geometry. While the effect of the gas/particle interaction is typically the focus of most cold spray research, it has become apparent that the size, shape, microstructure and quality of the feedstock powder have a large influence on the process efficiency. Hence, the effect of powder properties needs to be properly explored, understood, and considered in the powder selection process. This chapter aims to provide a complete reference on the effect of the feedstock particles on deposition quality in an additive manufacturing framework. It should provide

A. Nastic (✉) · D. MacDonald · B. Jodoin
Ottawa Cold Spray Laboratory, Department of Mechanical Engineering, University of Ottawa, Ottawa, ON, Canada
e-mail: Aleksandra.Nastic@uottawa.ca

D. MacDonald
e-mail: Daniel.MacDonald@uottawa.ca

B. Jodoin
e-mail: Bertrand.Jodoin@uottawa.ca

S. Pathak and G. C. Saha (eds.), *Cold Spray in the Realm of Additive Manufacturing*, Materials Forming, Machining and Tribology,
https://doi.org/10.1007/978-3-030-42756-6_3

the reader with a comprehensive resource for powder selection, pre-treatment, and storage. The effect of powder morphology will be presented. A descriptive analysis of the manufacturing processes used to produce particles will be included. The broad effect of powder size and shape on the particle velocity and resulting deposition will be discussed. Furthermore, the influence of the powder grain structure on particle distortion, dislocation generation, and recrystallization will be described on the basis of high strain rate deformation processes. Beyond the expected properties of the feedstock materials, it is also apparent that the "quality" of the powder is of great importance. The powder quality is thoroughly described by oxygen content and oxide layer type and thickness. This quality has been shown to greatly influence the process efficiency for some materials, and best practices for handling and storage of the powders is discussed. Finally, the status of powder recycling methods in cold spray will be considered along with the advantages of reprocessing in the field of additive manufacturing.

Keywords Feedstock powder · Grain size · Oxide layer · Additive manufacturing · Powder morphology

1 Introduction

In many popular metallic additive manufacturing (AM) processes, such as powder bed fusion or directed energy deposition, the initial state of the powder is not typically a focus of optimization as long as it presents adequate flowability and proper size distribution to achieve consistent and uniform melting [1]. Similarly, in cold spray AM (or CSAM), the focus of process refinement has mostly been in the modification of spray parameters such as gas temperature and pressure, gas nature, scanning strategies and speed, standoff distance, nozzle design, and powder feed rates [2–4]. However, the size, shape, composition, microstructure, and quality of feedstock powder has a significant effect on the process and resulting component properties. These powder characteristics can change the process outcome in several ways; by altering the gas-particle interaction during flight, by modifying the impact conditions and bonding requirements, and directly influencing the resulting chemical composition and microstructure of the produced AM part.

In CS, each individual particle has a minimum impact velocity that must be exceeded for bonding to occur, known as the critical velocity [5–10]. This velocity is influenced by many factors: the material properties of both the impacting particle and the material being impacted, particle size, particle geometry, particle impact temperature, and substrate temperature. This definition of the critical velocity is unambiguous; it is an exact velocity for a specific particle under prescribed conditions. This strict definition of critical velocity was explored by Hassani-Gangaraj et al. in their work using in situ observations of single impacts [11]. An example of these experimentally determined critical velocities can be found in Fig. 1 where it

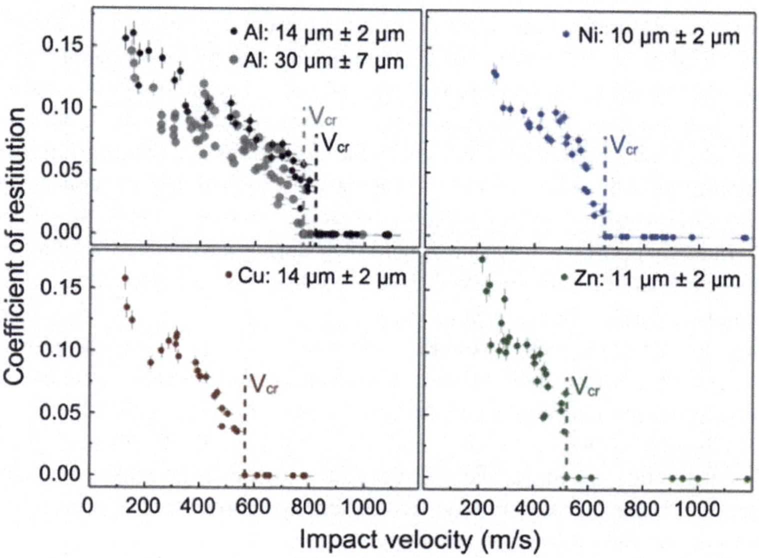

Fig. 1 Coefficient of restitution (ratio of particle rebound velocity to particle impact velocity) for several different metallic particles types and sizes showing a clear critical velocity [11]. Reproduced with permission of Elsevier

is clear that at a certain velocity, the particles no longer rebound from the substrate (coefficient of restitution of 0).

While this definition of critical velocity is straightforward, it is not always a practical parameter for cold spray applications. In the experimental work by Hassani-Gangaraj et al., each particle was fully characterized (microstructure, chemical composition, size, velocity, and temperature) and substrate conditions were known. However, the particles in a real CS gas plume do not experience such strict controlled conditions. Differences in particle size (distribution), shape, microstructure, chemical composition, and flight history (impact temperature) can occur—all of which can affect the bonding requirements. Therefore, measuring and defining a critical velocity for a specific powder type under CS conditions can sometimes be unclear or misleading. As a practical approach, authors often take the critical velocity as the mean velocity in which 50% of the particles adhere to the substrate [4, 12], or take it as the measured velocity of the largest particles to adhere [13–15]. For this work, the exact definition of the critical velocity is not crucially important, as the focus will be on how different powder characteristics influence critical velocity and not the critical velocity value itself.

The quality of CS AM components or coatings can be quantified in many ways, such as porosity, hardness, tensile strength, adhesion strength, particle flattening ratio, etc. [16, 17]. In addition to the quality of the deposit, an essential factor affecting the process economics is the deposition efficiency (DE). The DE is defined as the percentage of spent powder that consolidates into the final component, similar to the powder capture efficiency used in the directed energy deposition AM process [1]. The DE directly influences the deposition rate and process economics (spray time, gas consumption, electricity costs, capital costs, and labour costs) and wasted material (assuming the powder is not 100% recyclable). Fortunately, both coating quality and DE are related effects, where the conditions imposed to improve one typically improves the other. However, there are cases where DE and coating quality indicators are inversely correlated like in the work of Jenkins et al. [18], where they demonstrated that purposefully lowering the DE during the production of aluminum coatings decreased the porosity levels. Examples like this take place at very low DEs and are not typical for the CS AM process.

While the critical velocity provides crucial information on the minimum spray requirements, depositing under these conditions may not be ideal for manufacturing. Typically, as the particle impact velocity increases above the critical velocity, the quality of the deposit and the DE also tend to increase [17, 19, 20]. Particles travelling at higher velocities are more likely to develop zones of high plastic deformation, resulting in intimate contact of oxide-free surfaces promoting metallurgical bonding as well as interfacial mechanical anchors.

Conveniently, Assadi et al. [20] have defined the dimensionless parameter η, known as the particle impact velocity quotient;

$$\eta = \frac{v_{pi}}{v_{cr}} \tag{1}$$

where v_{pi} is the impact velocity and v_{cr} is the critical velocity. It has been demonstrated that an increase in the impact velocity quotient increases DE and coating quality (demonstrated through flattening ratio and tensile strength) [17, 19, 20]; an example can be seen in Fig. 2. This dimensionless parameter shows that to increase DE and coating quality, one should not only focus on expanding the particle impact velocity but also decreasing the critical velocity.

This chapter aims to provide a comprehensive resource on how the powder choice can influence the CS AM process. The chapter begins by outlining the major powder production methods and their effect on size, shape, microstructure and chemical makeup of the powders. Table 1 provides additional visual information concerning

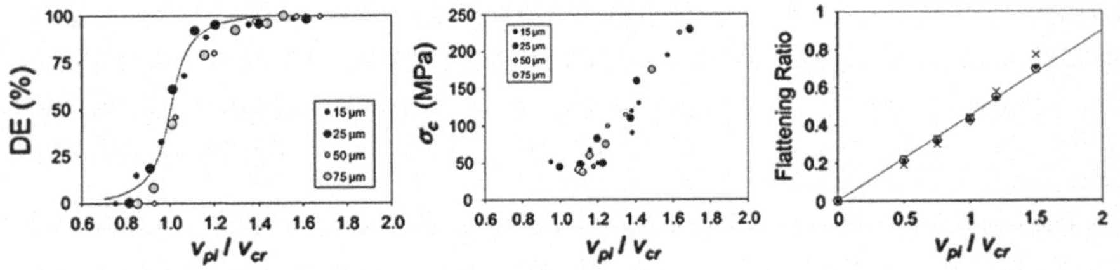

Fig. 2 DE, cohesive strength, and flattening ratio as a function of particle impact velocity quotient (η) for different copper particles sizes (indicated by different dots) under different deposition conditions [20]. Modified with permission from Springer Nature

Table 1 Powder particle geometry and composite powder features used in CS

a) **Spherical** (Plasma atomization/spheroidization)	b) **Spherical with Satellites** (Gas atomization)	c) **Irregular (Rounded)** (Gas atomization)	d) **Irregular (rod-like)** [21] (Water atomization)	e) **Irregular** (Water atomization)
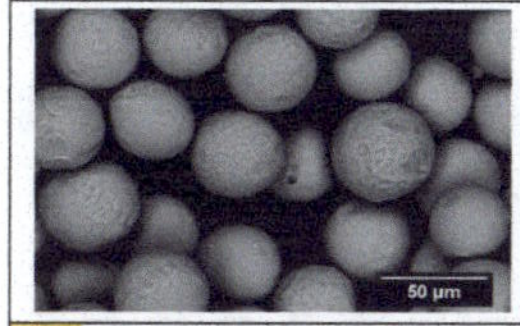	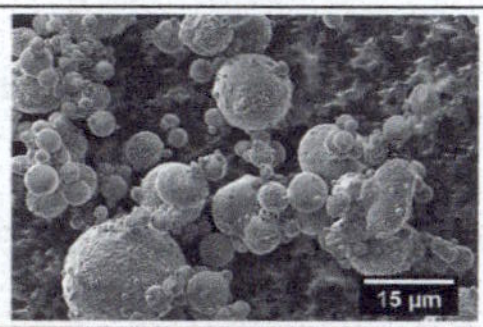	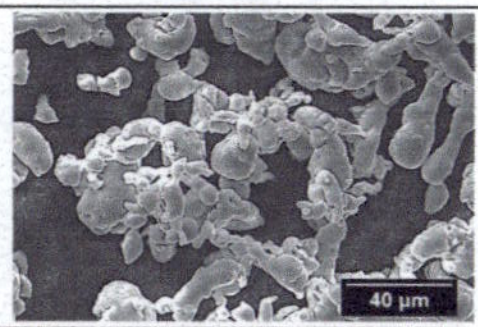		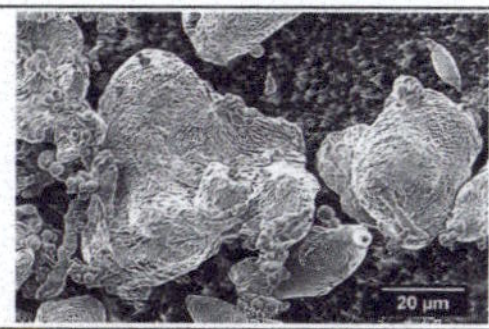
f) **Angular (Ti)** (HDH)	g) **Sponge (Ti)** [22] (Kroll process)	h) **Coral-like (Ti)** [23] (Armstrong)	i) **Dendritic** (Electrolysis)	j) **Film Coated** [24]
	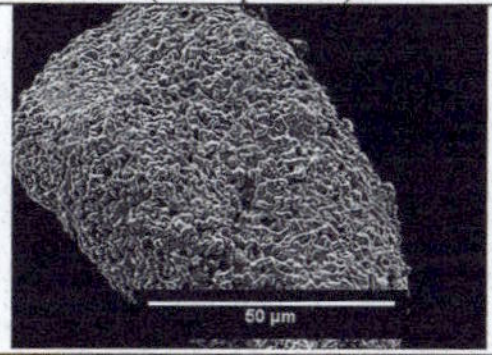	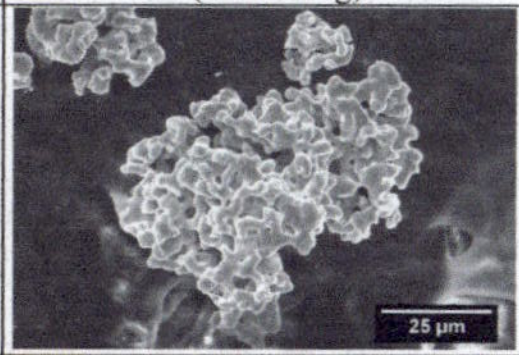	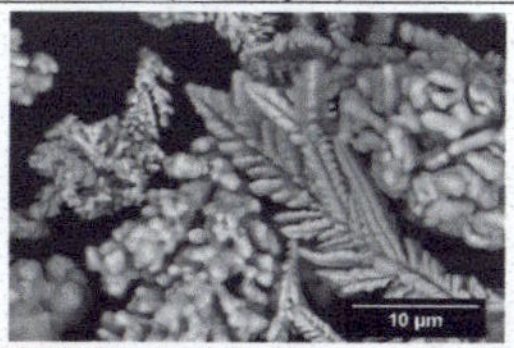	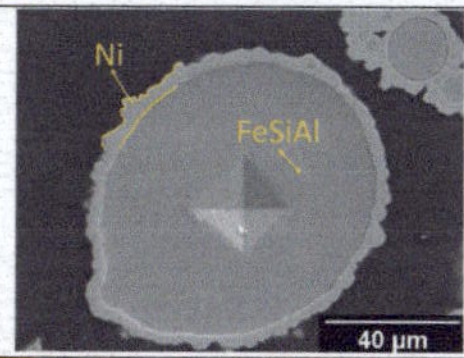
k) **Agglomerated Nano** (Spray drying) [25]	l) **Agglomerated and Sintered** (Spray drying) [26]	m) **Flaky** [27] (Milling)	n) **Lamellar Composite** [28] (Milling)	o) **Nano Composite** [29] (Milling)
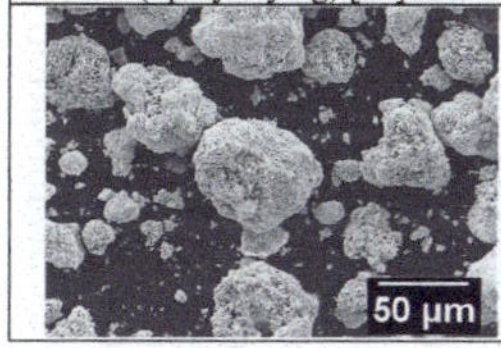	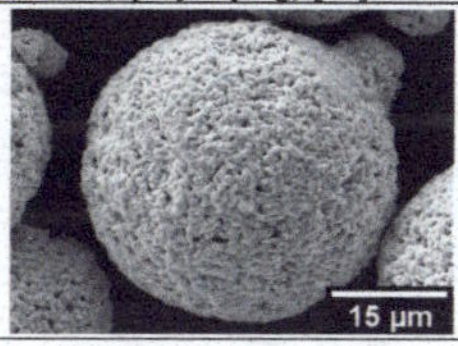		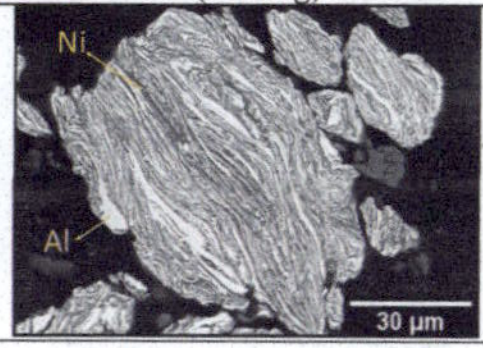	

the typical powder morphologies utilized in the CS field in support of the information contained in this chapter. Also, the effect of these powder characteristics on particle impact velocity quotient, and in turn, DE and deposition quality, will be thoroughly examined.

2 Powder Production

The principle powder characteristics—the shape, size, microstructure, and chemical composition—of the powders used in CS are a direct result of the powder manufacturing process. This section aims at reviewing the major metallic powder production methods used to produce powders utilized in CS and the influence these manufacturing processes have on resulting powder characteristics.

2.1 Atomization

The most common feedstock materials used in CS are spherical metallic powders produced through atomization [16, 30]. Atomization is the process of separating molten metal into droplets prior to solidification [16, 31–34]. It is suitable for most metals; however, it may be challenging for metals with melting temperature above 2000 K [16]. The atomized powders commonly utilized in CS are produced using inert gas atomization, plasma atomization, or water atomization. Other methods to atomize powder are not covered in this chapter, as they are uncommon in CS. The powders used for CS must be very fine compared to the powders produced with other methods, such as centrifugal force atomization, which results in particles in the range of 150–250 μm in diameter [31].

In inert gas atomization, the source material is first brought to a liquid state through conventional melting prior to being fed through a nozzle or by being melted in situ where the source material acts as an electrode. This continuous stream of molten metal is then broken into droplets by the impingement of a high-pressure jet of inert gas, often argon or helium. The metal solidifies during free fall inside the chamber and is collected at the bottom. The size of the droplets is critical, and getting a uniform and small particle size is difficult and requires the careful monitoring of many parameters [16, 31, 35]. These many parameters can include nozzle geometry, liquid properties such as surface tension and viscosity, pressure of the atomizing material, gas purity, gas jet geometry, melt temperature, etc. [16].

The powders produced via gas atomization are typically spherical (Table 1b); however, they can at times have a round, but irregular shape, as shown in Table 1c [16, 34, 36]. One major drawback of the process is the production of satellite particles. During the gas atomization process, the recirculation of the gas jet can cause fine particles to waft back and collide with partially melted larger particles [16, 34]. An example of a spherical aluminum alloy powder produced by gas atomization and having many satellite particles can be found in Table 1b. These small particles fuse with the larger particles, changing the flowability and dynamics during the CS process [16]. An additional issue that may arise during this production method is that inert gas can become trapped inside the particles, leaving small gas pores. This type of porosity that exists inside the powder, to begin with, can be a challenge to eliminate during consolidation by the CS process [16, 37].

Water atomization is a process very similar to gas atomization; however, it employs a jet of water instead of gas to atomize the stream of molten metal. This process results in irregularly shaped powder with higher levels of contamination compared to gas atomized [16, 34]. Examples of morphologies obtained from this method can be found in Table 1d, e. While the powder that is produced with this method is not ideal for CS, where chemical purity is often desirable, it is still a standard method of powder production as it has a much higher production rate compared to other atomization methods [16].

Plasma atomization was developed to produce fine and spherical powders without some of the drawbacks experienced in gas and water atomization. In the plasma atomization process, a pre-alloyed wire is fed directly into a plasma jet [35]. In general, the yield of fine powder from plasma atomization is higher than that of gas atomization, which is desirable for CS purposes [16]. Plasma atomization also typically results in highly spherical powders without the presence of satellite particles, as shown in Table 1a.

While both gas and plasma atomized are considered processes with low oxygen contamination, oxygen is still present in these powders. The amount of oxygen has significant effects on the resulting material properties, which is discussed in Sect. 6. This quantity of oxygen contamination can be influenced by both the processing parameters as well as particle size, as demonstrated in Fig. 3. In this plot, the different lines represent different processing parameters used in plasma atomization for the production of pure titanium powders. The difficulty in choosing a powder for CS AM can be seen in this plot; two powders of the same material, both plasma atomized with the same diameter, may have different chemical compositions based on the manufacturers operating conditions [35].

The other notable characteristic shown in Fig. 3 is that as particle size increases the oxygen content decreases. This is expected, as oxygen enters the particle at the surface and smaller particles will have a larger surface area to volume ratio.

Regardless of which process is used to atomize the powder, the droplets will experience rapid cooling and solidification. This quenching can force alloying elements into solution, induce high dislocation density, and result in a microstructure with fine, ultrafine, or even dendritic grains. Typically, these effects will change the particle hardness and its ability to deform upon impact. This effect will be further explored in Sect. 5. This microstructure will depend on the cooling rate, which is a function of the processing parameters and the particle size. Duflos et al. [38] presented an example of this, where gas atomized particles of different sizes produced at the same time had drastically different microstructures, as shown in Fig. 4.

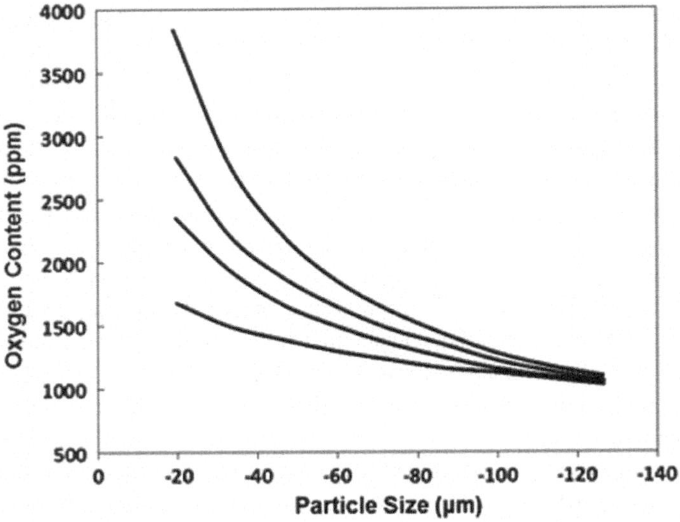

Fig. 3 Oxygen content of plasma atomized titanium powders as a function of particle size, with each line denoting different production parameters [35]. Reproduced with permission of Springer Nature

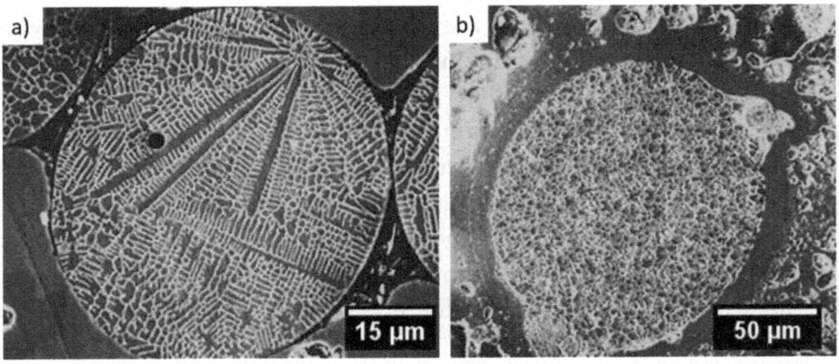

Fig. 4 Different grain structures **a** dendritic and **b** equiaxed seen in gas atomized Satellite 6 powders of different sizes as a result of different quenching rates [38]. Reproduced with permission of Springer Nature

2.2 *Mechanical Crushing*

Powders are sometimes produced through the crushing or grinding of raw material into a fine powder. In this process, energy in the form of impact, shear, wear, or pressure is applied to the materials, resulting in fracture. Powders produced through mechanical means typically have an irregular shape, as shown in Table 1f. This method works best for brittle materials since as particles get smaller, it becomes more challenging to apply the necessary forces on each particle required to reach fracture. In addition to the requirement of higher forces, re-bonding of particles may occur [16].

A unique form of titanium powder used in CS is produced through these mechanical means. Raw rutile is converted to titanium in a process known as the Kroll process. The resulting material is spongy and porous, but it can be mechanically crushed into a fine powder. This results in a powder with an irregular external shape capable of also maintaining its unique spongy characteristic (see Table 1g).

2.3 *Electrolysis*

Approximately 60 different metal powders, with Cu being the most popular in CS, have been successfully obtained via electrolysis, in which the applied electric current acts as the reducing agent. This process is very similar to electroplating, but with the conditions changed to produce a powder instead of a metallic layer [16, 39]. The primary characteristic of the produced powders is their resulting dendritic structure, as seen in Table 1i. These powders have a shallow oxygen content while presenting a high surface to volume ratio.

2.4 *Chemical Processes*

Some metal powders are manufactured through the reduction of metal compounds using a solid or gaseous reducing agent. Tungsten, molybdenum, and most importantly for CS, titanium powders are often produced this way. The Armstrong process is a metal halide reduction process in which a gaseous stream of titanium tetrachloride is injected into a stream of sodium metal. This reaction will result in a titanium metal and sodium chloride. This process can also work to produce pre-alloyed Ti-6Al-4V [40]. The material results in a high surface area to volume ratio and offers a unique coral-like morphology, as shown in Table 1h.

Another method for producing titanium powders is the hydride-dehydriding (HDH) process. HDH is a combined chemical and mechanical process in which commercially pure titanium powder is produced from bulk titanium at a much lower cost than atomization [35]. In this method, the titanium is heated under a hydrogen environment, resulting in the formation of TiH_2. This hydride is brittle and is easily mechanically crushed into a fine particle size. The fine particles then undergo a dehydride process where they are heated up under high vacuum, removing the hydrogen. This process produces irregular and angular powder, as shown in Table 1f [35].

2.5 Spray Drying (Agglomerated and Sintered)

Spray drying allows for particles too small to deposit using CS to be agglomerated into larger particles, including the agglomeration of multiple materials. Typically, in this process, a slurry is prepared and granulated by spraying into a fine mist in which the solvent dries, leaving only the powder behind. Following the granulation, particles are typically sintered to complete the bonding process. The powders produced by this technique are generally spherical, as shown in Table 11. However, depending on the materials, the spray drying conditions, and the sintering method, the powders can contain varying levels of internal porosity. Figure 5 shows two different cermet powders produced using this method but with different levels of internal porosity.

A type of spherical titanium powder has also been produced this way in a processed called the granulation-sintering-deoxygenation. In this process, Ti is milled to fine particles and then refined into spherical granules in the desired size range using spray-drying. The granules are then sintered, resulting in dense Ti particles with high oxygen content—which is later deoxygenated to industry standards. This method produces a low-cost powder due to its high yield and ability to use inexpensive starting materials; however, the resulting powder may have a higher oxygen content than the other covered methods [35].

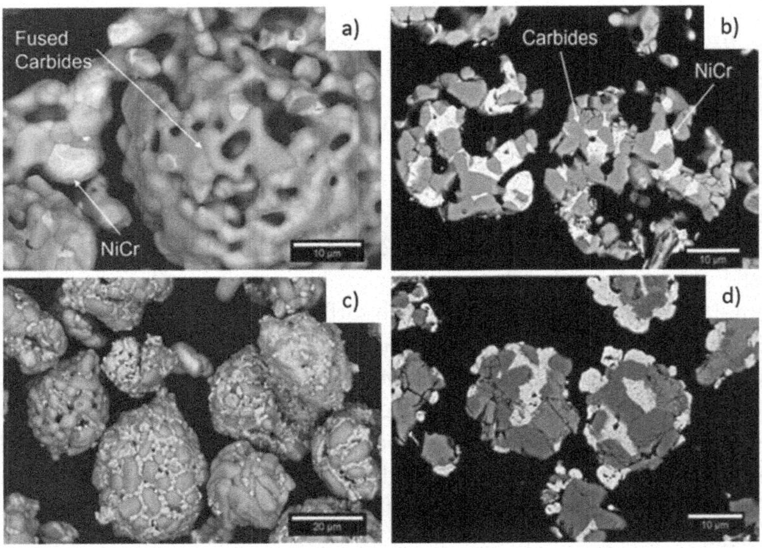

Fig. 5 Different CrC-NiCr powders with 25 wt% NiCr produced through agglomeration and sintering, showing different amounts of porosity [41]. **a, b** CRC-300 powder surface characteristics and cross-section and **c, d** Amperit 584 powder surface features and cross-section. Reproduced with permission of Springer Nature

2.6 *Powder Modification*

There are times when powder that results from a production method does not provide the desired characteristics for the CS AM application. In this case, several standard powder modification techniques may be used to provide different characteristics.

2.6.1 Plasma Spheroidization

Plasma spheroidization transforms particles of any shape and size into spherical particles. This process is very similar to plasma atomization, however in this case, instead of an alloyed wire being fed into the plasma jet, the powder is passed through the plasma jet. The plasma jet melts the powders, which then become spherical and rapidly solidify. Powders spheroidized by plasma spheroidization have similar properties to plasma atomized powders, but the size and chemical purity will be dependent on the source powder [16, 37].

2.6.2 Ball Milling

The shape and hardness of the powders can be modified by ball milling. In ball milling, the rotary motion of a drum forces the motion of very hard milling balls which impart high mechanical forces onto the powders. This results in deformation, fracture, and repeated re-welding of the particles. This method produces a powder that has small grains, is heavily work-hardened, and has an irregular, often flakey, morphology [16, 37], as shown in Table 1m. It is also used for mechanical alloying, in which more than one material is used. This process allows for a controlled, uniform distribution of a second phase inside a principal metallic matrix, as shown in Table 1n. It may also be tailored to create unique powders, like in the case of Table 1o, where nanosize copper particles have adhered to giant titanium powder through ball milling. Another process used to modify powders is known as the cryogenic ball milling, or cryomilling. This method, which is similar to the conventional milling, is done in a liquid nitrogen environment, therefore at a much lower temperature, resulting in powders with a nano grain structure [42, 43].

3 Effect of Powder Size

The relationship between the particle size and the CS process is complicated; not only does powder size affect the impact conditions (namely particle impact velocity and temperature), but it also directly influences the critical velocity. The majority of CS studies do not decouple these effects, resulting in the identification of optimum particle size for cold spray a challenging endeavour.

A simplified force balance approximation of the particle in flight can give insight into how particle size affects its velocity,

$$ma_p = \frac{1}{2}\rho V^2 C_d A_p \tag{2}$$

where m is the particle mass, a_p is the acceleration of the particle, ρ is the propellant gas density, V is the relative propellant gas velocity, C_d is the drag coefficient, and A_p is the cross-sectional area of the particle. The function can be further simplified since both the mass and cross-sectional area of the particle are functions of the diameter,

$$a_p = \frac{3\rho V^2 C_d}{\rho_p d} \tag{3}$$

where d is the particle diameter and ρ_p is the density of the particle materials. From Eq. 3, it becomes clear that for a particle travelling in a gas flow, the acceleration (and ultimately the velocity) of smaller particles should be higher than that of larger ones, with the maximum reachable particle velocity being the gas velocity. However, this simplified approximation may be misleading as the particle velocity upon impact with the substrate may differ from the in-flight velocity. The impingement of the gas jet upon the substrate surface gives rise to the formation of a bow shock. In the region beyond the bow shock, the gas velocity normal to the substrate is decreased substantially to subsonic speed, its density is largely increased, and its flow path becomes parallel to the substrate surface. These conditions will decelerate the particles in the direction normal to the substrate surface and accelerate them laterally. A full understanding of the interaction between the particles and the region after the bow shock is beyond the scope of this work, but a simplified view is that larger particles will be less affected (reduced deceleration) than smaller particles.

This leads to a complex relationship between impact velocity and particle size. Larger particles will experience less acceleration within the nozzle, but their deceleration within the stagnant region behind the bow shock will be reduced and often be negligible. Smaller particles acquire a higher velocity at the nozzle exit but may be greatly decelerated in the stagnation zone behind the shock. This relationship has been reported by many authors but is difficult to generalize since both the acceleration and deceleration of the particles depends significantly on the CS processing conditions such as nozzle design, gas stagnation properties, and standoff distance. However, the impact typically follows a curve similar to Fig. 6, where the solid line shows the effect of the deceleration through the bow shock [20].

The trend shown in Fig. 6 is crucially important for process optimization in CS. For materials under prescribed spray conditions, there should be a diameter that results in a maximum particle impact velocity. However, particle size is also known to affect the critical velocity, further complicating the process of choosing an optimal powder size.

It is generally agreed that increasing particle size decreases the critical velocity [4, 44]. In the work of Schmidt et al. [4], the critical velocity is based upon the onset of

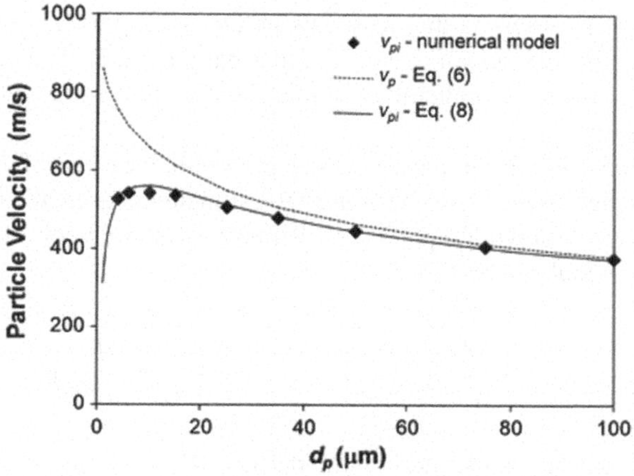

Fig. 6 Impact velocity (solid) and exit velocity (dotted) of copper particles of different diameters under set processing conditions [20]. Reproduced with permission of Springer Nature

adiabatic shear instability in the particle [13], determined through modelling. They propose that for small particle dimensions, the occurrence of shear instability (leading to the large deformation required to reach bonding) can be hindered by high cooling rates, which arise due to high-temperature gradients within the particle. This leads to the determination of a particle diameter at which adiabatic shear instability will never occur—the minimum critical diameter required for adhesion. Above this diameter, thermal diffusion is hypothesized to be slow and does not prevent localized shear instability from occurring at the particle surface, whereas particles smaller than this diameter would not reach adiabatic shear instability. A plot of the estimated minimum diameters for different metals is found in Fig. 7.

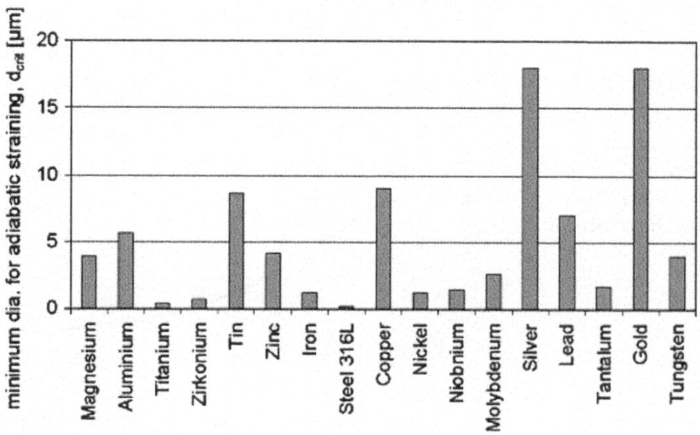

Fig. 7 Estimated critical (minimum) diameter required for bonding for different powder materials utilized in CS applications [4]. Reproduced with permission of Elsevier

From Fig. 7, it is observed that materials nature is a significant factor in determining the minimum particle size. For low thermal conductivity materials such as titanium, the minimal particle diameter is much smaller than for a high conductivity material such as copper.

While knowledge of minimum particle size is helpful, the effect of diameter on the onset of shear instability has also been used to estimate the critical velocity. Assadi et al. proposed the following equation to estimate critical velocity by matching a model to experimental results [20],

$$v_{crit} = k_1\sqrt{c_p(T_m - T_p) + 16\frac{\sigma}{\rho_p}\left(\frac{T_m - T_p}{T_m - 293}\right)} \tag{4}$$

where T_m is the melting temperature of the particle, T_p is the impact temperature of the particle, σ is the tensile strength of the particle at 293 K, and k_1 is a particle-size-dependent fitting parameter. The effect of particle size on k_1 is given in the equation,

$$k_1 = 0.64\left(\frac{d_p}{d_p^{ref}}\right)^{-0.18} \tag{5}$$

where d_p^{ref} is a reference particle diameter, a standard diameter used to determine the constants in the equation experimentally. Equations 4 and 5 indicate that an increase in particle size will decrease the critical velocity. This is attributed to the limited onset of shear instability in smaller particles due to dynamic effects such as; high-temperature gradients, higher strain rates and strain-rate hardening, and the higher viscous shear strength in the jetting region [4]. Figure 8 illustrates the critical velocity for copper particles at an impact temperature of 300 K computed using Eqs. 4 and 5 and compared to experimental measurements.

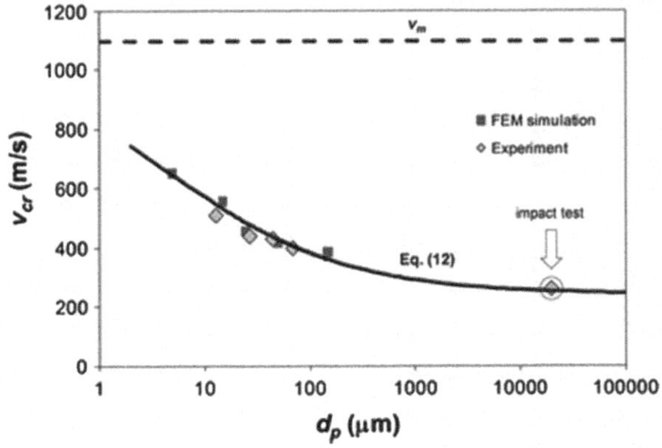

Fig. 8 Critical velocity (experimental and simulation) for copper particles of different diameters at 300 K [20]. Reproduced with permission of Springer Nature

While the onset of adiabatic shear instability being necessary for adhesion is an ongoing discussion topic [44–46], this criteria does match the empirical trend. From this criterion alone, one could conclude that using the largest particles possible would ensure the best results due to the lower critical velocity. However, as previously discussed, particle impact velocity is also a function of particle size. Plotting and comparing the particles critical velocity and the particle impact velocity assuming a specific set of spray conditions shows a range of suitable particle diameters, such as in Fig. 9. It also suggests that there should be an optimized diameter in which the particle velocity quotient is maximized. However, this diameter cannot be readily determined since it is a function of material properties and spray conditions. As such, it can only be determined using extensive experimentation and modelling efforts.

In all the work reported in this section, it has been assumed that the particle impact temperature was the same. However, particle size has a notable effect on particle impact temperature. The low particle temperatures and high particle velocities typical of CS have largely prevented experimental measurement of particle in-flight and impact temperature. Nastic and Jodoin have recorded, for the first time in CS, the temperatures of titanium particles with 150 μm diameter at the exit of the nozzle using a high-speed infrared camera. This new and innovative information was used to validate heat transfer models used in CS to predict particle temperatures [47]. Using these types of models, Schmidt et al. reported that when particles are injected into the high temperature (subsonic) region of the nozzle gas flow, they are heated prior to being cooled down in the supersonic region [48], as shown in Fig. 10. It also leads to the finding that tiny particles are heated up quickly in the hot subsonic section but are quickly cooled down in the supersonic section, revealing that larger particles can maintain their temperature over a longer distance [48, 49].

However, while larger particles can maintain their temperature longer into the cold section of the flow, Fig. 10 shows that particles that are above a specific diameter will have a reduction in temperature. This is due to larger particles requiring more energy to raise their temperature during preheating. This effect can be minimized by elongating the pre-chamber, allowing the particles to be in contact with the hot gas for a longer period of time [48, 49].

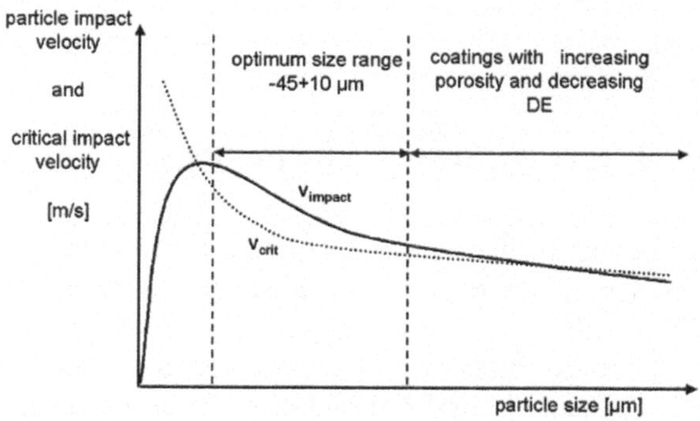

Fig. 9 An example of how critical velocity and particle impact velocity change with particle size [4]. Reproduced with permission of Elsevier

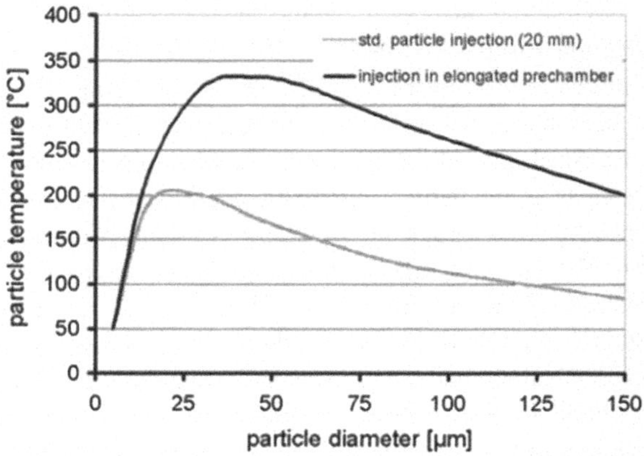

Fig. 10 Effect of particle diameter on particle temperature of copper particles inserted into the high-temperature subsonic region of the CS nozzle [48]. Reproduced with permission of Springer Nature

The particle impact temperature is crucial, as can be observed from the proposed equation for critical velocity (Eq. 4); as particle impact temperature is increased, the critical particle velocity decreases. Since particle size has a direct effect on impact temperature, it indirectly affects critical velocity through a completely different mechanism than previously discussed.

Beyond the immediate effects of particle size discussed in this section, microstructure and oxide contents must also be taken into consideration. As presented in Sect. 2, particles of different sizes are not exact scale replicas of themselves—the particle size has a relationship with microstructure and oxide content. This is often overlooked in studies that compare particle sizes but may explain some of the differences proposed in the models for critical velocity. Some studies even go as far as saying that particle size has no effect on critical velocity assuming the material properties and impact temperature could be kept the same [10, 50].

It should be clear to the reader that choosing an optimal particle size for CS AM is a challenging task. Particle size directly affects the in-flight velocity, the impact velocity, the impact temperature, and critical velocity. It is difficult to distinguish which effects are most significant, especially when you take into account differences in oxide content and microstructure. The relationship is convoluted, and typically, extensive experimentation or modelling is necessary to optimize the process.

4 Effect of Powder Morphology

The effect of the powder morphology on the in-flight particle velocity is consistent across literature; when comparing two particles of equal mass under the same CS conditions, non-spherical particles accelerate to higher velocities than spherical. Comparative studies have demonstrated this for milled [51], angular [19], coral-like [23, 52], dendritic [53, 54], and water atomized powders [55]. This is easily explained looking at the equation for particle acceleration (Eq. 2); where the drag coefficient and projected area are influential parameters for acceleration and are known to be lower for spherical shapes compared to any of the other powder shapes for the same particle mass [53, 55]. If the particle size is large enough to ensure the particle impact velocity is reasonably close to the in-flight velocity (see Sect. 3), then changing the particle shape from spherical to any non-spherical shape should increase the particle impact velocity quotient (η), assuming critical velocity does not change.

Attempts to calculate critical velocity (see Eq. 4) often do not consider powder morphology. This is because critical velocity is often considered an inherent material property. Quantifying the effect of different particle shapes on critical velocity adds substantial complexity. There have been numerous studies directly comparing different particle morphologies through experimental and numerical simulations that do demonstrate significant differences in deposition behaviour. However, these studies do not allow for the decoupling of the sole influence of particle morphology from other effects, such as impact velocity and temperature, microstructure, and chemical composition.

Irregular particles are usually found to travel faster than spherical particles under the same spray conditions; however, this does not always yield a better deposition. Yin et al. [56] showed that during impact, irregular shaped (elongated) particles that do not impact the substrate symmetrically would experience additional torque. While these irregular particles may have the highest value of plastic deformation in the model (PEEQ), this torque was shown to rotate the particle, detaching the bonded areas and deteriorating the coating quality. This can be observed in Fig. 11, where the particles impacting at 45° with the surface resulted in delamination.

This trend of deteriorating quality does not apply to all irregular powders. MacDonald et al. [57] showed that coral-like titanium powders manufactured using the Armstrong method produce components of higher density than would be seen

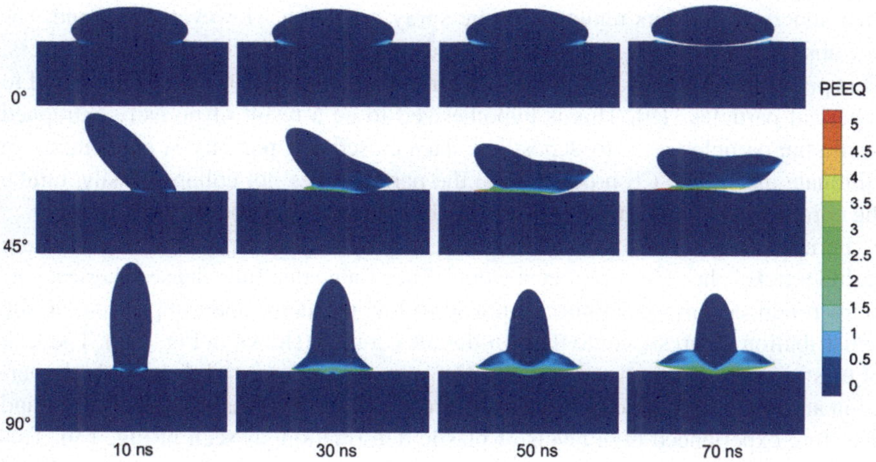

Fig. 11 Modelling results of irregular particles showing delamination in those that did not impact symmetrically [56]. Reproduced with permission of Springer Nature

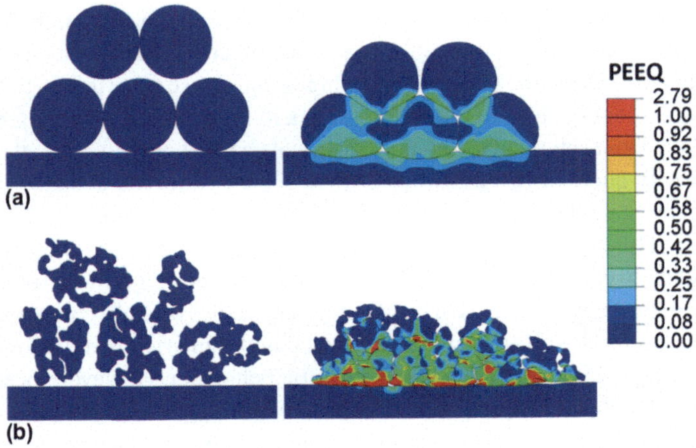

Fig. 12 Comparison of modelling results for the impact of spherical titanium particles to coral-like titanium particles, showing higher plastic deformation and lower porosity for the coral-like powders [23]. Reproduced with permission of Springer Nature

using spherical powders under the same conditions. This was attributed to the unique coral-like structure, allowing the particles to collapse while experiencing increased localized plastic deformation, as shown in Fig. 12. The same trend was observed by Munagala et al. with coral-like Ti-6Al-4V produced with the Armstrong process [52].

A similar trend was observed with dendritic powders. Luo et al. observed that dendritic nickel powders produced dense depositions while spherical powders would not adhere under the same spray conditions [53]. Similar results were seen by Ko et al. [54] with copper powders, where the DE of dendritic powders was higher than spherical powders under the same spray condition. However, this trend was not observed with spongy titanium powder manufactured with the Kroll process. This powder was shown to result in higher porosity depositions when compared to spherical particles [19]. This is hypothesized to be a result of porosity entrapped inside the particles prior to deposition. This closed cell porosity is challenging to eliminate during the CS process since the particle does not collapse easily, unlike the coral-like and dendritic powders, which have large open porosity.

Fernandez et al. [41] examined the influence of powder morphology for chromium carbide-nickel chromium cermet powders. They found that fully dense spherical particles produced through atomization lead to low plasticity upon impact due to the redistribution of stresses onto the non-ductile ceramic (shown in Fig. 13a). The lack of plasticity resulted in no deposition. Dense particles produced through agglomeration and sintering did not have the same redistribution of stresses upon impact and therefore experienced high levels of plastic deformation, as seen in Fig. 13b. This plasticity resulted in thick and dense deposits [41]. Porous particles, also produced through agglomeration and sintering, were found to build up in a similar manner; however, the preexisting porosity resulted in internal cracking and peeling of the depositions [41].

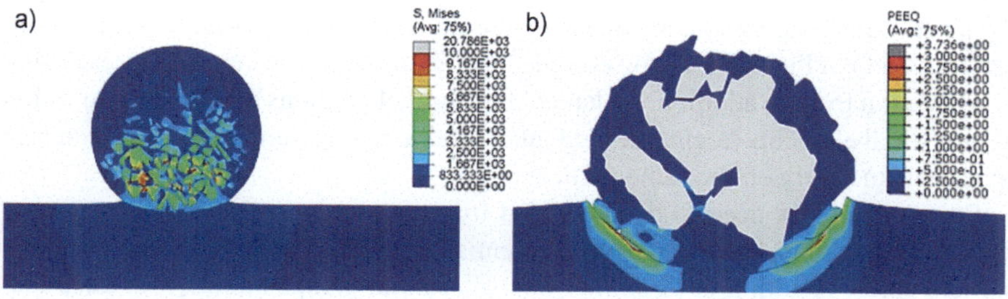

Fig. 13 Modelling results of chromium carbide-nickel chromium cermet powders with different morphologies. **a** Stress field upon the impact of dense spherical particles produced with atomization showing redistribution of stresses and low plastic deformation; **b** plastic equivalent strain of an agglomerated and sintered particle showing a high level of plastic deformation [41]. Reproduced with permission of Springer Nature

5 Effect of Powder Microstructure

In the CS AM process, the effect of powder microstructure and mechanical properties on the deposition process, subsequent build-up, and final material properties is of utmost importance. During atomization, the most popular method of CS powder production, high cooling rates induce a temperature gradient and lead to various microstructures, as shown in Fig. 14 and previously in Fig. 4. As an example, cooling rates of 100–1000 and 1000–10,000 °C/s are observed for plasma and gas atomization processes, respectively [58]. As depicted in Fig. 14a, in addition to microstructural variability between particles in a single powder batch, a single particle can also exhibit distinctive surface structural features and patterns. These distinct surface characteristics are indicative of directional solidification, heterogeneous nucleation

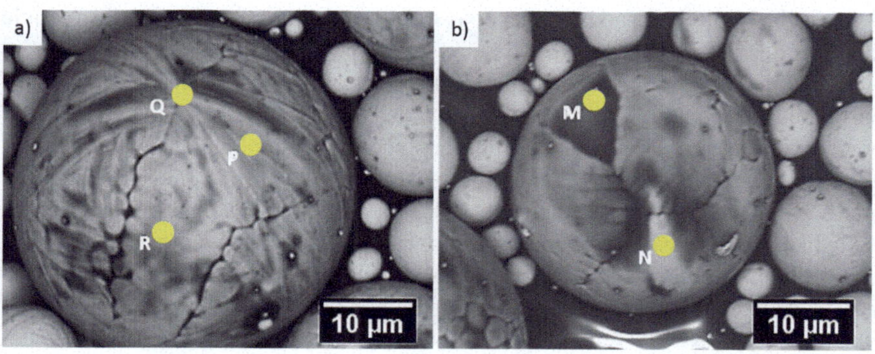

Fig. 14 SEM-BSD images of un-etched Ti-6Al-4V particles showing traits of powder solidification. **a** Particle showing a region of contact with a smaller particle that initiated solidification at region 'Q' of martensitic α' directionality, region 'R' highlighting the cell structure and 'P' zone showing rapid directional solidification along the principal arm. **b** Particle showing distinct surface feature over subsurface microstructure denoted as 'M' and tangential segregation and/or shrinkage within individual grains shown as 'N' [59]. Reproduced with permission of Elsevier

sites, and elemental segregation due to temperature gradient, particle to particle contact (region Q in Fig. 14a) during cooling, and interaction between structures during solidification [59]. In addition, as depicted in Fig. 14b, regions of high concentration of interstitial elements (region M) and intermetallic rich features (region N) can also appear distinctively on the same particle.

Exterior features need to be decoupled from interior bulk particle structures as corresponding solidification rates are potentially different at the surface and in the particle core—resulting in dissimilar microstructural compositions [59]. Such distinctive surface and subsurface microstructural differences have been detected during plasma atomization of Ti-6Al-4V powder [59]. The exterior shell solidifies rapidly reducing the internal thermal gradient and permitting the formation of internal homogeneous grains [58]. Birt et al. have linked the presence of the particle surface skin to a rapidly solidifying shell, which cracks from shrinkage due to thermal stresses upon further cooling [59]. If the surface skin is ultra-thin, the particle exterior is then transparent, providing a view into the internal microstructure, as depicted in region R of Fig. 14a. In addition, Birt et al. mentioned that the surface skin can also potentially delaminate during cooling, as seen from the relief in region M from Fig. 14b, although they claim that no evidence can be provided to support this occurrence. The presence of a delaminating, i.e. poorly adhered, surface shell, however, can affect the bonding process of particles during CS by impeding close subsurface contact between adjacent particles and introducing instabilities that reduce the energy left for particle deformation. In addition, if deposited, the surface skin, which has been demonstrated to be of a different structure than the particle interior, can introduce undesirable characteristics into the final manufactured component.

Analogously, for small particles with a diameter approaching 10 μm, diffusion and preferred growth orientation is limited due to accelerated cooling rates and high undercooling [60]. Consequently, they lack surface skin thermal stress separation due to their larger area-to-volume ratio and smaller overall mass [9], as seen in Fig. 14. As a result, featureless and homogeneous microstructures are present as the particle diameter decreases [59, 61–63].

In addition to size-dependant external skin features, powders can exhibit an internal cellular structure, referred to as Type I, which have been detected in gas atomized powder. This is attributed to rapid solidification and high cooling rates [59, 64], thermal equilibration and partial remelting of solid particles [65], and pre-solidified micro-droplets and dendrite fragments [64]. Type I powder displays a cellular dendritic like structure with localized segregation of alloying elements, as shown by the intercellular spacing in Fig. 15a [61, 66–74]. Hence, the cellular structure has been associated to undercooling processes (a function of powder size) which are achieved when the droplets cool well below the material's solidus temperature before nucleation and crystallization has had time to initiate—leading to thermodynamically unstable dendritic like microstructures [61, 75]. In addition, the cellular dendrites release heat during solidification under a phenomenon called recoalescence [62]. The combined effect of the temperature increase due to recoalescence and temperature decrease due to the cold surrounding gas also leads to this non-equilibrium microstructure. Type II particles have also been detected, which in contrast to

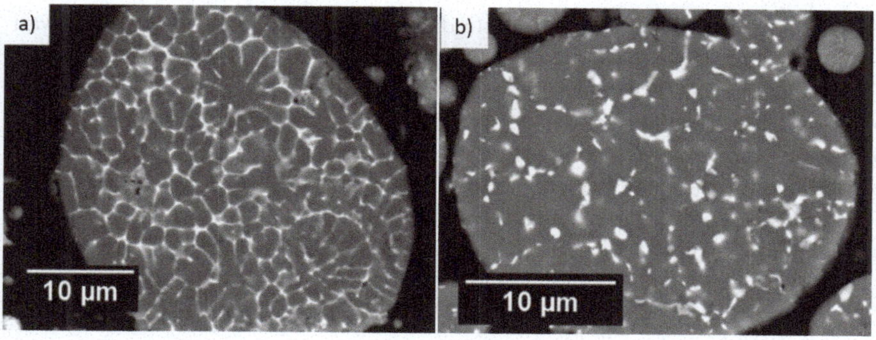

Fig. 15 SEM images of gas-atomized Al7075 particle cross-sections of **a** Type I internal grain structure and solute segregation and **b** Type II integral microstructure with grain boundary precipitates [76]. Reproduced with permission of Elsevier

Type I, exhibit coarser grains and extensive grain boundary precipitation by exposure to slower cooling rates during atomization [62, 76]. Type II powder cross-section is depicted in Fig. 15b.

The microstructural characteristics of particles are influenced by their initial composition, shape, and size and by the induced cooling rate based on the corresponding continuous cooling transformation diagram of the material. Figure 16 depicts the influence of the cooling rate on microscale particle properties for Ti-6Al-4V

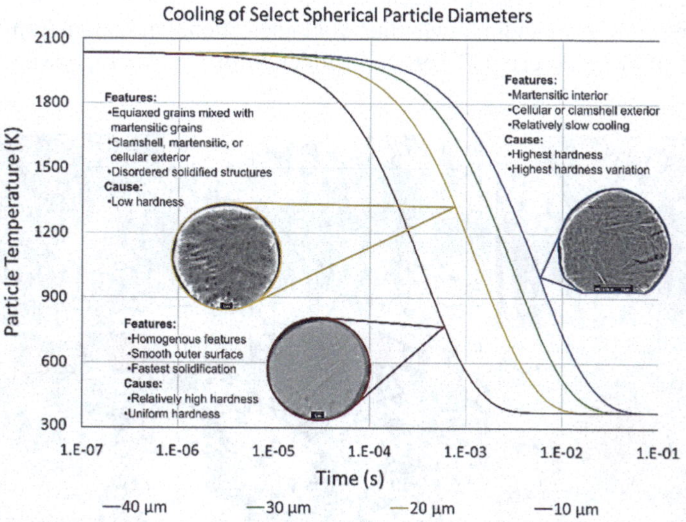

Fig. 16 Connection between resulting microstructure, cooling process and microscale particle properties. The particle reveals the presence of randomly distributed Martensitic α'. The martensitic α' occurs in a basketweave pattern and equiaxed α structures [59]. Reproduced with permission of Elsevier

[9]. As depicted, based on the particle diameter and resulting cooling rate, various microstructural features are observed on the surface and interior of the particle. These features, which include grain size and geometry, skin presence, crystal and phase structures, will have significant influences during CS deposition. The variety of features can result in a composite-like material during CS consolidation, which could produce favourable properties. However, due to the powder characteristics variability, quality control and repeatability are significant problems, which limit proper prediction of the deposit characteristics and generalization of the impact process.

5.1 Microstructural Characteristics

As presented previously, the grain structure of feedstock powder is dependant on multiple factors, which typically lead to the creation of non-uniform microstructures featuring both asymmetrical grain shapes and sizes, as depicted in Fig. 17a and Fig. 17d for Ti-6Al-4V. In addition, due to the high cooling rates during manufacturing, the particles are not fully relaxed and internal stress, and dislocations are present inside the grains. Rokni et al. [76] have shown through TEM observations that their Al7075 as-received powder contains a moderate density of dislocations and dislocation substructures prior to deposition.

Upon impact with a surface, the particles undergo deformation at high strain rate, between 10^3 and 10^9 s^{-1}, which induces the creation of a wide range of deformation microstructures. These features, such as dislocation nucleation, the formation of dislocation cell, shear banding, twinning, void nucleation, phase transformation and generation of point defects [77–79] are a function of the shock pressure reached.

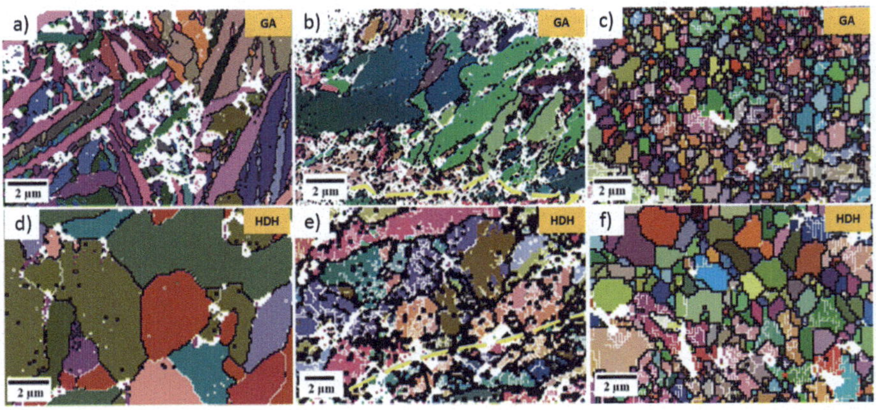

Fig. 17 EBSD Euler maps of **a**, **d** as-received, **b**, **e** as-deposited and **c**, **f** as-annealed Ti-6Al-4V powder manufactured through gas atomization (GA) and hydride de-hydride (HDH) process with an average diameter size of 33 ± 12 µm and 45 ± 15 µm respectively [81]. White regions represent unindexed features of the Euler pattern. Reproduced with permission of Springer Nature

The threshold shock pressure needed to induce secondary deformation features such as twinning and phase transformation is very high, i.e. deformation twinning in Cu sets at 17GPa [80]. Reported pressures reached in CS particles at critical velocities have been calculated to reach only 1.2 MPa for Al, 3.0 MPa for Cu and 3.9 MPa for Ni [80]. Nevertheless, there is evidence that the deformation process does affect the microstructural state of the deposition, as demonstrated in Fig. 17b, e for Ti-6Al-4V. This resulting microstructure is influenced by the microstructural state of the powder prior to deposition. As shown, the deposited gas atomized powder displays the presence of ultrafine grains (UFG) near contact boundaries which has been attributed, by many authors, to the dynamic recrystallization (DRX) processes [71, 76, 80–82]. On the other hand, the HDH deposited powder coating microstructure is markedly different; having an increased presence of low angle grain boundaries (LAGB) and an absence of clear UFGs.

The increase in DRX in the coating obtained from gas atomized powder has been linked to two significant factors: (1) lower critical strain for DRX formation due to the initial needle-like grain morphology [81, 83] and (2) lower impact velocities of the HDH powder [81]. Additionally, the deposition microstructure after annealing also differs, as shown in Fig. 17c, f. Hence, the initial powder feedstock microstructure affects the static recrystallization and recovery processes during annealing, both driven by the large strain energy present in the coatings. As such, much smaller grains are detected in the gas atomized powder deposits, as only a limited growth of the UFGs has had time to develop.

The formation of UFGs has been associated to the occurrence of continuous dynamic recrystallization (CDRX) [82], conversion of LAGBs to high angle grain boundaries (HAGBs) [84], geometric dynamic recrystallization (GDRX) [74, 83], and rotational dynamic recrystallization (RDX) [76]. Additionally, the presence of high local temperature (due to conversion of deformation work into heat) can lead to static recovery (SRV) and static recrystallization (SRX) in highly strained area. The activation energy required to induce these phenomena, both static and dynamic, depends on the material's stacking fault energy (SFE). The SFE, which is a material's intrinsic property governed by chemical composition, is related to the separation distance between two partial dislocations, called stacking fault, and to the energy associated to the generated dislocation sequence. Murr et al. have determined the following equation to describe the energy, γ_{SF}, related to this separation at equilibrium;

$$\gamma_{SF} = \frac{Gb^2}{2\pi d} \tag{6}$$

where G is the shear modulus, b is the Burgers vector of the partial dislocation and d is the separation distance of the partials.

Table 2 provides the SFE and associated melting temperatures of common powder feedstock material used in CS. Borchers et al. have suggested that the microstructural bonding features of cold sprayed Al, Cu and Ni differ considerably from each other due to their different intrinsic SFEs [80], although they all belong to the same isomechanical group. The high SFE feedstock Al powder presents low dislocation density

Table 2 Stacking fault free energies of commonly CS materials

Metal	Aluminum	Copper	Nickel	Steel	Titanium
SFE (mJ/m^2), γ	~166	~78	~128	~30	~15
T_m (K)	933	1356	1726	1643	1941

both prior and after CS deposition—exhibiting SRV rather than DRX or SRX. On the other hand, Ni and Cu, with medium and low SFE respectively, reach high dislocation density which allows DRX to occur along with submicron grain refinement [80].

Additionally, material thermal conductivities can affect the extent of recrystallization [84]. Bae et al. have observed that for materials such as Ti, which are poor thermal conductors, extensive local retention of transient thermal energy can induce substantial static recovery and static recrystallization noted by the presence of grain refinement for about half of the splat [85].

In addition to intrinsic materials properties, studies have shown that material behaviour under high strain rate can be affected by the grain size. Chen et al. have demonstrated that although aluminum is characterised by a relatively high SFE, ~110 to 135 mJ/m^2, twinning has been observed in nano-crystalline aluminum as well as in high purity aluminum [86]. Their study demonstrates that twinning becomes favourable and the preferred deformation mode when the grain size is reduced to the order of 10 nm. Deformation twinning strengthens the material effectively by decreasing the slip barrier spacing, which reduces the linear dimension over which the internal stress concentration can build-up—rising the applied stress required to initiate further plastic flow [42, 87, 88]. On the other hand, both Meyers et al. and Schmidt et al. have experimentally studied the grain size effect of Cu under high strain rate processes, which has shown to have a significant effect only when grain size was above 100 μm through profuse twinning [89, 90]. The proposed rationale is that plastic flow localization occurs readily in coarse grains while homogenous plastic deformation prevails in small Cu grain size specimen. The flow stress was shown to follow the Hall-Petch relationship, exhibiting significant hardening with grain size reduction [91]. Hence, the grain size has an influence on the twinning propensity of metals based on their SFE and dislocation slip processes and interactions [92].

Similarly, to effect of the grain size on deformation processes, alloying element addition has also been demonstrated to affect the SFE and consequently the material response to high strain rate loading processes. This was demonstrated by Millett et al. and their study of pure Ni and Nickel alloys [93, 94]. The pure Ni showed a nearly constant hardening rate with applied shock, while for the Ni-Co alloy the hardening process rapidly increased, which was associated to the shift from purely dislocation based mechanism to one which accommodates twinning.

Finally, in addition to deformation processes, one of the main features of the CS process is the strain localization during particle deformation. The quasi-adiabatic shear instability phenomenon has been related to the particle/substrate and particle/particle bonding and structure formation [50, 95–98]. The strain-to-failure of metals under high strain rate processes is related to the tendency of the strain to localize. The strain localization, also known as shear banding, is influenced by strain-hardening rate, temperature, strain-rate sensitivity of the flow stress, and presence of shear band initiation sites [98, 99]. Consequently, microstructural effects on deformation stability are evaluated based on their influence on strain localization. It has been shown that the alloy's initial content and microstructure, presence of dual-phase, and the precipitate content all influence and promote the strain localization during high strain rate deformation [90]. In the region of localization, the effect of grain size on mechanical properties is very complicated as the grain boundaries can provide strengthening, i.e. obstacles to dislocation slip, or positively contribute to material deformation leading to softening. The resulting effect of these two phenomena is influenced by the local temperature. The plastic deformation is controlled by the evolution of dislocation density, and as a result, the grain misorientation and lattice rotation play major roles in the deformation process of particles. During CS impact, localized intense lattice rotation and thermal softening have been found to be more pronounced for high angle grain misorientation. Guha et al. have demonstrated that material with high angle misorientation, subjected to high strain rate loading, show higher strength than material with low angle misoriented grains due to the greater difficulty of dislocation movement in slip systems across grain boundaries [100]. In addition, high angle misorientation can lead to high thermal softening and the generation of localized plastic strain zones [100]. Figure 18 provides an example of

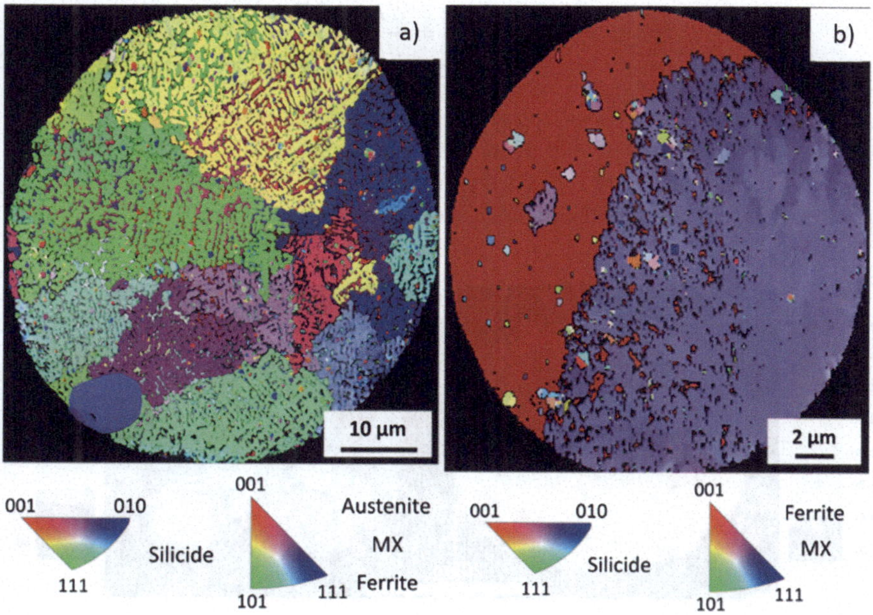

Fig. 18 EBSD derived IPFX orientation with crystallographic axes in the relevant stereographic triangles for a particle with a diameter **a** between 63 and 75 μm and **b** <20 μm [101]. Creative Commons Attribution License (CC BY)

atomized particles with different diameters and consequently different grain structure and orientation.

Hence, at high impact velocity, the particle experiences strain rate-dependent hardening and thermal softening processes, which dictate the final particle deformation. The strain rate developed during the impact of spherical particles has been approximated by Hutchings and related to particle properties [102];

$$\dot{\varepsilon} \approx 0.2\left(\frac{V_i^{1/2} H^{1/4}}{\rho^{1/4} r_p}\right) \tag{7}$$

where H is the indentation hardness, ρ is the particle density, V_i is the impact velocity and r_p is the particle radius. Hence, a 25 μm diameter copper and aluminium particles travelling at 100 m/s will experience strain rates in the order of 2.1×10^6 and $3.4 \times 10^6\ \mathrm{s}^{-1}$, respectively, which is in accordance to values obtained in FEM analysis of CS impact processes [102, 103]. It is clear from the previous equation that the particle properties, even if local, influence the particle deformation and subsequent deposition. A limited number of studies have measured the powder local mechanical properties using a nano-indentation mapping technique, as shown in Fig. 19. The material hardness, H, and reduced elastic modulus, E_r, are obtained from the unloading curves using the Oliver and Pharr method and following relations [104, 105];

$$H = P/A_c \quad \text{and} \quad S = \frac{dP}{dh} = \frac{2E_r\sqrt{A_c}}{\sqrt{\pi}} \tag{8}$$

where P is the maximum load applied, A_c is the contact area, $\frac{dP}{dh}$ is the unloading segment slope also referred to as the material stiffness, S, and E_r is the reduced elastic modulus. As depicted in Fig. 19, the Ti feedstock powder mechanical properties are not homogeneous, and the high hardness region is associated with the presence

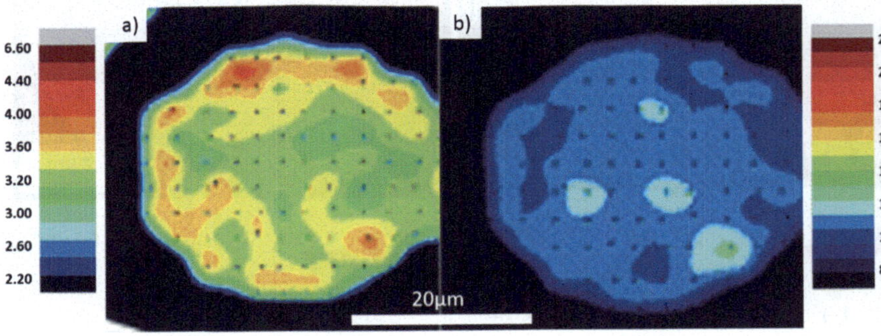

Fig. 19 **a** Nanohardness mapping (GPa) and **b** reduced modulus mapping through nanoindentation test (GPa) [107]. Reproduced with permission of Springer

of martensitic microstructure resulting from the surface tension during powder production. On the other hand, the reduced elastic modulus is fairly homogenous as the effect of material texture is less significant in a polycrystalline material [106].

Multiple studies have shown that the feedstock powder hardness has a direct impact on its deposition efficiency [51]. A high hardness limits the plastic deformation and flattening ratio of particles during impact. Jodoin et al. have shown that a substantial increase of particle impact velocity is necessary in order to compensate for the high mechanical strength of powders to promote deformation and eventual deposition [51].

5.2 *Amorphous Structure*

Amorphous metals, also known as metallic glasses (MGs) or glassy metals, lack long-range atomic order configuration and grain boundaries, which are common features of crystalline materials. MGs provide a unique combination of chemical and physical properties resulting in high hardness, high yield strength, high specific strength, excellent anti-wear characteristics, outstanding corrosion resistance, and good magnetic behaviour [96]. The formation of amorphous structures depends on the alloy composition and atomization conditions and is evaluated based on their glass-forming ability (GFA), and is possible due to the exceptionally high cooling rates during powder production. Inoue et al. have proposed three empirical rules for the manufacturing of amorphous alloys characterized by a wide supercooled liquid region and large GFA, i.e. (1) multicomponent system consisting of at least three alloying elements, (2) significantly different atomic size ratios above 13% and (3) optimal negative heats of mixing between alloying elements [96]. In addition, generating a multicomponent alloy near its eutectic point, minimizes the temperature difference between the glass transition and the liquidus, which stabilizes the liquid at low temperatures [96]. Hence, due to the high temperature involved in conventional manufacturing and powder metallurgy processes, the amorphous structure is hard to retain in the fabrication of bulk materials, which makes CS a promising alternative [108–111]. Multiple studies have shown that the amorphous microstructure is retained after CS deposition, i.e. Fe-based alloy [112, 113], $Cu_{50}Zr_{50}$ [108, 114], CuNiTiZr [110], NiTiZrSiSn [115] and FeSiCrBC [96]. Due to the particular thermal behaviour and mechanical deformation mechanisms involved during the impact of MG particles, the concept of critical velocity derived from dislocation based model and used for crystalline metals is inappropriate to reveal and understand the deposition mechanisms for MGs. Consequently, Concustell et al. [111] proposed a model based on the impact of liquid droplets and correlated experimental data with the Reynolds (Re) number, balancing inertia with viscous effects, and the Weber (We) number, balancing inertia with surface energy or capillary effects, of the MG particle before impact, as follows;

$$Re = \frac{\rho v_0 d}{\eta(T)}; \quad We = \frac{\rho v_0^2 d}{\sigma} \tag{9}$$

where ρ, v_0, d and η represent the density of the liquid, impacting velocity, particle diameter and viscosity of the liquid, respectively. The viscosity of metallic glass-forming liquids is given by the Vogel-Fulcher-Tamann (VFT) equation as follow;

$$\eta(T) = \eta_0 \exp\left(\frac{D \cdot T_0}{T - T_0}\right) \tag{10}$$

where T_0 is the VFT temperature and D is the fragility parameter. Multiple studies have shown that at low Re (low impact temperature and large viscosity), the low deformation features the presence of shear bands and phase transformations [109–111], as depicted in Fig. 20b. Under those conditions, similar to non-Newtonian liquids undergoing yielding, MGs are difficult to consolidate. However, above the glass transition, i.e. high Re and We, the deposition efficiency increases due to the homogeneous material flow during deformation activated by low material viscosity

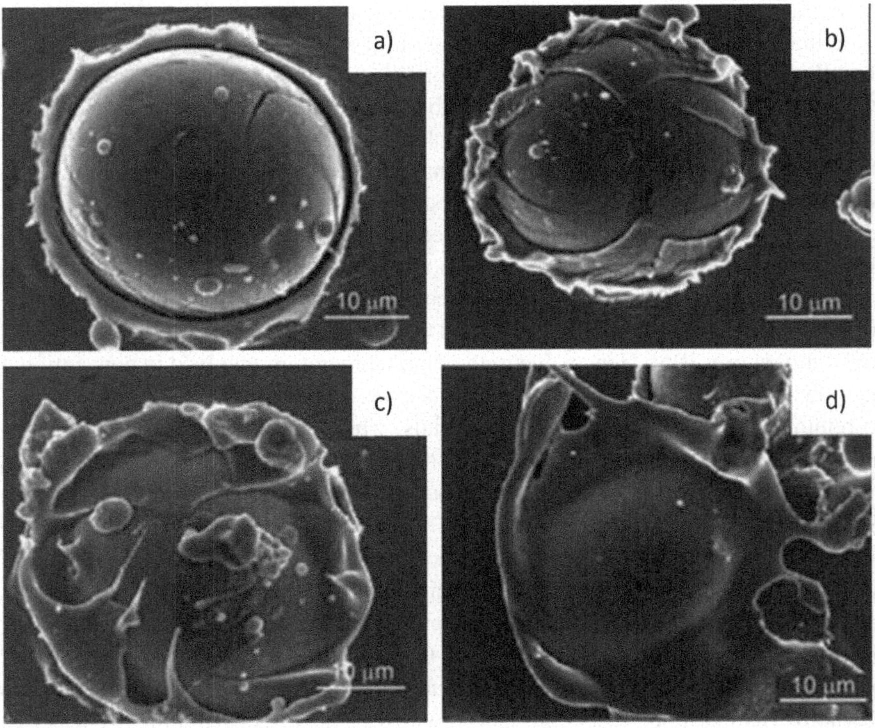

Fig. 20 Different deformation regime of Fe-based MG single particle impacts. **a** Elastic deformation, **b** shear banding formation, **c** mixed shear banding and homogeneous flow and **d** homogeneous flow [96]. Reproduced with permission of Springer

and inertial forces, as depicted in Fig. 20d. Under homogeneous flow and high strain rate deformation, shear-thinning processes can appear followed by Newtonian flow due to the nonstationary conditions under which the MGs deform upon impact, both of which increase the deposition efficiency and adhesion strength as they promote lateral viscous flow [108, 109, 111, 116].

For proper deposition, in addition to homogeneous deformation, crystallization phenomena need to be hindered in order to avoid restriction of viscous flow by atomic planes. It has been shown that high CS gas temperatures can induce and promote crystallization of particles during their flight, which reduces the resulting deposition efficiency [111]. Hence, high energy conditions and kinetics of crystallization need to be balanced appropriately to ensure proper MG deposition. Aiming to define a deposition window for MG materials, Concustell et al. have developed the following equation for critical velocity [111];

$$v_{crit} = \frac{0.004\eta(T)}{\rho D_p} \tag{11}$$

Therefore, unlike for crystalline materials, increasing the impact velocity of MGs does not necessarily lead to higher DE. Metallic glasses do deform homogeneously between their glass transition and crystallization temperatures, but their deformation process is also dependent on strain rate. Studies have shown that at high strain rate and temperature, inhomogeneous flow can occur and lead to poor deposition [117, 118]. Hence, due to the high strain rates experienced during the CS impact, the MG particles need to impact at temperatures above 1.15 to 1.3 T_g, approximately [108, 111].

5.3 *Multiphase*

The high cooling rates observed during atomization produces microstructures in powders that are fundamentally different from those found in bulk materials. Non-equilibrium microstructure deviates in relative phase fractions, solute contents, or phase morphologies. Borchers et al. have noticed the formation of metastable BCC structures in fine 316L powders, which has been associated with fast cooling rates from gas atomization [119]. The formation of metastable ferrite is favoured by its low surface energy in the surrounding liquid. After impact during CS deposition, it has been shown that the metastable BCC structure can undergo a phase transition to high-density equilibrium FCC austenite structure [119]. This diffusionless solid-state transformation is possible as propagating pressure waves of over 1GPa are commonly observed in impact models of the CS process, which is sufficient to overcome the activation energy of such transformation [103, 119]. Farinha et al. have reported that water atomized SS316 powders with a diameter close to 9 μm were almost entirely austenitic while powder with an average of 5 μm diameter was half ferritic [120]. Brewer et al. have studied the variability in austenitic stainless steel, SS304 and

SS316, feedstock powder effect on CS deposition. In their study, only the SS304 powder was purely austenite in phase while the three SS316 powders had ferrite content varying from 12 to 42%. The measured deposition efficiency was of 11.8% for the SS304 powder while it varied between 12.3 and 41.8% amongst the three tested SS316 powders. The variation in deposition efficiency was associated to the feedstock powder ferrite content, microhardness, crystallite size, and average grain orientation spread [121], which highlights the importance to report and analyze the powder features. In addition, as shown in Fig. 21, due to the irregularities in ferrite content of the feedstock SS316 tested powders, a significant difference is observed in the deposits ferrite fraction and consequently in resulting properties. Limited ferrite content is detected in the coating illustrated in Fig. 21b. In addition, the deposited coating displayed in Fig. 21a shows individual particles with a mixture of austenite and ferrite phases, while the coating in Fig. 21c also exhibits fully ferritic particles.

Similarly, in low-Co iron-based alloys such as Tristell 5183, produced by inert gas atomization, with a nominal composition of Fe-21%Cr-10%Ni-7.5%Nb-5%Si-2%C (in weight %), the high process cooling rates greatly influence the resulting particle microstructure, as shown in Fig. 22. This complex alloy system provides the perfect example of particle size effect on solidification phases and consequently demonstrates the importance of powder proper characterization prior to consolidation. As depicted in Fig. 22a, particles in the range of 63–75 μm diameter show a largely austenitic (FCC) dendritic iron-based matrix, a silicide isostructural interdendritic phase ($Fe_5Ni_3Si_2$) and a micron-sized primary carbonitride. In Fig. 22b, an entirely different phase composition is detected for particles with a diameter less

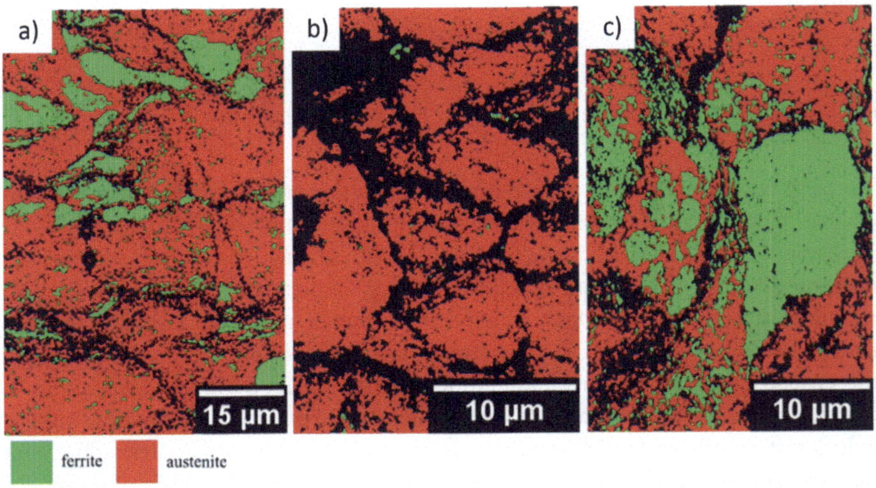

Fig. 21 Austenite (FCC-red) and ferrite (BCC-green) mapping after deposition of three SS316 powders acquired from different manufacturers [121]. Initial corresponding average feedstock powder size prior to deposition have been measured to **a** 45.6 μm, **b** 17.2 μm and **c** 61.3 μm. Black regions refer to features that have not been indexed with reliability using EBSD. Reproduced with permission of Elsevier

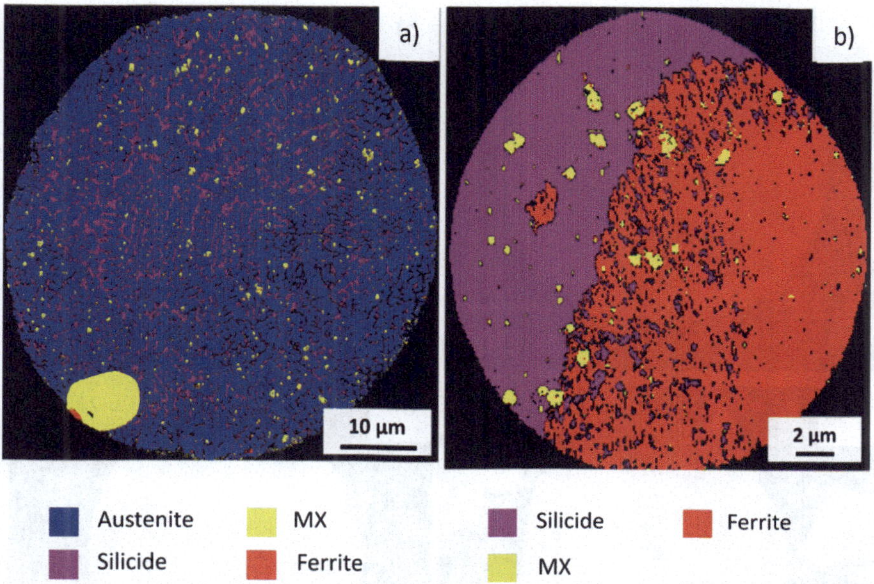

Fig. 22 EBSD derived phase orientation maps of powder cross-section consistent with **a** 63–75 μm diameter and **b** <20 μm diameter [101]. Creative Commons Attribution License (CC BY)

than 20 μm. A ferritic (α-Fe) phase conforming to an irregular dendritic morphology occupies half of the particle while the other half is taken by the silicide phase. In addition, micron-sized carbonitride particles are dispersed throughout the particle. The increased undercooling prior to nucleation induces the formation of the metastable phases in small diameter particles.

In addition to inconsistent phase distribution and content, the effect of the composition on microstructure and elemental distribution has a major role in the final properties of the fabricated material. Liu et al. have studied a series of Al-Cu binary alloy powders, from 2 to 5 wt% Cu, which presented a cellular structure with θ (Al_2Cu) phase at the cell boundaries, as shown in Fig. 23a, b [122]. With the increase of Cu content, the solidification structure becomes slightly dendritic, as seen in Fig. 23b. Due to the non-equilibrium gas atomization process, the resultant phase content differed significantly from common lever rule and basic Gulliver Scheil model, which describe the composition, amount of specific phases, and solute redistribution. The measured Cu content in the α aluminum was also above the equilibrium solubility limit of 0.77 wt%, which indicates that any post-annealing processes would initiate Cu-rich phase precipitation. As particle strength increases with θ content, the powder deformation upon impact during CS process would decrease.

A study conducted by Coddet et al. showed that the proportion of Ag-rich phases in Cu-23.7Ag was in good agreement with values predicted by Cu-Ag eutectic equilibrium phase diagram [123], despite the high cooling rates of the atomization processes. As depicted in Fig. 23c, d, the powders exhibit a dendritic solidification structure

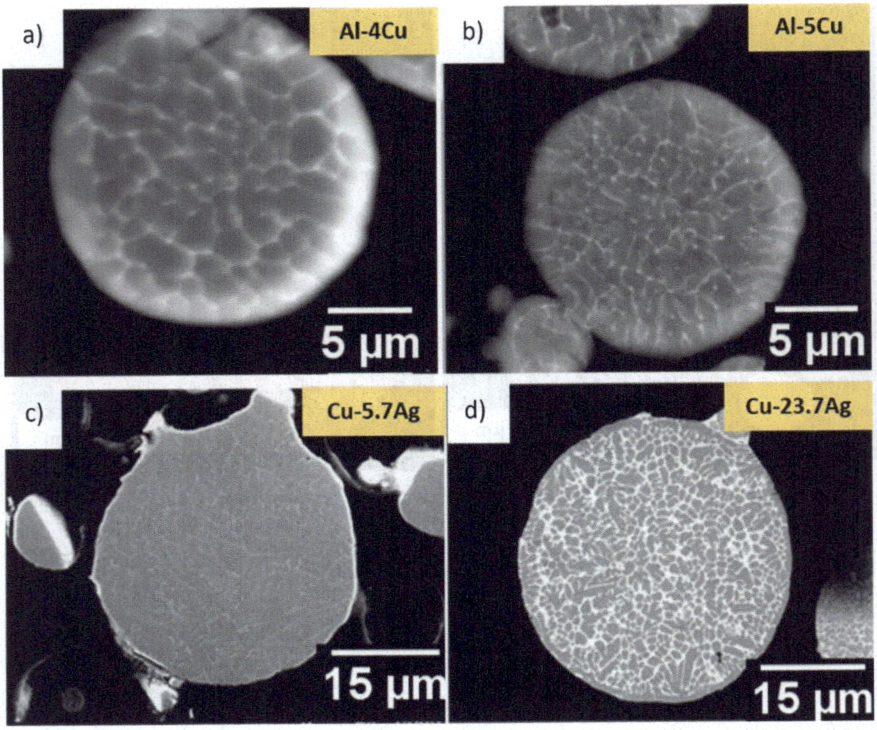

Fig. 23 Cross-section images of as-atomized powders. **a** cellular structure Al-4Cu, **b** cellular and dendritic structure Al-5Cu, **c** Cu-5.7Ag and **d** Cu-23.7Ag [122, 123]. Reproduced with permission of Elsevier

similar to the Al-Cu binary alloy with increasing Cu content, shown in Fig. 23b. In addition to the powder characterization, Coddet et al. have demonstrated that a post-heat treatment at 400 °C of the CS deposits can generate sufficient thermal energy, under the driving force of the reduction of the interfacial energy, to reorganize the Ag-rich phases homogeneously through the deposits. Temperature and duration of the heat treatment need to be correctly selected to avoid the depletion in Ag of the Cu solid solution and subsequent decrease in mechanical properties [123]. The heat treatment procedure does not affect the mechanical properties obtained from the second phase precipitation but does influence the additional properties obtained from the cold working process occurring during CS deposition. In fact, Coddet et al. have shown that the cold work effect reduction during heat treatment is more pronounced when the Ag content is increased [123].

5.4 Heat Treatment

In addition, to post-heat treatments of CS deposits to suit the mechanical properties to the specific application, a solution heat treatment of the powder prior to spraying can be considered. In some cases, a post-heat treatment can be affected by the microstructural state of the coating and result in unforeseen material properties, which would explain the use of powder pre-treatment. As an example, Rokni et al. have stated that the grain growth during post-spray annealing is limited and irregular due to the presence of grain boundary solute segregation in Al7075 alloy CS microstructure [66]. High-temperature powder heat treatment could eliminate the alloying element segregation and/or work hardening of the deposited CS material, but it is sometimes considered undesirable in multi-material applications [61]. As shown earlier in this section, solute atoms tend to segregate along cell boundaries during the atomization process, which decreases the quantity of solute available for strengthening precipitates in precipitation-strengthened alloys. The segregated cell boundaries form a brittle intermetallic network affecting the overall mechanical properties in a variety of materials, such as Fe-based alloys [124] and Al alloys [125]. In addition, the inconsistent intermetallic network affects the powder formability and any post-deposition ageing heat treatment [126]. Hence, a solution heat treatment of the powder can be performed to reduce solute segregation, increase ductility and offer the potential of a post-deposition heat treatment via controlled constituent rearrangement. For the heat treatment to be effective multiple factors need to be considered, such as (1) gas used based on particle flammability, (2) solutionizing temperature to avoid sintering and (3) cooling rate to avoid precipitation and oxidation [127].

As shown in Fig. 24, Story et al. have removed the cellular grain network in Al7075 powder through heat treatment using a novel furnace designed to eliminate sintering of powders. The homogenisation of the powder, even if some precipitates

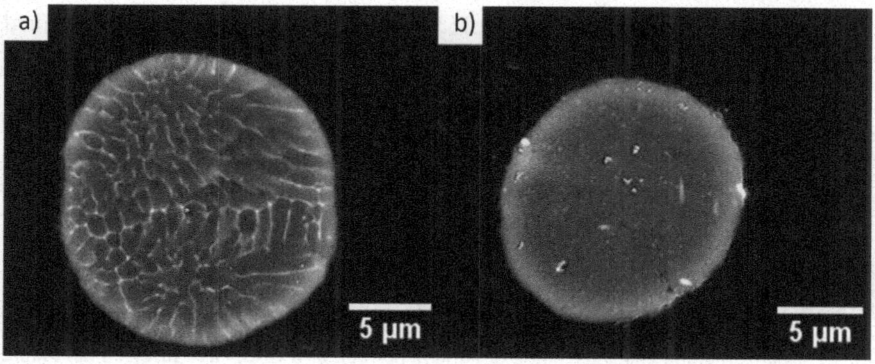

Fig. 24 BSD images of Al7075 **a** as-received powder and **b** after heat treatment, a homogeneous microstructure is revealed with the presence of coarse needle-shaped precipitates. Bright phase segregated along cell boundaries represents zinc-rich intermetallics [127]. Reproduced with permission of Springer

remain after cooling, have shown an approximate 10% increase in DE for all three aluminum alloys, i.e. Al2024, Al6061 and Al7075, tested in their study [127]. Similar results have been obtained by Sabard et al., which have shown that the solution heat-treatment of Al7075 has decreased the powder hardness and consequently increased particle deformation, i.e. jetting, during impact generating an improved deposition efficiency and bonding strength [61]. Their study showed that under identical spraying parameters, the heat-treated homogenized Al7075 coating reaches a thickness of 300 μm while only 40 μm is reached for un-treated powder [61].

In addition to eliminating the cellular network, the solution heat treatment also homogenizes the grain size distribution and decreases the dislocations/lattice defects. The presence of LAGBs, composed of an array of dislocations substructure, in the feedstock powder, has been associated with potential residual stress in the droplet after atomisation [82]. The heat treatment process reduces the quantity of LAGBs during recrystallization, as shown in Fig. 25b, and replaces the dendritic asymmetrical microstructure and micron-sized grains by fully relaxed equiaxed grains [61, 95]. During impact, the decreased dislocations/lattice defects facilitate the accumulation of localized dislocations phenomena at the impact zone during deformation leading to enhanced RDX.

It is, however, important to note that the microstructure that evolves during recrystallization annealing of powders is related to the SFE, as discussed in Sect. 5a. In low stacking fault energy FCC materials, such as copper, α-brass, austenitic iron and nickel, annealing twins can also develop, as shown in Fig. 26b [128, 129].

Annealing twin generation during recrystallization is mainly dependent on the local lattice orientation and dislocation density distribution resulting from packing sequence defects in the original grains, such as stacking faults [128]. Resulting grain growth is dependent on twinning processes, which leads to varying microstructures after heat treatment, as shown in Fig. 26. A study conducted by Li et al. has demonstrated that although the heat-treated copper particles include annealed twins structures, the deposited coating microstructure showed extensive particle deformation, metal jetting, and particle interlocking [130]. Similarly, Ning et al. have confirmed

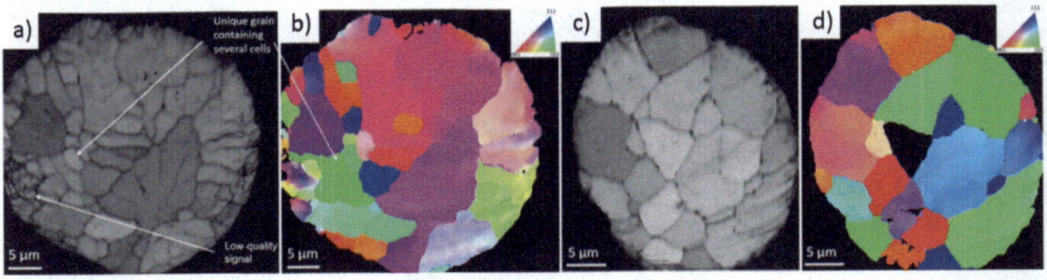

Fig. 25 Microstructural characterisation of Al6061 powder. **a, b** EBSD and Inverse Pole Figure showing irregular grain size distribution and shape of as-received powder. **c, d** Solution heat treated resulting microstructure based on EBSD and Inverse Pole Figure analysis showing large equiaxed grains and homogenised overall grain size distribution. Creative Commons Attribution License (CC BY)

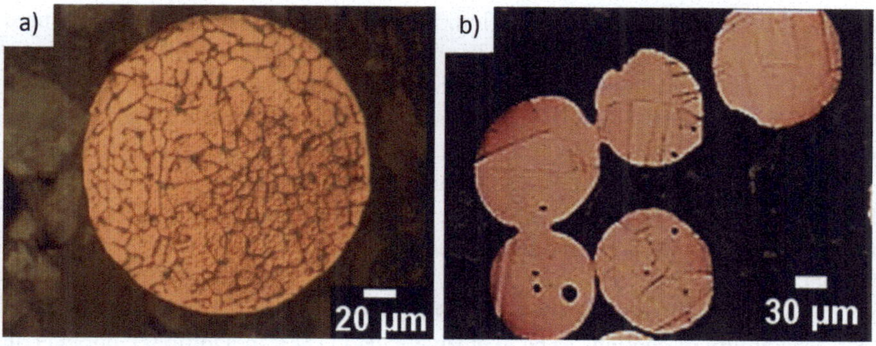

Fig. 26 Cross-section of **a** as received gas atomized Cu particle and **b** annealed Cu particle. Annealing twins observed in the annealed powder [130]. Reproduced with permission of Elsevier

that vacuum annealing at 500 °C of copper particles did decrease the critical velocity and increased deposition efficiency by reducing the microhardness [131].

In addition to solutionizing, other powder heat treatment preprocessing can lead to improved deposition. Rokni et al. have applied a degassing process at 400 °C in nitrogen for 6 h on gas atomized Al5056 feedstock powder. The preprocessing was followed by cooling to room temperature under continuous nitrogen flow, which transported away gases and moisture. Their study revealed that powder preprocessing led to the homogeneous distribution of Mg solute, softer powders, increased particle deformation upon impact, improved cohesion, and enhanced tensile strength and ductility of the deposit [120].

Powder heat treatment prior to deposition eliminates microstructural and elemental inconsistencies that reduce the coating reproducibility and homogeneity. Although the mechanical properties differ in deposits with and without heat treating of the feedstock powder, various post-heat treatment can be applied to the CS material, i.e. low-temperature ageing for the development of excellent dispersion strengthening precipitates to adjust the final general mechanical properties [61]. Additionally, it is important to note that post-deposition annealing typically results in lower strength but higher ductility due to the decrease and annihilation of dislocation density generated by the cold work effect of the CS process [132, 133].

6 Effect of Oxygen

6.1 Oxygen Content

The mechanical properties of materials strongly depend on the presence of interstitial elements such as oxygen, nitrogen or carbon. A loss in ductility is caused by oxide dispersion and interstitial hardening, which reduce the extent of deformation during the impact of the particle in CS deposition. The oxygen content of powder is measured using the inert gas fusion method, which provides the total content of contamination, i.e. surface and solid solution content. From the measurements, the surface oxide can be deduced once the oxide thickness is measured using high-resolution imaging techniques.

Powder with large interstitial residual oxygen content will deposit with a reduced deposition efficiency from a lack of particle ductility. Subsequently, the particles upon impact are prone to cracking and failure due to the combined effect of residual stress, thermal shock, thermal expansion and oxygen embrittlement. In addition, increased oxygen content will result in lower deposit ductility even after heat treatment and decreased ability of coatings to be explosively clad [134]. In addition, Conrad et al. have shown that high oxygen and nitrogen content in solution leads to a higher rate of strain hardening in titanium [135]. Oxygen is the most common contaminant in Ti and Ti alloy powders due to its high solubility, up to 34 at.% in alpha titanium, and affinity for titanium material. In titanium, oxygen occupies octahedral sites, which increases lattice parameters and induces lattice strain. The solid solution oxygen interacts with both hydrostatic and shear stress fields of dislocation, which hinders plastic deformation motion, increases powder hardness and decreases its ductility, which as a result influences the material deposition process [136, 137].

Similarly, Barnett et al. noted that reduced interstitial oxygen content in tantalum, tungsten and their alloys enables to increase powder ductility, which promotes plastic flow during impact and larger material consolidation [134]. In addition to mechanical properties, the solid solution oxygen content can substantially affect the electrical conductivity of copper by scattering sites for electrons on atomic scales [120].

The oxygen content in particles is commonly known to be detrimental in many ways; however, the presence of second-phase metal-oxide inclusion can also have a profound effect on the particle microstructural evolution under high strain and high strain rate deformation processes. In dynamic recrystallization, the size of the hard second-phase particles directly affects the grain refinement rate by inhibiting recovery, grain boundary migration and development of HAGBs [138, 139]. As shown in Fig. 27, the presence of second-phase oxygen inclusions initiates the particle-simulated nucleation of recrystallization (PSN), and as a result, local lattice rotation is created. This is demonstrated by a larger refined zone in the two-phase splat in comparison to the single-phase splat. Whether the second phase particles can generate sites of recrystallization nucleation depends on the dislocation structure around them. Brown et al. have proposed a dislocation plasticity model, which states that the transition from laminar to rotational motion surrounding a particle of diameter, d, occurs if the following dislocation density is reached [140];

$$\rho = \frac{b^2}{d^4}\left(\frac{\sqrt{2\alpha\mu}}{\sigma_f}\right)^2 \tag{12}$$

where b is the Burgers vector, α is a constant of ~0.5, μ is the shear modulus, σ_f is the friction stress in the matrix. Zhang et al. have used this correlation in their study and

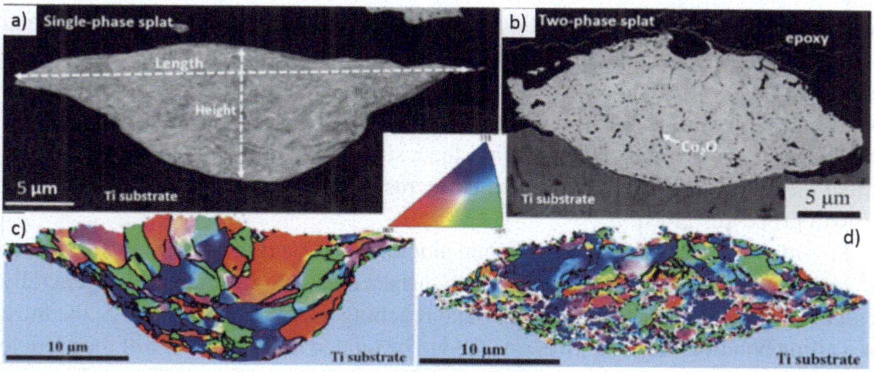

Fig. 27 BSE images after CS impact process deformation of **a** pure Cu and **b** Cu-0.21wt.%O on Ti substrate. EBSD mapping of **c** single phase particle and **d** two-phase particle, where high angle boundaries are represented by black lines (>15°) [141]. Reproduced with permission of Springer

for the case shown in Fig. 27, this dislocation density has been calculated to be $5.4 \times 10^{13}\ \mathrm{m}^{-2}$, which is easily reached at high strain rate deformation processes resulting from CS particle impact [120]. Hence, submicron second-phase oxygen-containing particles can be used to increase the recrystallization process and creation of ultrafine grains in CS deposition processes.

In addition to the influence that oxygen has on microstructural evolution during the high strain rate deformation occurring at the impact of particles in CS, it also has an effect on crystallization kinetics of MG particles. Oxygen is known to destabilize the amorphous structure of MG particles and affect the GFA by promoting crystallization [96]. Due to the high oxygen affiliation energy of common alloying elements in MG materials, such as Ti and Zr, based on the Ellingham diagram, sublayer crystallization occurs due to inward diffusion of oxygen and elemental component oxidation [110]. Oxygen content dramatically increases the necessary cooling rate for glass formation, i.e. contamination of 0.5% increase the cooling rate by two orders of magnitude [142], which consequently renders the production of MG particles difficult and eventual use in CS challenging.

6.2 *Oxide Shells*

Oxide layer adhesion to the metallic particle is associated with multiple interfacial features such as electrostatic interactions from space charge development, chemical bonding, van der Waals forces and mechanical interactions from non-planar boundaries [143]. The adhesion relates to the interfacial energy, γ_i, through;

$$W_{ad} = \gamma_i - \gamma_m - \gamma_{ox} \tag{13}$$

where W_{ad} is the adhesion work, γ_m and γ_{ox} are the surface energies of the metal and oxide phases separated, respectively [143]. The oxide/metal interface can be robust on an atomic scale based on the ionization transfer process allowing chemical bond and electron orbital linking of the two phases across the interface. Growth mechanisms, evolving metal/oxide interface, stress development and phase changes affect the adhesion state and complicate the resulting interfacial relations and oxide growth processes [143].

Provided a source of oxygen, the total amount of oxide film formed on the surface of a metallic particle is proportional to its surface area, and its thickness is generally irregular around the particle periphery and reaching a few nm [144]. In CS, the most prevailing bonding mechanism is assumed to be related to the breakage of the particle and substrate surface natural oxide films to allow intimate conformal contact between clean metal surfaces under high local pressure [8, 145–147]. Hassani-Gangaraj et al. have studied the effect of oxide film layer on critical velocity by conducting particle impact experiments with three types of powder differing in their affinity to oxygen; gold on gold, silver on silver and aluminum on aluminum [148]. Their results, shown in Fig. 28a, demonstrate that the critical velocity decreases with the material decrease to oxygen affinity. In addition to demonstrating the effect of oxide film on bonding, their study validates that besides the breakage and ejection of the native oxide film, particles also require material jetting to induce close contact, i.e. the deposition of gold noble metal particles requires a critical velocity of 253 ± 7 m/s.

Li et al. have shown that the increase of oxygen content in Cu, 316L steel and Monel alloy powders increased their critical velocity to almost the same value, i.e. near 600 m/s, despite having significantly different critical velocities at low oxygen content. The authors state that these results suggest that the critical velocity will be dominated by oxide scale thickness once reaching a certain degree of oxidation rather than the simple oxygen or oxide content of the materials [15]. Additional studies

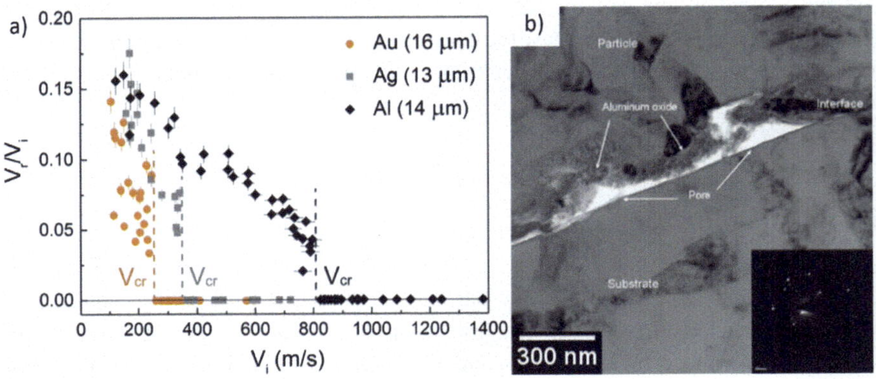

Fig. 28 **a** Coefficient of restitution and critical velocity for gold, silver and aluminum particles sprayed on Au, Ag and Al substrates respectively [148] and **b** HRTEM with SAED insert of the interface between aluminum particle with the oxygen content of 0.045% [150]. Reproduced with permission of Elsevier

have shown the importance to evaluate the oxygen content as it alludes directly to the relative oxide scale thickness. As an example, Hassani-Gangaraj et al. have summarized the work of many on the effect of oxygen on the Cu deposition process and resulting critical velocity. In their report, they use the measured oxygen wt% of Cu feedstock powder and calculate an equivalent oxide thickness layer by assuming that the entire oxygen content is occupied by the film layer. At a measured oxygen content of 0.336 and 0.02% and corresponding calculated oxide layer thickness of 16.7 and 2.8 nm, the obtained Cu critical velocity is 640 m/s and 310 m/s, respectively. This significant decrease in critical velocity by up to 50% cannot be justified using size or temperature effects under their experimental conditions [148].

In addition to affecting the bonding process and critical velocity, improper removal of the oxide film upon impact can leave significant residual oxide fragments in the consolidated part, as depicted in Fig. 28b. Such oxide debris entrained as interparticle boundaries in the deposited material can impede the ability of the consolidated structure to adopt equiaxed grain after annealing. Mccune et al. have found that the residual oxides after CS deposition of direct-reduction copper powder show coalescence of Cu_2O phases into 1 μm particulate, which effectively pin the grain structure during high-temperature annealing [149].

6.3 Powder Storage and Handling

The oxidation behaviour of powders during storage is of great concern due to the evident effect of oxygen on the deposition process in CS applications. A general oxidation reaction that occurs at the surface of a metal (M) can be written as [151];

$$M(s) + O_2(g) = MO_2(s). \tag{14}$$

Thermodynamically, the formation of the oxide film, $MO_2(s)$, is controlled by the standard free energy of formation ΔG°, which can be expressed as;

$$\Delta G^\circ = RT \ln pO_2(g) \tag{15}$$

Here $pO_2(g)$ is the partial pressure of oxygen at a given temperature T and R is the gas constant.

Ellingham/Richardson diagrams provide the information about the required partial pressure of oxygen needed to form an oxide at any temperature [151]. However, these diagrams do not include the oxidation kinetics, which provides the time required for a possible reaction to occur. Moreover, if there are possibilities that more than one type of oxide forms at the surface, the reaction kinetics of the formation of different oxides need to be considered and analysed. The oxidation of metal particles depends on numerous factors such as temperature, surface preparation, oxygen pressure, and metal pre-treatment [151].

Aluminum, a thermodynamically unstable metal with respect to its oxide and hydroxide in air, rapidly oxidizes to form a 2–10 nm stable natural self-healing amorphous layer. Hence, after a few minutes or hours, the oxidation kinetics decreases drastically to very low or negligible values unless cracking of the protective film occurs. This oxidation growth process occurs at room temperature for aluminum while for metals such as copper, iron and barium, the oxide layer, reaching 4–5 nm, forms similarly at temperatures as low as −195 °C [144]. Cabrera and Mott suggested that this oxidation process is possible due to an electric field (contact potential difference) set up through the oxide film between the metal and absorbed oxygen, which enables the metal ions to move through the oxide layer without much help of temperature, i.e. kinetic energy [144]. The oxidation process follows a logarithmic law:

$$1/X = A - B \ln t \tag{16}$$

where X is the thickness, t is the time, A is an integration constant, and B is the rate constant. In addition to the induced field mechanism proposed by Cabrera and Mott, which is valid under uniform passivation film growth, Yang et al. have shown that the passivation of Cu is rather dominated by surface and interface diffusion [152].

At intermediate temperatures, the oxide film will continue to grow only if the film is thin enough to induce a strong electric field and lead to metal ion movement or if the temperature is high enough to induce crystallization, which enables grain boundary diffusion [144, 153]. At sufficiently high temperatures, the oxidation behaviour is different, the rate is comparatively faster, and thicker films are formed. The oxidation at high temperatures conforms to the parabolic law [144];

$$x^n = Kt \tag{17}$$

where x is the oxide thickness, K and n are constant, and t is the time. In this regime, the oxidation mechanism involves the migration of electrons and cations into the reacting zone. For copper, it has been shown using radioactive tracers, that it is the metal that diffuses and not the oxide [152]. The metal ion migration occurs by the formation of vacant cationic sites at the oxide-oxygen interface after their diffusion to the metal-oxide interface. Hence, the reacting element is soluble in the oxide in the form of interstitial metal, vacancies or other point defects, which makes the oxide thickness proportional to the concentration gradient and follows the given parabolic growth law.

Needless to say that the oxidation process involves complex interface phenomena, which renders the actual oxide film growth mechanism hard to precisely identify. Fujita et al. have summarized this complexity by plotting the relationship between temperature, oxide thickness and oxide rate law of copper in the low-temperature regime, as shown in Fig. 29.

The oxidation of copper is very complex as it includes many processes such as electron migration across the metal/oxide interface, external/internal diffusion of O_2, chemical reactions and recrystallization of the pseudomorphic oxide layer [154,

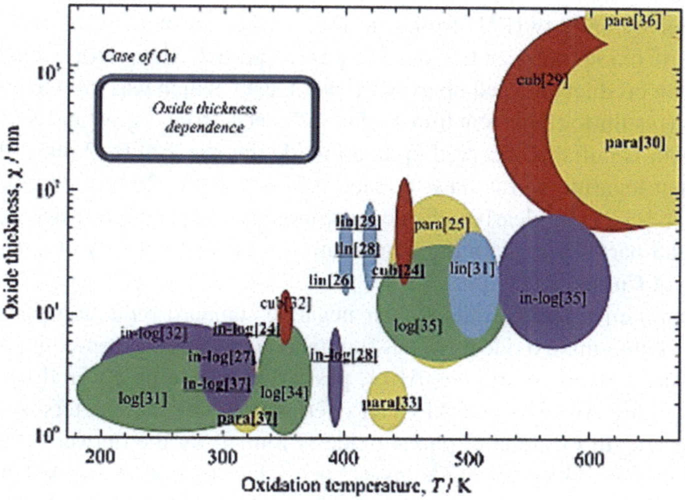

Fig. 29 Copper oxidation process reported by various studies conducted at low temperature. The relationship between, oxide thickness, oxidation temperature and growth rate law are shown. "lin", "para", "cub", "log" and "in-log" indicate linear law, parabolic law, cubic law, logarithmic law and inverse-logarithmic law, respectively [154]. Collected data have been taken from various sources shown in brackets; please refer to the original paper for provided references [154]. Reproduced with permission of Elsevier

155]. Feng et al. have studied the oxidation of copper powder in oxygen and in dry and humid air. Their results agree with observations previously reported in multiple studies conducted with copper wafers and films [156]. As shown in Fig. 30, increasing

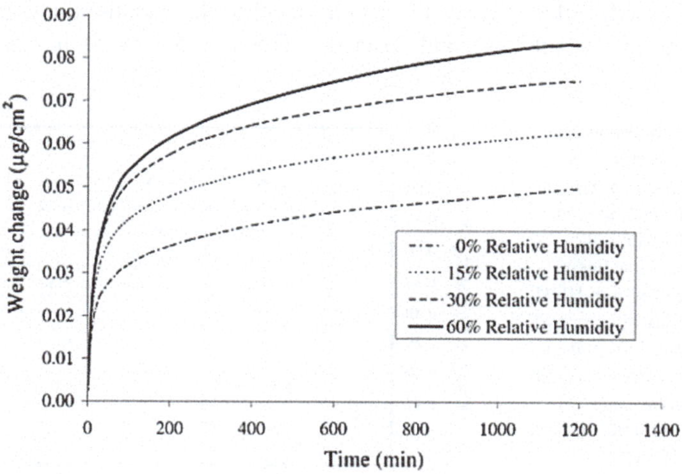

Fig. 30 Oxidation of 1 μm diameter copper particle in the air at 30 °C as a function of relative humidity [156]. Reproduced with permission of Springer Nature

the air relative humidity (RH) from 0 to 60% causes an increase of about 50% in the extent of oxidation over the same exposure period. Using XRD analysis, the only reaction product detected up to 90 °C was Cu_2O, which was used to convert the weight gain results to equivalent film thickness. Hence, from Fig. 30 and XRD results, linear growth is initially observed up to an oxide thickness of 0.13 nm followed by a subsequent logarithmic rate law to reach 0.74 nm within 20 h of exposure in dry air [156]. These results clearly indicate the importance of proper powder storage and handling and partially explain the large scatter of critical velocity obtained for the deposition of Cu in CS [148].

For aluminum powders, due to their negative standard reduction potential, the formation of aluminum oxide proceeds spontaneously in an oxygen-containing environment. The formed amorphous Al_2O_3 passive layer can be covered by a porous water-containing $Al(OH)_3$ and ALOOH overlay of a few nanometers in thickness [157]. The growth of the amorphous oxide is limited by the outward diffusion of aluminum cations. Once the oxide layer reaches a critical thickness value, with an increase in temperature, it becomes thermodynamically unstable, and phase transition occurs to γ-Al_2O_3 for which the inward oxygen diffusion becomes the limiting growth factor. In addition to oxide increase with temperature, Gobard [158] has shown the increase in oxidation with air humidity. His results show an oxide thickness of 22 nm and 170 nm after exposure to air at 52% and 100% humidity, respectively, over a period of 5 years.

Bariel et al. have studied the effect of storage on the oxygen pick-up of commercially pure Ti and Ti-6Al-4V powder over a period of 2.5 years in normal conditions, i.e. placed in an unsealed steel container filled with an air of uncontrolled humidity varying between 20 and 60% RH [159]. Their results, illustrated in Fig. 31, show great resistance to further increase in oxide content of the titanium powders. The passive oxide layer formed on the surface of titanium is very thin, i.e. 5–10 nm, and is composed of three layers: (1) TiO adjacent to the metallic titanium surface (2) intermediary layer of Ti_2O_3 and (3) anatase TiO_2 in contact with the environment

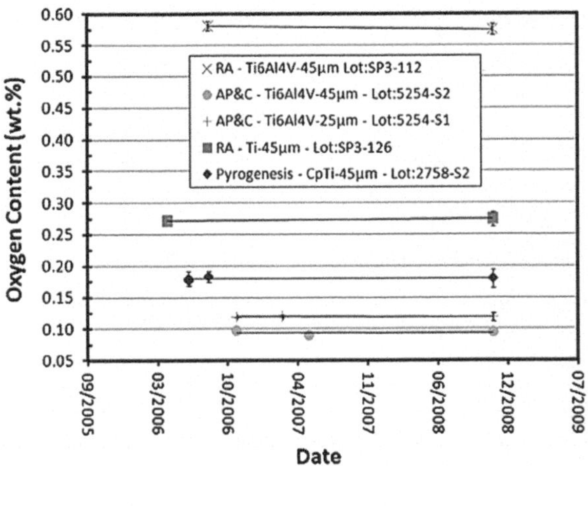

Fig. 31 Titanium powders oxygen content stored in unsealed steel containers in the air under uncontrolled conditions. Reading alloy (RA) powders have been manufactured through HDH: Hydride-Dehydride process and AP&C and Pyrogenesis have been generated through plasma atomization [159]. Reproduced with permission of Taylor & Francis

[160]. The TiO_2 layer protects the metal from further oxidation in various media and environments. Even under a saturated water vapour pressure at 120 °C oxidation is prevented [161].

In addition to the temporal oxygen content, Fig. 31, also demonstrates the initial oxygen content in Ti powder based on the particle manufacturing process. As shown, both Ti and Ti-6Al-4V manufactured through HDH show significantly higher oxygen than the plasma atomized powders. Similarly, Wong et al. have detected a higher oxygen content, both on the surface and in solid solution, in irregular and sponge Ti powder than in spherical feedstock [19]. Venkata et al. have, however, shown that although HDH powder exhibited 75% higher oxygen content than plasma atomized powder, their nano hardness was the same and attributed to the difference in microstructure [58].

Hence, it is of high importance to understand the growth mechanism of the oxide film under various atmospheric conditions in order to comprehend its effect during CS deposition. As the CS process relies on particle energy upon local impact, a thicker stronger and well-adhered oxide layer will significantly affect both the bonding process and particle resulting plastic flow due to the energy it takes away from the particle in order to fracture. Yin et al. have simulated the behaviour of Al_2O_3 oxide film layer around Al6061-T6 particles and demonstrated the importance of particle surface cleaning by the ejection of oxide debris for proper creation of metallurgical bonding [162].

Three materials have been presented in this section, namely copper, aluminum and titanium, based on their dissimilar behavior at normal room conditions and popularity in the AM field. The complexity of the subject due to the numerous variables that influence oxide growth restricts, however, any generalization of oxidation processes.

7 Powder Recycling

Using recycled powder in additive manufacturing processes can minimize the cost of production. In powder bed fusion processes such as selective laser melting (SLM), only a small portion of the feedstock powder is melted and used to build the required component [163]. The remaining of the powder is partially melted, highly heated and/or unmelted [163]. Hamed et al. have shown that recycled AlSi10Mg powder in SLM processes provides similar particle average size, microstructure, morphology, composition and mechanical properties, which confirms the usage of recycled powder to minimize cost [163]. On the other hand, Strondl et al. have shown that the ductility and toughness decrease in recycled Ti-6Al-4V powder due to a possible variation in oxygen content. Hence, although numerous studies have investigated the effect of powder recycling, more fundamental research is necessary to understand the effect of recycling on final part properties better and to validate the method for industrial applications and industrial relevant environments before extended commercial use [164].

Fig. 32 Impact of a single Cu particle **a** as-received and **b** reclaimed. Inserts are showing the powder geometry and size prior to deposition [165]. Reproduced with permission of Springer

In CS applications, the recycled powder properties differ primarily from those reclaimed in conventional AM processes due to the undeniable differences encountered during particle consolidation. The un-deposited powder in CS is composed of several different powder states. One of which is large particles that have not achieved critical velocity for a proper deposition but did undergo plastic deformation upon impact. Another being particles that have partially bonded to the substrate and debonded during the elastic rebound process. Lastly, particles that have experienced multiple particle impacts during flight. Perry et al. have conducted a study on the deposition efficiency of recycled Cu for pin array heat sinks produced using CS AM [165]. Figure 32 compares the deformation behaviour of as-received and reclaimed 20 μm powder, which demonstrates successful deposition and similar peripheral phenomena occurring at the substrate/particle interface and consequently shows the viability of recycling processes in CS applications.

In addition, their study showed that the deposition of a mixture of recycled and as-received powder leads to a similar fin component microstructure, a reduction of DE limited to 4% and low porosity levels, which provides a production cost reduction of 30%. Unlike other powder AM processes, the reclaimed powder did show lower flowability and average particle velocity, which has been associated to the selective content nature and deformed geometry of the reclaimed powder in the CS process [166, 167]. Due to the very limited studies of powder recycling in the CS field, powder reclamation is not currently used industrially.

8 Conclusion

The CS process benefits immensely from its solid-state deposition principle, low thermal stress, compressive residual stress, low oxidation, and high work hardening of the material. However, it is crucial to understand the influence of feedstock powder characteristics on the deposition process in order to appropriately tailor process parameters to each AM component. Emphasis has been put in this chapter on various facets of feedstock powder features. This chapter is a tool to help understand and recognize the importance of the particle's effect on the deposition, to offer a throughout summary of important controllable particle features prior to part fabrication, and to illustrate and explain the options available to the user of the CS process in the AM framework.

References

1. Gibson, I., Rosen, D., & Strucker, B. (2015). *Additive manufacturing technologies*. New York: Springer.
2. MacDonald, D., Nastic, A., & Jodoin, B. (2018). Understanding adhesion. In *Cold-spray coatings recent trends and future perspectives* (pp. 421–450). Berlin: Springer.
3. Van Steenkiste, T. H., et al. (1999). Kinetic spray coatings. *Surface and Coatings Technology, 111*(1), 62–71.
4. Schmidt, T., Gärtner, F., Assadi, H., & Kreye, H. (2006). Development of a generalized parameter window for cold spray deposition. *Acta Materialia, 54*(3), 729–742.
5. Stoltenhoff, T., Kreye, H., & Richter, H. J. (2002). An analysis of the cold spray process and its coatings. *Journal of Thermal Spray Technology, 11*(4), 542–550.
6. Dykhuizen, R. C., & Smith, M. F. (1998). Gas dynamic principles of cold spray. *Journal of Thermal Spray Technology, 7*(2), 205–212.
7. Kosarev, V. F., Klinkov, S. V., Alkhimov, A. P., & Papyrin, A. N. (2003). On some aspects of gas dynamics of the cold spray process. *Journal of Thermal Spray Technology, 12*(2), 265–281.
8. Grujicic, M., Zhao, C., Tong, C., DeRosset, W., & Helfritch, D. (2004). Analysis of the impact velocity of powder particles in the cold-gas dynamic-spray process. *Materials Science and Engineering: A, 368*(1–2), 222–230.
9. Klinkov, S. V., Kosarev, V. F., & Rein, M. (2005). Cold spray deposition: Significance of particle impact phenomena. *Aerospace Science and Technology, 9*(7), 582–591.
10. Li, C.-J., Li, W.-Y., & Liao, H. (2006). Examination of the critical velocity for deposition of particles in cold spraying. *Journal of Thermal Spray Technology, 15*(2), 212–222.
11. Hassani-Gangaraj, M., Veysset, D., Nelson, K. A., & Schuh, C. A. (2018). In-situ observations of single micro-particle impact bonding. *Scripta Materialia, 145*, 9–13.
12. Gilmore, D. L., Dykhuizen, R. C., Neiser, R. A., Smith, M. F., & Roemer, T. J. (1999). Particle velocity and deposition efficiency in the cold spray process. *Journal of Thermal Spray Technology, 8*(4), 576–582.
13. Assadi, H., Gärtner, F., Stoltenhoff, T., & Kreye, H. (2003). Bonding mechanism in cold gas spraying. *Acta Materialia, 51*(15), 4379–4394.
14. Gärtner, F., Stoltenhoff, T., Schmidt, T., & Kreye, H. (2006). The cold spray process and its potential for industrial applications. *Journal of Thermal Spray Technology, 15*(2), 223–232.
15. Li, C. J., et al. (2009). Influence of spray materials and their surface oxidation on the critical velocity in cold spraying. *Journal of Thermal Spray Technology, 19*(1–2), 342–347.
16. Villafuerte, J. (2015). *Modern cold spray*. Ontario: Springer.
17. Assadi, H., Kreye, H., Gärtner, F., & Klassen, T. (2016). Cold spraying—A materials perspective. *Acta Materialia, 116*, 382–407.
18. Jenkins, R., Yin, S., Aldwell, B., Meyer, M., & Lupoi, R. (2018). New insights into the in-process densification mechanism of cold spray Al coatings: Low deposition efficiency induced densification. *Journal of Materials Science and Technology, 35*(3), 427–431.
19. Wong, W., et al. (2013). Effect of particle morphology and size distribution on cold-sprayed pure titanium coatings. *Journal of Thermal Spray Technology, 22*(7), 1140–1153.

20. Assadi, H., et al. (2011). On parameter selection in cold spraying. *Journal of Thermal Spray Technology, 20*(6), 1161–1176.
21. Bastwros, M., et al. (2014). Effect of ball milling on graphene reinforced Al6061 composite fabricated by semi-solid sintering. *Composites Part B: Engineering, 60,* 111–118.
22. Sun, Y. Y., et al. (2015). Manipulation and characterization of a novel titanium powder precursor for additive manufacturing applications. *JOM Journal of the Minerals Metals and Materials Society, 67*(3), 564–572.
23. MacDonald, D., Fernández, R., Delloro, F., & Jodoin, B. (2017). Cold spraying of armstrong process titanium powder for additive manufacturing. *Journal of Thermal Spray Technology, 26*(4), 598–609.
24. Chen, C., et al. (2018). Cold spraying of thermally softened Ni-coated FeSiAl composite powder: Microstructure characterization, tribological performance and magnetic property. *Materials and Design, 160,* 270–283.
25. Sundberg, K., Champagne, V., McNally, B., Helfritch, D., & Sisson, R. (2015). Effectiveness of nanomaterial copper cold spray surfaces on inactivation of influenza A virus. *Journal of Biotechnology & Biomaterials, 5*(4), 205.
26. Gao, P.-H., Li, C.-J., Yang, G.-J., Li, Y.-G., & Li, C.-X. (2008). Influence of substrate hardness on deposition behavior of single porous WC-12Co particle in cold spraying. *Surface and Coatings Technology, 203*(3–4), 384–390.
27. Hall, A. C., Brewer, L. N., & Roemer, T. J. (2008). Preparation of aluminum coatings containing homogenous nanocrystalline microstructures using the cold spray process. *Journal of Thermal Spray Technology, 17*(3), 352–359.
28. Zhang, Q., et al. (2008). Formation of NiAl intermetallic compound by cold spraying of ball-milled Ni/Al alloy powder through postannealing treatment. *Journal of Thermal Spray Technology, 17*(5–6), 715–720.
29. Tria, S., et al. (2011). Deposition and characterization of cold sprayed nanocrystalline NiTi. *Powder Technology, 210*(2), 181–188.
30. Jeandin, M., Rolland, G., Descurninges, L. L., & Berger, M. H. (2014). Which powders for cold spray? *Surface Engineering, 30*(5), 291–298.
31. Lagutkin, S., Achelis, L., Sheikhaliev, S., Uhlenwinkel, V., & Srivastava, V. (2004). Atomization process for metal powder. *Materials Science and Engineering: A, 383*(1), 1–6.
32. Ünal, A. (1989). Liquid break-up in gas atomization of fine aluminum powders. *Metallurgical Transactions B, 20*(1), 61–69.
33. Yule, A., & Dunkley, J. (1994). *Atomization of melts for powder production and spray deposition*. Michigan: Clarendon Press.
34. Datta, B. K. (2014). *Powder metallurgy: An advanced technique of processing engineering materials*. New Delhi: PHI Learning.
35. Sun, P., Fang, Z. Z., Zhang, Y., & Xia, Y. (2017). Review of the methods for production of spherical Ti and Ti alloy powder. *JOM Journal of the Minerals Metals and Materials Society, 69*(10), 1853–1860.
36. Macdonald, D., Rahmati, S., & Jodoin, B. (2018). An economical approach to cold gas dynamic spraying using in-line nitrogen-helium blending. In *International Thermal Spray Conference*, Orlando, Florida, May 7–10, 2018.
37. Fauchais, P., Montavon, G., & Bertrand, G. (2010). From powders to thermally sprayed coatings. *Journal of Thermal Spray Technology, 19*(1–2), 56–80.
38. Duflos, F., & Stohr, J. F. (1982). Comparison of the quench rates attained in gas-atomized powders and melt-spun ribbons of Co- and Ni-base superalloys: Influence on resulting microstructures. *Journal Materials Science, 17*(12), 3641–3652.
39. Peissker, E. (1991). Production and handling of electrolytic powders. *Metal Powder Report, 46*(4), 20–25.
40. Araci, K., Mangabhai, D., & Akhtar, K. (2015). Production of titanium by the Armstrong Process. In *Titanium powder metallurgy*. Amsterdam: Elsevier.
41. Fernandez, R., & Jodoin, B. (2017). Effect of particle morphology on cold spray deposition of chromium carbide-nickel chromium cermet powders. *Journal of Thermal Spray Technology, 26*(6), 1356–1380.

42. Ghelichi, R., et al. (2014). Fatigue strength of Al alloy cold sprayed with nanocrystalline powders. *International Journal of Fatigue, 65,* 51–57.
43. Ajdelsztajn, L., Zúñiga, A., Jodoin, B., & Lavernia, E. J. (2006). Cold-spray processing of a nanocrystalline Al-Cu-Mg-Fe-Ni alloy with Sc. *Journal of Thermal Spray Technology, 15*(2), 184–190.
44. Hassani-Gangaraj, M., Veysset, D., Champagne, V. K., Nelson, K. A., & Schuh, C. A. (2018). Adiabatic shear instability is not necessary for adhesion in cold spray. *Acta Materialia, 158,* 430–439.
45. Hassani-Gangaraj, M., Veysset, D., Champagne, V. K., Nelson, K. A., & Schuh, C. A. (2019). Response to Comment on "Adiabatic shear instability is not necessary for adhesion in cold spray". *Scripta Materialia, 162,* 515–519.
46. Assadi, H., Gärtner, F., Klassen, T., & Kreye, H. (2019). Comment on "Adiabatic shear instability is not necessary for adhesion in cold spray". *Scripta Materialia, 162,* 512–514.
47. Nastic, A., & Jodoin, B. (2018). Evaluation of heat transfer transport coefficient for cold spray through computational fluid dynamics and particle in-flight temperature measurement using a high-speed IR camera. *Journal of Thermal Spray Technology, 27*(8), 1491–1517.
48. Schmidt, T., Gaertner, F., & Kreye, H. (2006). New developments in cold spray based on higher gas and particle temperatures. *Journal of Thermal Spray Technology, 15*(4), 488–494.
49. Sova, A., Grigoriev, S., Kochetkova, A., & Smurov, I. (2013). Influence of powder injection point position on efficiency of powder preheating in cold spray: Numerical study. *Surface and Coatings Technology,* 1–6.
50. Grujicic, M., Zhao, C., DeRosset, W., & Helfritch, D. (2004). Adiabatic shear instability based mechanism for particles/substrate bonding in the cold-gas dynamic-spray process. *Materials and Design, 25*(8), 681–688.
51. Jodoin, B., et al. (2006). Effect of particle size, morphology, and hardness on cold gas dynamic sprayed aluminum alloy coatings. *Surface and Coatings Technology, 201*(6), 3422–3429.
52. Munagala, V. N. V., Akinyi, V., Vo, P., & Chromik, R. R. (2018). Influence of powder morphology and microstructure on the cold spray and mechanical properties of Ti6Al4V coatings. *Journal of Thermal Spray Technology, 27*(5), 827–842.
53. Luo, X. T., Li, Y. J., & Li, C. J. (2016). A comparison of cold spray deposition behavior between gas atomized and dendritic porous electrolytic Ni powders under the same spray conditions. *Materials Letters, 163,* 58–60.
54. Ko, K. H., Choi, J. O., & Lee, H. (2015). Characteristics of cold sprayed dendritic Cu coatings. *Surface Engineering, 32*(9), 650–654.
55. Fukanuma, H., Ohno, N., Sun, B., & Huang, R. (2006). In-flight particle velocity measurements with DPV-2000 in cold spray. *Surface and Coatings Technology, 201*(5), 1935–1941.
56. Yin, S., He, P., Liao, H., & Wang, X. (2014). Deposition features of Ti coating using irregular powders in cold spray. *Journal of Thermal Spray Technology, 23*(6), 984–990.
57. MacDonald, D., Leblanc-Robert, S., Fernández, R., Farjam, A., & Jodoin, B. (2016). Effect of nozzle material on downstream lateral injection cold spray performance. *Journal of Thermal Spray Technology, 25*(6), 1149–1157.
58. Bhattiprolu, V. S., Johnson, K. W., Ozdemir, O. C., & Crawford, G. A. (2018). Influence of feedstock powder and cold spray processing parameters on microstructure and mechanical properties of Ti-6Al-4V cold spray depositions. *Surface and Coatings Technology, 335,* 1–12.
59. Birt, A. M., Champagne, V. K., Sisson, R. D., & Apelian, D. (2015). Microstructural analysis of Ti–6Al–4V powder for cold gas dynamic spray applications. *Advanced Powder Technology, 26*(5), 1335–1347.
60. Jones, H. (1984). Microstructure of rapidly solidified materials. *Materials Science and Engineering, 65*(1), 145–156.
61. Sabard, A., de Villiers Lovelock, H. L., & Hussain, T. (2018). Microstructural evolution in solution heat treatment of gas-atomized Al alloy (7075) powder for cold spray. *Journal of Thermal Spray Technology, 27*(1–2), 145–158.
62. Kong, C. J., Brown, P. D., Harris, S. J., & McCartney, D. G. (2007). Analysis of microstructure formation in gas-atomised Al–12 wt.% Sn–1 wt.% Cu alloy powder. *Materials Science and Engineering: A, 454–455,* 252–259.

63. Wang, J., Yang, H., Ruan, J., Wang, Y., & Ji, S. (2019). Microstructure and properties of CoCrNi medium-entropy alloy produced by gas atomization and spark plasma sintering. *Journal of Materials Research,* 1–11.
64. Wang, F., Xiong, B., Zhang, Y., Liu, H., & He, X. (2009). Microstructural development of spray-deposited Al–Zn–Mg–Cu alloy during subsequent processing. *Journal of Alloys and Compounds, 477*(1–2), 616–621.
65. Annavarapu, S., & Doherty, R. D. (1993). Evolution of microstructure in spray casting. *International Journal of Powder Metallurgy (1986), 29*(4), 331–343.
66. Rokni, M. R., Widener, C. A., Champagne, V. K., & Crawford, G. A. (2015). Microstructure and mechanical properties of cold sprayed 7075 deposition during non-isothermal annealing. *Surface and Coatings Technology, 276,* 305–315.
67. Ajdelsztajn, L., Jodoin, B., & Schoenung, J. M. (2006). Synthesis and mechanical properties of nanocrystalline Ni coatings produced by cold gas dynamic spraying. *Surface and Coatings Technology, 201*(3–4), 1166–1172.
68. Ajdelsztajn, L., Zúñiga, A., Jodoin, B., & Lavernia, E. J. (2006). Cold gas dynamic spraying of a high temperature Al alloy. *Surface and Coatings Technology, 201*(6), 2109–2116.
69. Bérubé, G., et al. (2012). Phase stability of Al-5Fe-V-Si coatings produced by cold gas dynamic spray process using rapidly solidified feedstock materials. *Journal of Thermal Spray Technology, 21*(2), 240–254.
70. Trivedi, R., Jin, F., & Anderson, I. E. (2003). Dynamical evolution of microstructure in finely atomized droplets of Al-Si alloys. *Acta Materialia, 51*(2), 289–300.
71. Zhang, Y. Y., & Zhang, J. S. (2011). Recrystallization in the particles interfacial region of the cold-sprayed aluminum coating: Strain-induced boundary migration. *Materials Letters, 65*(12), 1856–1858.
72. Zambon, A., Badan, B., Norman, A. F., Greer, A. L., & Ramous, E. (1997). Development of solidification microstructures in atomized Fe-Ni alloy droplets. *Materials Science and Engineering: A, 226–228,* 119–123.
73. Kalay, Y. E., Chumbley, L. S., Anderson, I. E., & Napolitano, R. E. (2007). Characterization of hypereutectic Al-Si powders solidified under far-from equilibrium conditions. *Metallurgical and Materials Transactions A: Physical Metallurgy and Materials Science, 38*(7), 1452–1457.
74. Rokni, M. R., Widener, C. A., Crawford, G. A., & West, M. K. (2015). An investigation into microstructure and mechanical properties of cold sprayed 7075 Al deposition. *Materials Science and Engineering: A, 625,* 19–27.
75. Hariprasad, S., Sastry, S. M. L., & Jerina, K. L. (1996). Undercooling and supersaturation of alloying elements in rapidly solidified Al-8.5% Fe-1.2% V-1.7% Si alloy. *Journal of Materials Science, 31*(4), 921–925.
76. Rokni, M. R., Widener, C. A., & Crawford, G. A. (2014). Microstructural evolution of 7075 Al gas atomized powder and high-pressure cold sprayed deposition. *Surface and Coatings Technology, 251,* 254–263.
77. Kestenbach, H. J., & Meyers, M. A. (1976). The effect of grain size on the shock-loading response of 304-type stainless steel. *Metallurgical Transactions A, 7*(12), 1943–1950.
78. Sevsek, S., et al. (2019). Strain-rate-dependent deformation behavior and mechanical properties of a multi-phase medium-manganese steel. *Metals (Basel), 9*(3), 344.
79. Holian, B. L., Hammerberg, J. E., & Lomdahl, P. S. (1998). The birth of dislocations in shock waves and high-speed friction. *Journal of Computer-Aided Materials Design, 5*(2/3), 207–224.
80. Borchers, C., Gärtner, F., Stoltenhoff, T., & Kreye, H. (2004). Microstructural bonding features of cold sprayed face centered cubic metals. *Journal of Applied Physics, 96*(8), 4288–4292.
81. Bhattiprolu, V. S., Johnson, K. W., & Crawford, G. A. (2019). Influence of powder microstructure on the microstructural evolution of as-sprayed and heat treated cold-sprayed Ti-6Al-4V coatings. *Journal of Thermal Spray Technology, 28*(1–2), 174–188.
82. Zou, Y., et al. (2009). Dynamic recrystallization in the particle/particle interfacial region of cold-sprayed nickel coating: Electron backscatter diffraction characterization. *Scripta Materialia, 61*(9), 899–902.

83. Humphreys, F. J., & Hatherly, M. (2004). *Recrystallization and related annealing phenomena*. Amsterdam: Elsevier.
84. Rokni, M. R., Nutt, S. R., Widener, C. A., Crawford, G. A., & Champagne, V. K. (2018) Structure–properties relations in high-pressure cold-sprayed deposits. In *Cold-spray coatings* (pp. 143–192). Cham: Springer.
85. Bae, G., Kang, K., Kim, J.-J., & Lee, C. (2010). Nanostructure formation and its effects on the mechanical properties of kinetic sprayed titanium coating. *Materials Science and Engineering: A, 527*(23), 6313–6319.
86. Chen, M, et al. (2003). Deformation twinning in nanocrystalline aluminum. *Science (80-), 300*(5623), 1275–1277.
87. Nabarro, F. R. N., & Duesbery , M. S. (2007). *Dislocations in solids* (Vol. 13). North Holland: Elsevier.
88. Zerilli, F. J., Armstrong, R. W. (1997). Dislocation mechanics based analysis of material dynamics behavior: Enhanced ductility, deformation twinning, shock deformation, shear instability, dynamic recovery. *Le Journal de Physique IV, 07*(C3), C3-637–C3-642.
89. Meyers, M. A., et al. (2003). Laser-induced shock compression of monocrystalline copper: Characterization and analysis. *Acta Materialia, 51*(5), 1211–1228.
90. Schmidt, C. G., Caligiuri, R. D., Giovanola, J. H., & Erlich, D. C. (1991). Effect of grain size on high strain rate deformation of copper. *Metallurgical Transactions A, 22*(10), 2349–2357.
91. Lasalmonie, A., & Strudel, J. L. (1986). Influence of grain size on the mechanical behaviour of some high strength materials. *Journal Materials Science, 21*(6), 1837–1852.
92. Rohatgi, A., Vecchio, K. S., & Gray, G. T., III. (2001). A metallographic and quantitative analysis of the influence of stacking fault energy on shock-hardening in Cu and Cu–Al alloys. *Acta Materialia, 49*(3), 427–438.
93. Millett, J. C. F., Meziere, Y. J. E., & Bourne, N. K. (2007). Shear stress measurement in nickel and nickel–60 wt% cobalt during one-dimensional shock loading. *Journal Materials Science, 42*(15), 5941–5948.
94. Millett, J. C. F., Bourne, N. K., & Gray, G. T. (2008). The behavior of Ni, Ni-60Co, and Ni3Al during one-dimensional shock loading. *Metallurgical and Materials Transactions A: Physical Metallurgy and Materials Science, 39*(2), 322–334.
95. Sabard, A., & Hussain, T. (2018). Bonding mechanisms in cold spray deposition of gas atomised and solution heat-treated Al 6061 powder by EBSD. Cornell University.
96. Henao, J., & Sharma, M. M. (2018). Characterization, deposition mechanisms, and modeling of metallic glass powders for cold spray. In *Cold-spray coatings* (pp. 251–272). Cham: Springer.
97. Legoux, J. G., Irissou, E., & Moreau, C. (2007). Effect of substrate temperature on the formation mechanism of cold-sprayed aluminum, zinc and tin coatings. *Journal of Thermal Spray Technology, 16*(5–6), 619–626.
98. Fressengeas, C., & Molinari, A. (1987). Instability and localization of plastic flow in shear at high strain rates. *Journal of the Mechanics and Physics of Solids, 35*(2), 185–211.
99. Peirce, D., Asaro, R. J., & Needleman, A. (1983). Material rate dependence and localized deformation in crystalline solids. *Acta Metallurgica, 31*(12), 1951–1976.
100. Guha, R. D., Sharma, A. J., Diwan, P., & Khanikar, P. (2017). Effect of grain orientation on high strain-rate plastic deformation. *Procedia Engineering, 173,* 1048–1055.
101. Carrington, M. J., et al. (2019). Microstructural characterisation of Tristelle 5183 (Fe-21%Cr-10%Ni-7.5%Nb-5%Si-2%C in wt%) alloy powder produced by gas atomisation. *Materials & Design, 164,* 107548.
102. Yildirim, B., Yang, H., Gouldstone, A., & Müftü, S. (2017). Rebound mechanics of micrometre-scale, spherical particles in high-velocity impacts. *Proceedings of the Royal Society A: Mathematical, Physical and Engineering Sciences, 473*(2204), 20160936.
103. Nastic, A., Vijay, M., Tieu, A., Rahmati, S., & Jodoin, B. (2017). Experimental and numerical study of the influence of substrate surface preparation on adhesion mechanisms of aluminum cold spray coatings on 300 M steel substrates. *Journal of Thermal Spray Technology, 26*(7), 1461–1483.

104. Oliver, W. C., & Pharr, G. M. (2004). Measurement of hardness and elastic modulus by instrumented indentation: Advances in understanding and refinements to methodology. *Journal of Materials Research, 19*(01), 3–20.
105. Nastic, A., et al. (2015). Instrumented and vickers indentation for the characterization of stiffness, hardness and toughness of zirconia toughened Al2O3 and SiC armor. *Journal of Materials Science and Technology, 31*(8), 773–783.
106. Goldbaum, D., Chromik, R. R., Brodusch, N., & Gauvin, R. (2015). Microstructure and mechanical properties of Ti cold-spray splats determined by electron channeling contrast imaging and nanoindentation mapping. *Microscopy and Microanalysis, 21,* 570–581.
107. Goldbaum, D., Chromik, R. R., Yue, S., Irissou, E., & Legoux, J.-G. (2011). Mechanical property mapping of cold sprayed Ti splats and coatings. *Journal of Thermal Spray Technology, 20*(3), 486–496.
108. List, A., et al. (2014). Cold spraying of amorphous Cu50Zr50 alloys. *Journal of Thermal Spray Technology, 24*(1–2), 108–118.
109. Henao, J., et al. (2016). Influence of the substrate on the formation of metallic glass coatings by cold gas spraying. *Journal of Thermal Spray Technology, 25*(5), 992–1008.
110. Yoon, S., Xiong, Y., Kim, H., & Lee, C. (2009). Dependence of initial powder temperature on impact behaviour of bulk metallic glass in a kinetic spray process. *Journal of Physics D: Applied Physics, 42,* 5.
111. Concustell, A., et al. (2015). On the formation of metallic glass coatings by means of Cold Gas Spray technology. *Journal of Alloys and Compounds, 651,* 764–772.
112. Ajdelsztajn, L., Jodoin, B., Richer, P., Sansoucy, E., & Lavernia, E. J. (2006). Cold gas dynamic spraying of iron-base amorphous alloy. *Journal of Thermal Spray Technology, 15*(4), 495–500.
113. List, A., Gärtner, F., Schmidt, T., & Klassen, T. (2012). Impact conditions for cold spraying of hard metallic glasses. *Journal of Thermal Spray Technology, 21*(3–4), 531–540.
114. Kwon, O.-J., et al. (2007). Thermal and mechanical behaviors of Cu–Zr amorphous alloys. *Materials Science and Engineering: A, 449–451,* 169–171.
115. Yoon, S., Lee, C., Choi, H., & Jo, H. (2006). Kinetic spraying deposition behavior of bulk amorphous NiTiZrSiSn feedstock. *Materials Science and Engineering: A, 415*(1–2), 45–52.
116. Demetriou, M. D., & Johnson, W. L. (2004). Modeling the transient flow of undercooled glass-forming liquids. *Journal of Applied Physics, 95*(5), 2857–2865.
117. Chen, M. (2008). Mechanical behavior of metallic glasses: Microscopic understanding of strength and ductility. *Annual Review of Materials Research, 38*(1), 445–469.
118. Greer, A. L., Cheng, Y. Q., & Ma, E. (2013). Shear bands in metallic glasses. *Materials Science and Engineering: R: Reports, 74*(4), 71–132.
119. Borchers, C., Schmidt, T., Gärtner, F., & Kreye, H. (2008). High strain rate deformation microstructures of stainless steel 316L by cold spraying and explosive powder compaction. *Applied Physics A, 90*(3), 517–526.
120. Gärtner, F., et al. (2006). Mechanical properties of cold-sprayed and thermally sprayed copper coatings. *Surface and Coatings Technology, 200*(24), 6770–6782.
121. Brewer, L. N., Schiel, J. F., Menon, E. S. K., & Woo, D. J. (2018). The connections between powder variability and coating microstructures for cold spray deposition of austenitic stainless steel. *Surface and Coatings Technology, 334,* 50–60.
122. Liu, T., Leazer, J. D., Menon, S. K., & Brewer, L. N. (2018). Microstructural analysis of gas atomized Al-Cu alloy feedstock powders for cold spray deposition. *Surface and Coatings Technology, 350,* 621–632.
123. Coddet, P., Verdy, C., Coddet, C., & Debray, F. (2015). Effect of cold work, second phase precipitation and heat treatments on the mechanical properties of copper–silver alloys manufactured by cold spray. *Materials Science and Engineering: A, 637,* 40–47.
124. Li, K., Song, C., Zhai, Q., Stoica, M., & Eckert, J. (2014). Microstructure evolution of gas-atomized Fe–6.5 wt% Si droplets. *Journal of Materials Research, 29*(04), 527–534.
125. Rokni, M. R., Widener, C. A., & Champagne, V. R. (2014). Microstructural evolution of 6061 aluminum gas-atomized powder and high-pressure cold-sprayed deposition. *Journal of Thermal Spray Technology, 23*(3), 514–524.

126. Rokni, M. R., Zarei-Hanzaki, A., & Abedi, H. R. (2012). Microstructure evolution and mechanical properties of back extruded 7075 aluminum alloy at elevated temperatures. *Materials Science and Engineering A, 532,* 593–600.
127. Story, W. A., & Brewer, L. N. (2018). Heat treatment of gas-atomized powders for cold spray deposition. *Metallurgical and Materials Transactions A: Physical Metallurgy and Materials Science, 49*(2), 446–449.
128. Field, D. P., Bradford, L. T., Nowell, M. M., & Lillo, T. M. (2007). The role of annealing twins during recrystallization of Cu. *Acta Materialia, 55*(12), 4233–4241.
129. Borchers, C., Gärtner, F., Stoltenhoff, T., Assadi, H., & Kreye, H. (2003). Microstructural and macroscopic properties of cold sprayed copper coatings. *Journal of Applied Physics, 93*(12), 10064–10070.
130. Li, W., Huang, C., Yu, M., & Liao, H. (2013). Investigation on mechanical property of annealed copper particles and cold sprayed copper coating by a micro-indentation testing. *Materials & Design, 46,* 219–226.
131. Ning, X.-J., Jang, J.-H., & Kim, H.-J. (2007). The effects of powder properties on in-flight particle velocity and deposition process during low pressure cold spray process. *Applied Surface Science, 253*(18), 7449–7455.
132. Hall, A. C., Cook, D. J., Neiser, R. A., Roemer, T. J., & Hirschfeld, D. A. (2006). The effect of a simple annealing heat treatment on the mechanical properties of cold-sprayed aluminum. *Journal of Thermal Spray Technology, 15*(2), 233–238.
133. Qiu, X., et al. (2017). Effect of heat treatment on microstructure and mechanical properties of A380 aluminum alloy deposited by cold spray. *Journal of Thermal Spray Technology, 26*(8), 1898–1907.
134. Barnett, B., Trexler, M., & Champagne, V. (2015). Cold sprayed refractory metals for chrome reduction in gun barrel liners. *International Journal of Refractory Metals and Hard Materials, 53,* 139–143.
135. Conrad, H. (1981). Effect of interstitial solutes on the strength and ductility of titanium. *Progress in Materials Science, 26*(2–4), 123–403.
136. Champagne, V. K. (2007). The *cold spray materials deposition process: Fundamentals and applications*. Lisle, IL: Woodhead.
137. Oh, J.-M., et al. (2011). Oxygen effects on the mechanical properties and lattice strain of Ti and Ti-6Al-4V. *Metals and Materials International, 17*(5), 733–736.
138. Robson, J. D., Henry, D. T., & Davis, B. (2011). Particle effects on recrystallization in magnesium–manganese alloys: Particle pinning. *Materials Science and Engineering: A, 528*(12), 4239–4247.
139. Apps, P. J., Berta, M., & Prangnell, P. B. (2005). The effect of dispersoids on the grain refinement mechanisms during deformation of aluminium alloys to ultra-high strains. *Acta Materialia, 53*(2), 499–511.
140. Brown, L. M. (1997). Transition from laminar to rotational motion in plasticity. *Philosophical Transactions of the Royal Society of London. Series A: Mathematical, Physical and Engineering Sciences, 355*(1731), 1979–1990.
141. Zhang, Y., et al. (2017). The effect of submicron second-phase particles on the rate of grain refinement in a copper-oxygen alloy during cold spray. *Journal of Thermal Spray Technology, 26*(7), 1509–1516.
142. Lin, X. H., Johnson, W. L., & Rhim, W. K. (1997). Effect of oxygen impurity on crystallization of an undercooled bulk glass forming Zr-Ti-Cu-Ni-Al alloy. *Materials Transactions JIM, 38*(5), 473–477.
143. Wood, G. C., & Stringer, J. (1993). The adhesion of growing oxide scales to the substrate. *Le Journal de Physique IV, 03*(C9), C9-65–C9-74.
144. Cabrera, N., & Mott, N. F. (1949). Theory of the oxidation of metals. *Reports on Progress in Physics, 12*(1), 308.
145. Chen, C., et al. (2018). On the role of oxide film's cleaning effect into the metallurgical bonding during cold spray. *Materials Letters, 210,* 199–202.

146. Watanabe, Y., Yoshida, C., Atsumi, K., Yamada, M., & Fukumoto, M. (2014). Influence of substrate temperature on adhesion strength of cold-sprayed coatings. *Journal of Thermal Spray Technology, 24*(1–2), 86–91.
147. Ichikawa, Y., Tokoro, R., Tanno, M., & Ogawa, K. (2019). Elucidation of cold-spray deposition mechanism by auger electron spectroscopic evaluation of bonding interface oxide film. *Acta Materialia, 164,* 39–49.
148. Hassani-Gangaraj, M., Veysset, D., Nelson, K. A., & Schuh, C. A. (2019). Impact-bonding with aluminum, silver, and gold microparticles: Toward understanding the role of native oxide layer. *Applied Surface Science, 476,* 528–532.
149. McCune, R. C., Donlon, W. T., Popoola, O. O., & Cartwright, E. L. (2000). Characterization of copper layers produced by cold gas-dynamic spraying. *Journal of Thermal Spray Technology, 9*(1), 73–82.
150. Kang, K., Yoon, S., Ji, Y., & Lee, C. (2008). Oxidation dependency of critical velocity for aluminum feedstock deposition in kinetic spraying process. *Materials Science and Engineering: A, 486*(1–2), 300–307.
151. Khanna, A. S. (2012). High temperature oxidation. *Handbook of Environmental Degradation of Materials,* 127–194.
152. Casati, R., Fabrizi, A., Tuissi, A., Xia, K., & Vedani, M. (2015). ECAP consolidation of Al matrix composites reinforced with in-situ γ-Al2O3 nanoparticles. *Materials Science and Engineering: A, 648,* 113–122.
153. Mimura, K., Lim, J.-W., Isshiki, M., Zhu, Y., & Jiang, Q. (2006). Brief review of oxidation kinetics of copper at 350 °C to 1050 °C. *Metallurgical and Materials Transactions A: Physical Metallurgy and Materials Science, 37*(4), 1231–1237.
154. Fujita, K., Ando, D., Uchikoshi, M., Mimura, K., & Isshiki, M. (2013). New model for low-temperature oxidation of copper single crystal. *Applied Surface Science, 276,* 347–358.
155. Liang, Y., et al. (2017). Oxidation-resistant micron-sized Cu–Sn solid particles fabricated by a one-step and scalable method. *RSC Advances, 7*(38), 23468–23477.
156. Feng, Z., Marks, C. R., & Barkatt, A. (2003). Oxidation-rate excursions during the oxidation of copper in gaseous environments at moderate temperatures. *Oxidation of Metals, 60*(5/6), 393–408.
157. Bauer, D. M., et al. (2017). Investigations of aging behaviour for aluminium powders during an atmosphere simulation of the LBM process. *Powder Metallurgy, 60*(3), 175–183.
158. Godard, H. P. (1967). Oxide film growth over five years on some aluminum sheet alloys in air of varying humidity at room temperature. *Journal of the Electrochemical Society, 114*(4), 354.
159. Baril, E., Lefebvre, L. P., & Thomas, Y. (2011). Interstitial elements in titanium powder metallurgy: Sources and control. *Powder Metallurgy, 54*(3), 183–186.
160. Pouilleau, J., Devilliers, D., Garrido, F., Durand-Vidal, S., & Mahé, E. (1997). Structure and composition of passive titanium oxide films. *Materials Science and Engineering: B, 47*(3), 235–243.
161. Hiromoto, S., Hanawa, T., & Asami, K. (2004). Composition of surface oxide film of titanium with culturing murine fibroblasts L929. *Biomaterials, 25*(6), 979–986.
162. Yin, S., Wang, X., Li, W., Liao, H., & Jie, H. (2012). Deformation behavior of the oxide film on the surface of cold sprayed powder particle. *Applied Surface Science, 259,* 294–300.
163. Asgari, H., Baxter, C., Hosseinkhani, K., & Mohammadi, M. (2017). On microstructure and mechanical properties of additively manufactured AlSi10Mg_200C using recycled powder. *Materials Science and Engineering: A, 707,* 148–158.
164. Ardila, L. C., et al. (2014). Effect of IN718 recycled powder reuse on properties of parts manufactured by means of selective laser melting. *Physics Procedia, 56,* 99–107.
165. Perry, J., Richer, P., Jodoin, B., & Matte, E. (2019). Pin fin array heat sinks by cold spray additive manufacturing: Economics of powder recycling. *Journal of Thermal Spray Technology, 28*(1–2), 144–160.

166. Tang, H. P., et al. (2015). Effect of powder reuse times on additive manufacturing of Ti-6Al-4V by selective electron beam melting. *JOM Journal of the Minerals Metals and Materials Society, 67*(3), 555–563.
167. Strondl, A., Lyckfeldt, O., Brodin, H., & Ackelid, U. (2015). Characterization and control of powder properties for additive manufacturing. *JOM Journal of the Minerals Metals and Materials Society, 67*(3), 549–554.

Cold Spray Deposition on Polymeric and Composite Substrates

Antonello Astarita, Luca Boccarusso, Luigi Carrino, Massimo Durante, Alessia Serena Perna, and Antonio Viscusi

Abstract The aim of this chapter is to bring together, summarize and explain the latest development in the field of the cold spray deposition on polymeric and composite substrates. Although the deposition on metallic substrates has been widely studied and there are interesting books reporting the phenomena occurring and describing the process in detail, the deposition on polymeric substrates is a relatively new branch of cold spraying and to date there are no review papers or books focused on this topic. Several original research papers dealing with this topic have been published in the last years so it can be assumed that the available material is enough to allow the writing of a book chapter that summarizes what was done and highlights what more needs to be done. First it is worth to notice why this topic is of interest. Polymer matrix composites (PMCs) are widely used in the aerospace industry and in the military because of their low density, high specific strength and stiffness, and other unique properties such as ease formation and machining, well size stability. At the same time the use of polymeric components in aircraft, automobile, and power sector has gradually increased due to the development of new polymers with enhanced mechanical and physical properties. Some special characteristics such as electrical conductivity, thermal conductivity, electromagnetic shielding, erosion and radiation protection have to be improved to further widen the fields of application of these materials. Therefore, the surface metallization of a PMC or polymeric substrate is considered to be an effective technique to enhance the above-mentioned surface properties. To date there are several conventional techniques for the polymer metallization. The most common metallization techniques include vapor phase deposition, thermal spray deposition, electro deposition, electroforming and electroless deposition. However, these techniques have disadvantages. High equipment and processing costs, size limitations of workpiece produced by vapor phase deposition-based techniques (deposition in atomic dimension and workpiece should be placed in a limited

A. Astarita (✉) · L. Boccarusso · L. Carrino · M. Durante · A. S. Perna · A. Viscusi
Department of Chemical, Materials and Industrial Production Engineering, University of Naples Federico II, Piazzale Tecchio 80, 80125 Naples, Italy
e-mail: antonello.astarita@unina.it

A. S. Perna
University of Bergamo, Bergamo, Italy

S. Pathak and G. C. Saha (eds.), *Cold Spray in the Realm of Additive Manufacturing*, Materials Forming, Machining and Tribology,
https://doi.org/10.1007/978-3-030-42756-6_4

cavity). Low adhesive force, poor stability in the electroplating case and the consequent pollution it generates, long production cycle and high cost of the mold for electroforming, distortion of the substrate surface resulting from the molten particles, and high-temperature flames in thermal spraying (TS). Recently, the cold spray technique, with suitable modification, has been increasingly used for the metallization of polymers and PMCs, especially for the advance polymer composite such as carbon fiber-reinforced polymer (CFRP). Compared to the techniques mentioned in the previous indent the cold spray offers some intriguing advantages when spraying metallic particles on polymeric substrates: (i) the deposition is made possible by the significant plastic deformation of the particles and of the substrate, some polymers (the thermoplastic ones) can experience an huge amount of plastic deformation so this type of interaction easily occurs; (ii) no chemical reaction are required to take place between the particles and the substrate, so it is not a problem the fact that due to their different chemical nature it would be almost impossible to observe a chemical bonding between a metal and a polymer; (iii) compared with thermal spray processes a less heat input is required in cold spray, therefore, heat effects such as surface distortion, oxides, void, phase transformation, and residual stresses are considerably reduced in the coatings; on this premise cold spray appears to be a suitable technique to process temperature-sensitive materials, such as polymers; (iv) in addition, CS has a technical advantage (similar to the TS)—the spraying gun can be held by a robot arm or the workpiece can be installed onto a numerical control working platform to prepare large-sized or complex surface parts with flexibility and compatibility, and this can be a big competitive advantage in the realm of additive manufacturing. In this chapter it will be provided a comprehensive overview of the deposition of metallic coatings on polymeric substrates through cold spray. The deposition behavior of the coatings and the adhesion mechanisms between the substrate and the particles will have discussed to set a foundation for the practical application of cold spray on polymers. The microstructure and the properties of the coatings will be as well reported to provide a useful guideline for practical applications. Moreover, a section will be devoted to the study of the tailored design of the PMCs substrates and to the study of proper chemical formulations of the polymers to facilitate the cold spray deposition. The main topics that are covered in this chapter are below reported:

- Different experimental set up adopted;
- Various substrate/coating configurations that can be produced;
- Mechanisms that rule the adhesion between the particles and the substrate;
- Deposition behavior and influence of the primary process parameters;
- Mechanical, tribological, and physical properties of the coatings, including microstructure;
- Modeling of the process;
- Tailored design and production of the PMCs to facilitate the deposition;
- Tailored chemical composition of the polymers to facilitate the deposition; and
- Direction of further investigation.

Keywords Cold spray · Additive manufacturing · Carbon fiber-reinforced polymer · Polymer matrix composite · Numerical modeling

1 Introduction

This chapter aims to bring together, summarize, and explain the latest development in the field of the Cold Gas-Dynamic Spray (CGDS) deposition on both polymeric and fibres reinforced plastics (FRPs) substrates. Although the deposition on metallic substrates has been widely studied, the deposition on polymeric, with or without reinforcement, substrates is a relatively new branch of cold spraying; however, it is important to highlight that the bonding mechanism is different from the one that occurs when a metallic substrate is considered. Indeed, in recent years these materials are increasingly using in several sectors of engineering such as automotive, aerospace and construction replacing metals for different applications. However, if on one hand these materials offer advantages such as lightness, high strength to weight ratio and flexibility in designing shapes and forms, on the other hand it could be useful to improve some of their properties such as the electrical ones, electromagnetic shielding capabilities, thermal conductivity, flame resistance, and erosion and radiation protection, in order to further widen the fields of application of these materials. In this regard, the surface metallization is considered to be an effective technique to enhance the above-mentioned surface properties and then expand their engineering application fields.

To date, a lot of conventional techniques were developed for the metallization; for instance, the vapour phase deposition, the thermal spray deposition, the electrodeposition, the electroforming and electro-less deposition are the most used for research applications. Unfortunately, the abovementioned technologies suffer from a series of disadvantages that can limit their use in some exceptional cases. The vapour phase deposition based techniques are characterized by relatively high equipment and processing costs and limitations on the maximum dimensions of the processable work-pieces. At the same, the electroforming deposition processes are characterized by low bonding force, time-consuming production cycle, and expensive moulds; finally, the thermal spraying techniques (TS) can lead to degradation of the substrates, especially for the thermal-sensitive ones, due to the molten state of the impacting particles. Several studies were carried out with the scope to investigate the possibility to lay down metallic particles on a target surface at temperatures well below the melting point of the material, overcoming the abovementioned issues [1]. On these premises, CGDS technology appears to be a suitable technique to process temperature-sensitive materials [2].

Compared to other techniques, the cold spray offers some intriguing advantages when spraying metallic particles on polymeric or FRPs substrates: (i) the deposition is made possible by only mechanical interlocking mechanism, and no chemical reactions are required to take place between the particles and the substrate that would be almost impossible between a metal and a polymer; (ii) suitability of powder granulometry's is less restrictive than for conventional thermal spraying. Then, cold spraying covers a broader range of powder size, i.e., from micron down to submicron; (iii) compared with thermal spray processes, less heat input is required in cold spray so that the powders can retain their primary properties during deposition. Therefore, the typical cold sprayed coatings are characterized by reduced oxide phenomena and surface distortion, absence of voids and phase transformation, and weak residual stresses, which are emphasized in TS coatings. On these premises, cold spray appears to be a suitable technique to process temperature-sensitive materials, such as polymers and polymeric composite without reaching the melting or degradation temperature of the substrate material; (iv) furthermore, one of the most essential advantages of cold spray is that the spray gun can be connected to a robotic arm, or the sample can be placed on a numerically controlled system, so that complex or free shape surfaces can be easily prepared. This can be a significant competitive advantage in the realm of additive manufacturing [3–5].

In this chapter, a comprehensive overview of the metallic coating depositions on polymeric and FRPs substrates through cold spray will be provided. The deposition behavior of the coatings and the adhesion mechanisms between the substrate and the particles will be examined in order to understand the phenomena better so that the practical applications of cold spray on polymers can be extended. The microstructure and the properties of the coatings will be reported to provide a useful guideline for practical applications.

2 Bonding Mechanism

Polymers are classified, according to their response when they are subjected to heat, in thermoplastics and thermosets. Thermoplastics, or simply called as plastics, consist of linear or branched polymer chains that can become soft when they are heated and hard if they are cooled; that means the thermoplastic materials are workable almost indefinitely. Different is the case for thermosets, which are materials containing polymer chains organized in a network structure formed during the curing process; this structure is the result of an irreversible chemical reaction meaning that it is no possible leading the material in its original state because the solid polymer degrades upon the application of heat.

The bonding mechanisms for polymeric materials are strongly dependent on the type of matrix chosen: while the thermoplastics, being more ductile, are more suitable for the process, the thermosets are subject to breakages during impact due to their fragile behavior.

Concerning the fibre-reinforced plastics, different fabrics, such as carbon, glass, basalt, kevlar, and so on, can be used as reinforcement leading to a considerable number of different composites. Regarding the cold spray deposition on composite substrates, the adhesion is mainly ruled by the properties of the polymer used as a matrix. On these premises, it was decided to discuss the bonding separately on thermosets and thermoplastic polymers, while the contribution of the reinforcement is not taken into account.

2.1 Bonding Mechanism for Thermoplastic Substrates

It is widely accepted that the bonding between the target surface and the impacting metal particles is due to the adiabatic shear instability phenomenon on the local interface. Due to this phenomenon, the particles impacting the substrate destroy the surface oxide layer causing a close contact between the surfaces of the powders and the substrate undergoing robust plastic deformation. It seems clear that this mechanism cannot be effective for the formation of coatings for thermoplastic polymeric materials whose bonding mechanism is, to date, only partially understood. In fact, due to the different nature of the materials involved, it could be difficult, if not impossible, that a strong adhesion strength can occur between the metal particles and the relatively softer polymeric substrate. There are two main theories that attempt to explain how bonding can take place. The former confirms the presence of the adiabatic shear instability phenomenon; however, in this case, it leads to the formation of a metallic-nonmetallic bond, causing the adhesion of particles on the substrate. This theory has been carried on by Assadi et al. [6] and Grujicic et al. [7] that analyzed the beneficial effects on the adhesion strength due to adiabatic shear instability occurring at the particle/substrate or particle/particle interfaces. The second theory, recently observed by Hussain et al. [8], proposed instead the idea that a superficial portion of the metal particles could melt and fill the micro-pores on the rough surface of the substrate forming a mechanical interlock favoured by the high pressures. The latter theory is also reflected in the beneficial effect on the deposition of the addition of hard particles, such as Al_2O_3, as these lead to an increase in the roughness of the target surface. The authors [9] also found exciting results by spraying TiO_2 hard particles onto a polymeric based substrate; in particular, the relatively hot carrier gas was proved to lead the softening of the surface that can embed the impacting particle. Under these conditions, when TiO_2 particles penetrate the softer substrate, a little portion of the materials is squeezed out and extruded from the substrate itself, which can act as binder among the particles and the surface to make the coating. It is noticeable that it can be challenging, if not impossible, the coating build-up under this bonding process and the formation of a thicker coating was proved to be very hard to achieve. In Fig. 1, it is conceptually shown how the flow of the polymer around the particle prevents it from falling out from the cavity created by the impact.

Just like it is observed when spraying on a metal substrate, the particles must exceed the critical velocity threshold, v_c, below which particles rebound while hitting the substrate. The critical velocity depends on the type of particle used; however, the threshold values have been estimated only with regard to the adhesion of metal powders on substrates composed of the same material. The adhesion of the particles on a different substrate takes place with different speeds, which can be both higher

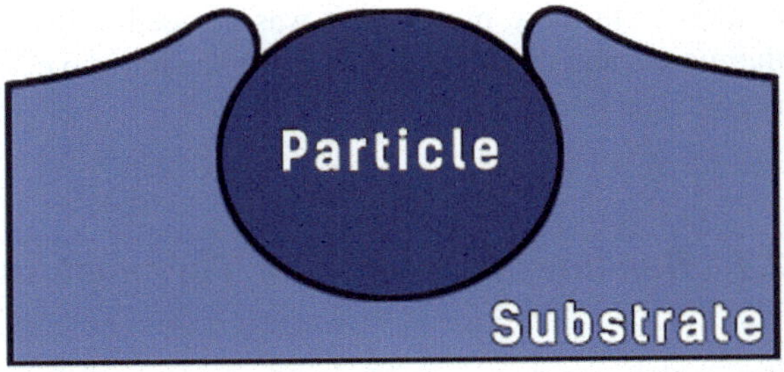

Fig. 1 Example of an embedded particle in a thermoplastic substrate. The flow of the polymer prevents the particle from falling out

than lower depending on the substrate/particle combination, and this value remains valid until a sufficient amount of spray material has been deposited to minimize the substrate effect. When the substrates used are softer and/or less dense than the particles, the embedment of the sprayed material occurs, as the particles are captured by the deformation of the softer substrate. In the case of polymers, the flow surrounds the particles preventing them from falling out.

It is clear that with the increase of the temperature of the carrier gas, it can be observed not only an increase of the speed of the particles but also an extension of the localized molten area with a consequent reduction in the stored elastic energy. In this way, the particles can more easily penetrate into the substrate rather than bounce above, leading to an increase in deposition efficiency. However, for thermoplastic polymers, when the substrate is heated to or above its glass transition temperature, the latter passes to a soft rubbery state and loses its stiffness. In fact, the long polymeric chains can slide on one another, and therefore, in this state, the contact of the impacting particles with the surface is not very strong, the particles are not well anchored to the substrate and can simply be torn away from the substrate leading to a decrease in deposition efficiency.

2.1.1 Influence of Glass Transition Temperature

For thermoplastic polymers, it is necessary to consider the effect of the local thermal softening of the substrate. In fact, when it is exposed to the flow of particles at a temperature close to its glass transition temperature (T_g), it softens, allowing the metallic particles to penetrate. During cooling, powders remain mechanically interlocked in it. It is for this reason that, unlike the conventional cold spray, the particles do not undergo significant plastic deformation when they impact. In Fig. 2, it can be seen as a copper particle that penetrates the PVC substrate, maintaining its round morphology, showing the typical wave shape phenomenon taking place at the coating interface. At temperatures much lower than the glass transition temperature, the polymer is relatively hard and brittle, so that at low impact speed, the particles cannot deform the substrate by sticking, while at higher speeds, only the substrate erosion is achieved. On the other hand, at temperatures significantly higher than T_g due to the drastic lowering of polymer strength, the deposition becomes complicated again as

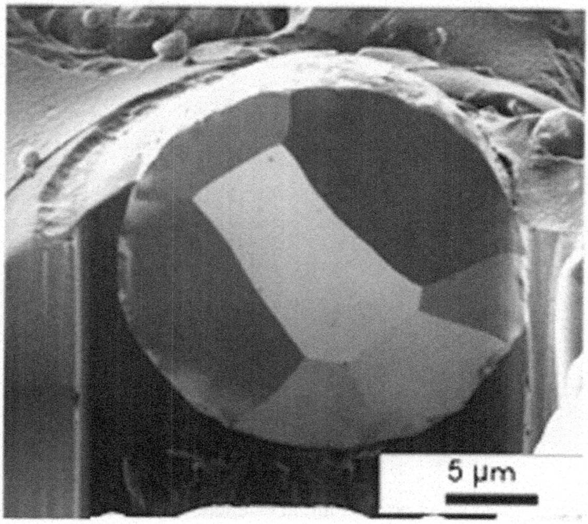

Fig. 2 SEM image of a copper sprayed particle on a PVC substrate

it is easy to erode the substrate. The possibility of carrying out successfully depositions is, therefore, strictly dependent on the substrate's ability to maintain excellent resistance to high temperatures; in this sense, the high-performing polymers, like PEEK, are favored.

Experimental tests have proved these phenomena; an example is the cold sprayed copper particles on ABS (T_g equal to 105 °C) substrate by setting a carrier gas temperature at 425 °C. Under these conditions, the particles impacting the surface can reach an impact temperature that exceeds 200 °C, which is far above the glass transition temperature of the polymer. The result is an extensive deterioration of ABS strength and erosion of the substrate itself. On the contrary, successfully results can be obtained with the relatively more performant PEEK and PEI materials if they are coated with copper particles at 425 °C.

The effectiveness of the cold spray deposition depends on the powder type. On this subject, Lupoi et al. [10] carried out experimental activities for cold spray (CS) applied on polymer substrates, and the found results were summarized in the chart shown in Fig. 3. It was found experimentally that the copper material, when is sprayed onto the substrate, produces an extensive erosion of the surface; differently, if aluminum particles are used to coat the polymer substrate, slight erosion of the target surface was observed without obtaining a good deposition. Better results were found for tin particles that can be efficiently sprayed onto the substrate without showing significant erosion of the impacted surface. The reasons are well explained in Fig. 3 reporting the particle impact energy ($E = 0.5\,mv^2$) against the powder material. Three different zones were identified on the chart which was defined from experimental observations. In fact, if the impact energy value is higher than given threshold values, the severe contact stresses generated can produce erosion of the polymer substrate. By looking at Fig. 3, the relatively high density and strong copper material needing high impact energy for the deposition, due to its thermal and mechanical properties, was

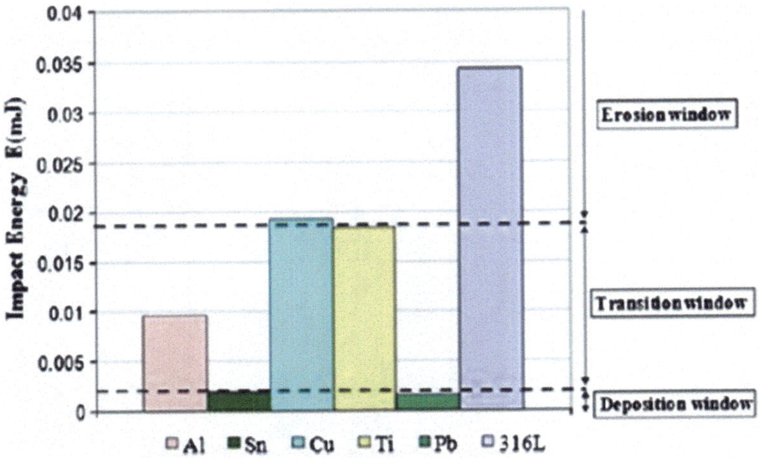

Fig. 3 Cold spray process initial characterization chart for polymer substrates

proved to overcome the erosion limit. On the contrary, the lower strength materials, such as tin, can stick on the substrate for lower impact velocities resulting in impact energy values that do not produce a damage of the surface, so that the coating can be efficiently formed; it can be noticed that a particle of copper can generate about 10.7 times more energy at impact than tin. The slight aluminum particles with the high deposition velocity are characterized by greater impact energy values than tin, for which the damage of the substrate is not substantial like copper, as the impact energy is relatively low. As for stainless steel material (316L), the correspondent deposition impact energy is relatively high, and the predominant effect is most likely erosion. Similar considerations can be done for both lead and titanium: in this case, the former can be easily cold sprayed on polymer substrate due to the low levels of impact energy; more difficulties can be found for titanium.

2.2 Bonding Mechanism for Thermosetting Substrates

Regarding the deposition on thermosetting polymers, the classic bonding mechanisms are not directly applicable. In fact, it has been experimentally verified that the particles do not deposit or undergo strong plastic deformations even when they impact the substrate with higher velocities than the critical one. It is therefore clear that the adiabatic shear instability mechanism cannot be considered valid when trying to deposit metal powders on thermosetting substrates. The reason why this happens is due to the low erosion resistance and fragile behavior of the thermosets. In fact, the critical velocities of most metals are so high that the particles cause severe erosion of the substrates without achieving the bonding.

To achieve an effective adhesion of the coating, the bonding strength of the first layer must be strong enough to avoid erosion of the substrate caused by the shoot peening effect. The deposition of the first layer, in fact, can affect the coating grow-up due to the fact that if the particles impacting on the polymer substrate are not firmly anchored on the surface, the upcoming particles can destroy the first layer making it impossible to make a thick coating.

The phenomenon that rules the deposition of the first layer is only the mechanical interlocking since there is no metallurgical bond. This implies that the particles have to reach a sufficiently high velocity, defined as v_{int}, without exceeding the substrate erosion velocity, $v_{ero,sub}$.

However, it is necessary that a partial erosion of the substrate occurs, in order to form micro-cracks on it. Metallic powders, in fact, settle into those craters, although the bonding is so weak that the particles merely attach. An example of this phenomenon is shown in Fig. 4. For some metals, there is a considerable erosion of the substrate before the deposition can take place ($v_{ero,sub} < v_{int}$): for those ones, it is no possible to carry out the deposition on thermosetting substrates.

After the first layer has been successfully deposited, it is possible to carry out metal deposition on the first layer referring to the conventional cold spray process. In fact, from this point on, the phenomenon that regulates the process is adiabatic shear instability. The velocity to be achieved in this case is the same critical velocity necessary for metal-on-metal deposition. It is, however, important that the bond of the first layer with the substrate is strong enough to resist at the shoot peening effect of the particles having a $v > v_{crit}$. If, in fact, this bond is not strong enough, the incoming particles will cause the removal of the first layer and the erosion of the substrate, making the build-up of the coating impossible to achieve.

Fig. 4 SEM micrograph of a metallic particle embedded in a thermoplastic substrate

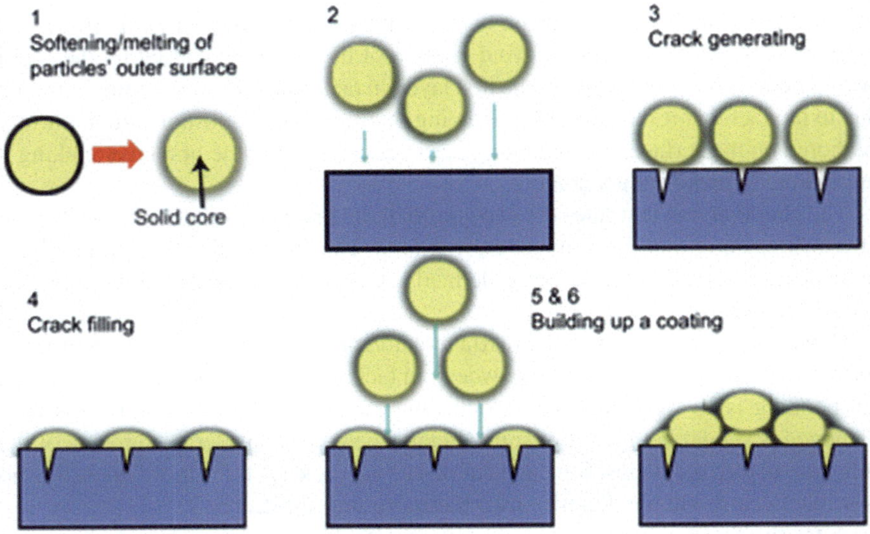

Fig. 5 A schematic crack filling process on an epoxy substrate

A new erosion velocity, v_{ero}, was then defined, which depends on the intensity of the bond between the first layer and the substrate and is ideally equal to that of metal-on-metal deposition if the bond with the first layer is sufficiently strong. It seems clear that the deposition window on thermosets is narrow, making it impossible to create a coating with most of the commonly used metals. A schematic representation of this process is shown in Fig. 5.

Some metals such as copper cannot be deposited on a thermosetting substrate [6] as they cause erosion without gaining mechanical interlocking. On the contrary, tin particles manage to obtain the mechanical anchorage by deforming during the impact with the substrate because of their softness. In this case, the partial fusion of powders plays a predominant role. In fact, deposition can take place only at relatively high temperatures (greater than 300 °C [7]). This temperature is higher than the tin melting point, but the particles stay in the hot carrier gas only for a short time, making it impossible to obtain the melting. When those powders impact the substrate, the molten part of the particles penetrates the micro-cracks achieving the interlocking and protecting the surface from further erosion.

Furthermore, tin is the only metal which makes the coating growth possible, because of the sufficiently low critical velocity. In fact, despite v_{ero} is quite low due to the weak bond strength reached with the mechanical interlocking, the critical velocity is usually still so low to make the deposition of the subsequent layers possible.

3 Improvements in Deposition Efficiency: The Interlayer Concept

From the literature review, the concept of the interlayer is often present. It is useful to deposit a ductile layer on a polymeric substrate to protect it and favour the adhesion of the successive layers [11]. Tin is usually used as a coupling metal, thanks to its great ability to adhere to surfaces without eroding it, due to its low critical speed. In fact, the impact energy of metal powders is usually so high (in the order of E = 0.02 mJ for copper), to cause severe damages to the polymer, contrary to what happens with tin, where all the impact energy, lower than an order of magnitude, is exploited to deform the powders and adhere to the substrate, as shown in Fig. 6. The high deposition efficiency of tin is also favoured by its ability to spread on the substrate, forming a dense and homogeneous coating. In fact, the degree of spreading (*D*/*d*) results to be directly proportional to the velocity and inversely proportional to the yield stress (σ_y) of the considered powders. The relation is shown in the modified Madejski's equation (Eq. 1):

$$(D/d) \propto (\rho, V_d, d) \propto^{-1} (\sigma_y) \tag{1}$$

where D, d, ρ, σ_y, and V_d are the diameter of the particle after impact, the diameter of the particle before impact, the density, the yield stress, and the particle impact velocity, respectively. From the previous equation, it is clear that the yielding tension strongly influences the deformation, which is why the tin particles become more deformed and have a higher deposition efficiency.

The aid in achieving a thick coating thanks to a softer interlayer is confirmed by several studies. Ganesan et al. [12] stated that coating a PVC substrate with a first layer of tin increases its hardness especially if dendritic copper particles are then sprayed on the top surface; in fact, those powders are able to deform themselves and develop a thick coating (800–1000 μm).

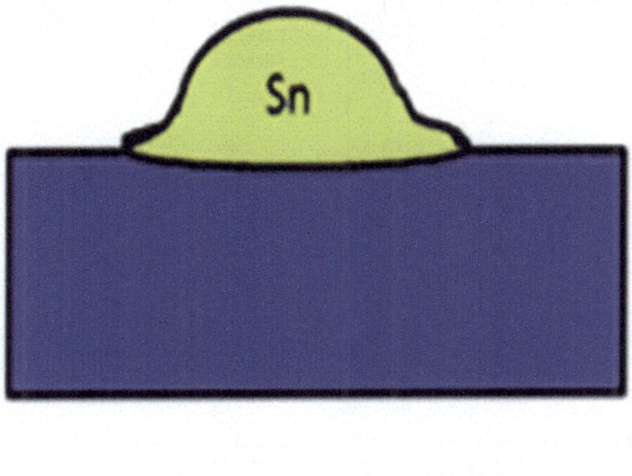

Fig. 6 Tin deformation on the substrate

4 Influence of Process Parameter on Thermoplastics and Thermosetting Substrates

Ganesan et al. [12] carried out experimental activities by spraying a variety of selected metal powders on both thermoplastic (PVC) and thermosetting (epoxy) polymer substrates, in order to analyze the influence of the gas pressure and temperature on the CS deposition efficiency (DE). They used a standoff distance of 30 mm, a traverse speed of 50 mm/min, and a powder feed rate of 5 rpm. The powder chosen were spherical and irregular copper to further analyze the effect of the shape of the powders in the deposition. The trend of DE respect to carrier gas temperature is shown in Fig. 7 for three different gas pressure values and sprayed powders. The pictures prove that the variation of DE with the gas temperature is more appreciable than that obtained by varying the gas pressure, which leads to similar conditions (see Fig. 7a–c).

As concerning the PVC substrate, the maximum values of DE were observed for spherical copper particles, in all the tested conditions; moreover, it can be seen that

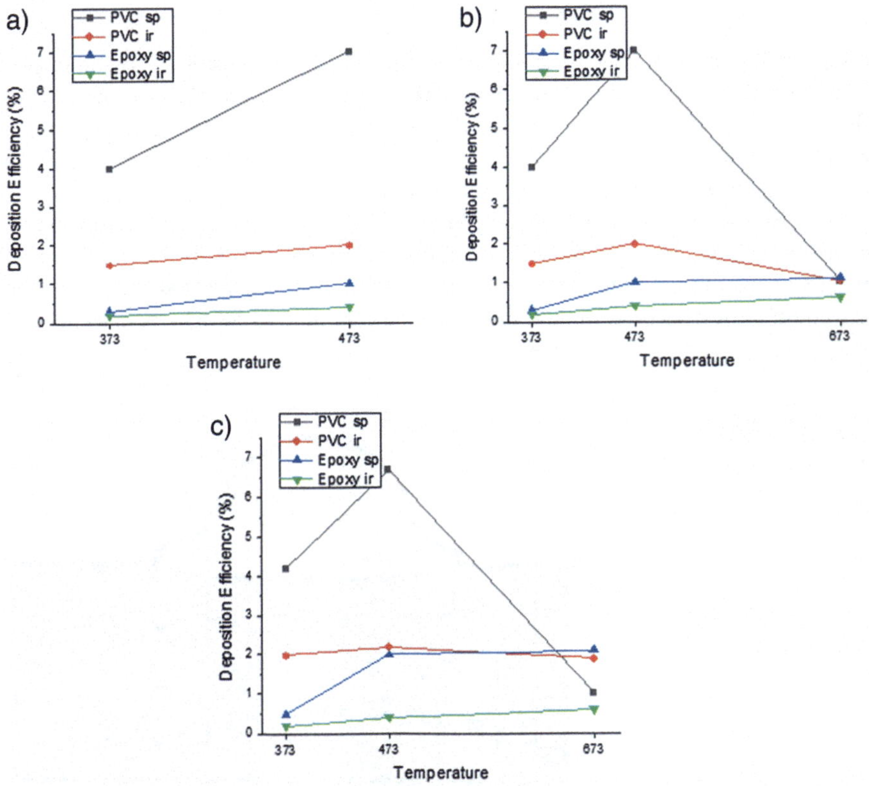

Fig. 7 Deposition efficiency versus carrier gas temperature for three different gas pressure values and sprayed powders: 1 MPa (**a**), 2 MPa (**b**), 3 MPa (**c**)

DE tends slightly to increase with the increase of gas temperature, up to 473 K, above which DE shows a sudden drop because of the glass transition temperature of the polymeric substrate (T_g = 353 K for PVC). The deposition efficiency of dendritic particles is the second highest. In both cases, the plastic nature of the substrate leads to the particles to attach and deeply penetrate the impacted surface without showing any significant rebounding phenomena (Fig. 8). Under these conditions, the polymer erosion is minimized, and DE can reach then the highest values. It can be seen in Fig. 8a that for gas temperature and pressure of 373 K and 20 bar, respectively, the particles do not have the right adhesion energy for bonding, explaining the reasons for which they do not attach on the substrate. When the gas temperature increases and reaches the value of 473 K, the temperature of the impacted surface also increases approaching T_g of the plastic substrate. In this case, the particles can adhere to the substrate leading to higher DE values (Fig. 8b). However, with the increase of the gas temperature to 673 K, namely above the substrate glass transition temperature, the polymer substrate begins slightly to lose its mechanical rigidity and its state changes to a soft rubbery state. Under these conditions, the bonding contact between the impacting particles and the substrate is not strong enough to ensure a dense and compact coating; the result is that the upcoming particles can easily destroy the layer leading to poor DE values (Fig. 8c).

Regarding the impacts on the epoxy surface, for both spherical and dendritic copper particles, DE slightly increases up to 673 K (Fig. 9). The reason is that the

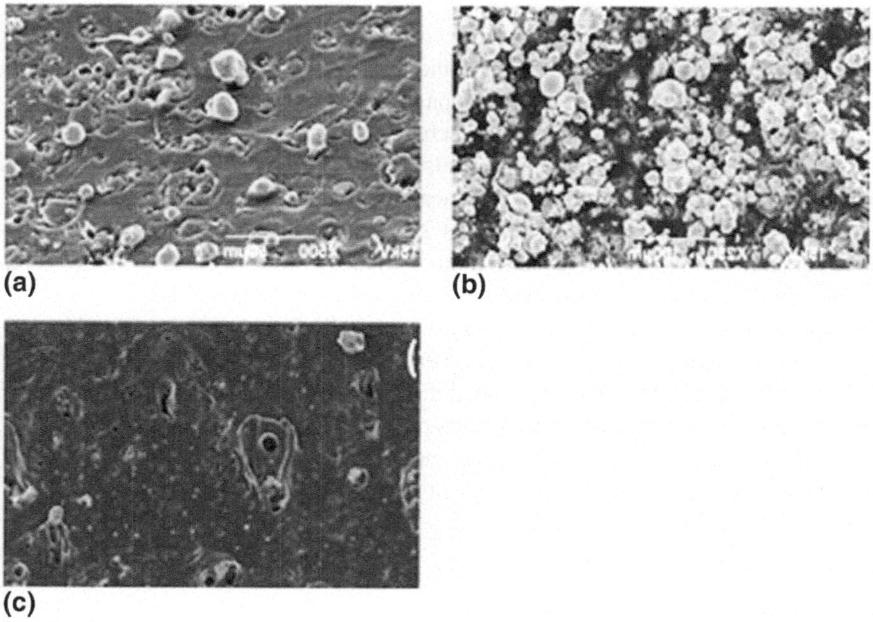

Fig. 8 Micrograph images of spherical copper particles sprayed on PVC substrate at three different temperature conditions: 373 K (**a**), 473 K (**b**), 673 K (**c**)

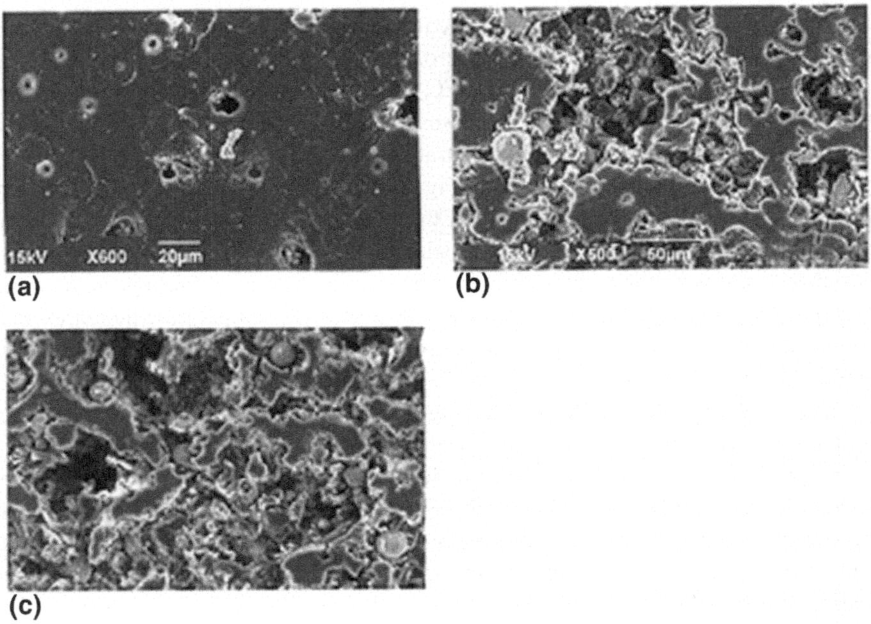

Fig. 9 Micrograph images of spherical copper particles sprayed on the epoxy substrate at three different temperature conditions: 373 K (**a**), 473 K (**b**), 673 K (**c**)

brittle behavior of epoxy is responsible for the formation of voids and damages experiencing on the target surface upon the particle's impact at high velocity. It is likely that during the impact process, the particles can merely fasten inside the voids and then firmly attach to the substrate, increasing DE. When the temperature of the gas, namely of the substrate, increases, also the velocity of the particles increases being a function of temperature. That means that the higher the temperature, the higher the particle velocity, the greater the substrate destruction (Fig. 9). Under this working condition, the particles have a more chance to fill the craters and the voids created by the particle bombardment, resulting in higher values of DE.

As a validation of the previously mentioned bonding mechanisms, the cold spray deposition efficiency was proved to be higher for the PVC substrate than for the epoxy cause of the ability of the PVC to embed the particle. The lower value of DE of the epoxy is due to the brittle behavior of this polymer that, under the bombardment of high-velocity particles, can be damaged and destroyed, leading to the formation of voids and craters on the surface. The result is poor DE.

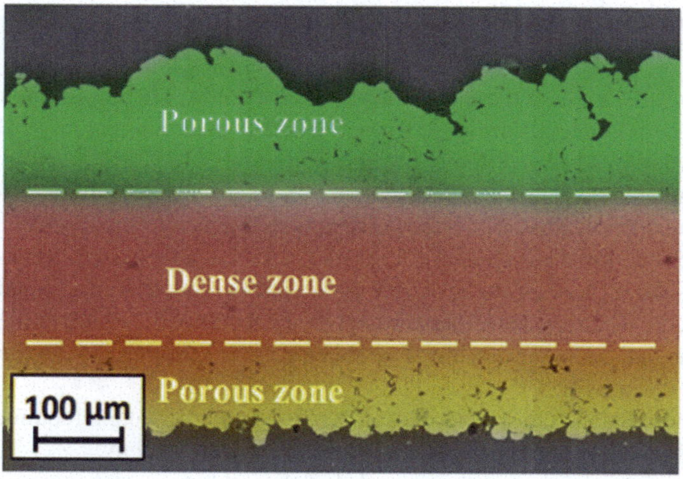

Fig. 10 SEM image of an Al coating on a thermoplastic substrate

5 Microstructure Analysis

Cold sprayed coatings on polymer substrates show a multi-layered microstructure, different than the one found on metallic materials. In fact, in the latter, the structure presents a gradual increase in porosity from the substrate/coating interface to the surface. This is due to the peening effect of the subsequently sprayed particles, which favours the densification of the coating.

When deposition occurs on a polymer, it can be instead observed a porous area at the substrate/coating interface. This peculiar structure is formed because the thermoplastic polymer acts as a shock absorber, causing the impacting particles to lose part of their kinetic energy. Due to the lower energy, the particles do not deform sufficiently to create a dense and cohesive structure. However, when the coating grows to reach a critical thickness, the shock-absorbing effect of the substrate no longer affects the adhesion. From this point on, the coating begins to thicken.

Finally, the last layer presents a porous structure again, due to the absence of the peening effect of the impinging particles. It is, therefore, clear that the polymeric substrate influences the microstructure for a discrete portion of the coating thickness. The whole structure formed is shown in Fig. 10.

6 Experimental Achievements and Case Studies

In this section, some main achievements regarding the deposition on both polymers and fibre reinforced polymers will be briefly discussed and summarized.

6.1 Spray Tests on Polymeric Substrates

In the literature, some of the metals most used for cold spray deposition have been deposited on polymeric substrates. These include copper, aluminum, iron, and tin, which were sprayed onto different polymeric substrates.

Among the various metals, however, only tin has been successfully deposited on a greater number of substrates, once the right nozzle geometry has been chosen. The thicknesses deposited with this metal vary from 45 μm up to nearly 100 μm.

The results of various studies show that copper produces a massive amount of impact energy when it reaches the critical velocity necessary to ensure the deposition. This energy may cause the erosion of the substrate rather than the effective deposition and limits the growth of the coating.

Aluminum, on the other hand, possessing a low specific weight requires to reach higher temperatures and pressures to achieve the critical velocity.

From literature review, it was found that the cold spray technology was used to process different materials, such as aluminum (Al), copper (Cu) and tin (Sn); in particular, micron-sized particles of these materials were sprayed on ABS blend (PC/ABS), polystyrene, polyamide-6, polypropylene, carbon PEEK450CA30, polyvinyl chloride (PVC) and resin epoxy substrates to prove the feasibility of the process.

6.1.1 Properties of the Substrates

- *Polycarbonate–ABS blend (PC/ABS)*

This blend of PC and ABS exhibits an excellent balance of properties, most notably high-impact resistance (even at cold temperatures), rigidity, dimensional stability, excellent creep resistance, low moisture absorption, and good heat resistance.

- *Polystyrene*

Polystyrene is a rigid and transparent thermoplastic, which is present in solid or glassy state at normal temperature. But, when heated above its glass transition temperature, it turns into liquid form that flows and can be easily used for extrusion. It becomes solid again when it cools off. It is resistant to many aqueous chemicals, and it is also an excellent electrical insulator. Crystal forms of the polymer polystyrene have low impact strength and get degraded on exposure to sunlight, due to photo-oxidation, which affects its mechanical properties.

- *Polyamide-6*

Better-known, like nylon, is a semi-crystalline polymer belonging to the polyamide family. Like all semi-crystalline polymers, it belongs to high chemical resistance. The acronym PA is almost always followed by a number that identifies the number of carbon atoms that make up the main polymeric chain and which can give polyamide different characteristics. PA6 achieve very high performance and, in exceptional cases, are metal substitutes.

- *Polypropylene*

Some of the most significant properties of polypropylene are low density, good thermal and abrasion resistance, chemical resistance, elasticity and toughness, fatigue resistance and insulation.

- *PEEK450CA30*

Polyether ether ketone (PEEK) is an organic thermoplastic polymer having a semi-crystalline structure able to retain excellent mechanical and chemical resistance properties under high-temperature conditions, i.e. close to the melting point of the material (343 °C).

- *Polyvinyl chloride (PVC)*

PVC has good mechanical properties and resistance to abrasion, wear and ageing, to chemical agents and to the attack of fungi and bacteria, it is a source of light, it is water-repellent, it is hardly inflammable and self-extinguishing.

- *Epoxy*

Epoxy resins are easily and quickly cured at any temperature from 5 to 150 °C, depending on the choice of curing agent. High adhesive strength and high mechanical properties are also enhanced by high electrical insulation and good chemical resistance. The epoxy resin has very good stiffness, toughness and heat resistant properties.

PC/ABS

Lupoi and O'Neill [10] reported the results obtained by spraying spherical copper, prismatic aluminum, and spherical tin particles on PC/ABS substrates.

Copper

The macrographs top view of the cold sprayed spherical copper particles are reported in Fig. 11. The coatings were obtained onto PC/ABS substrates by setting only two values for the pressure of the carrier gas, namely, 5 and 30 bar.

The process parameters set to spray the copper particles under the two abovementioned conditions are reported in Table 1.

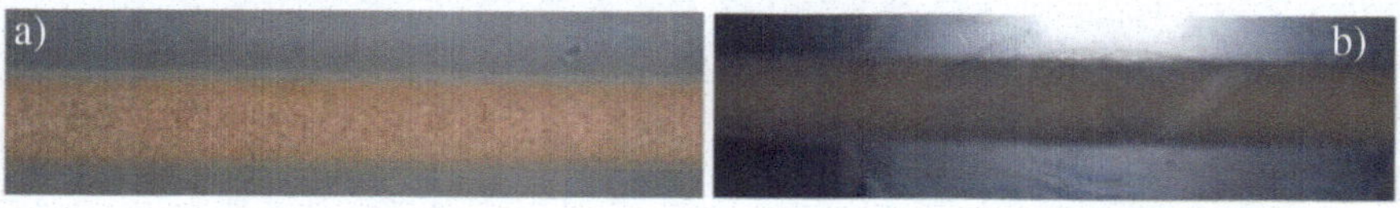

Fig. 11 Macrograph top view of the cold sprayed spherical copper particles onto PC/ABS substrates; the gas pressure was set to 5 bar (**a**) and 30 bar (**b**)

Table 1 Process parameters set to spray the copper particles

(a)	(b)
Standoff distance SoD: 40 mm Substrate transverse speed: 8.3 mm/s Rotational speed: 15 rpm Spraying pressure: 5 bar Spraying temperature: room temperature	Standoff distance SoD: 40 mm Substrate transverse speed: 16.6 mm/s Rotational speed: 18 rpm Spraying pressure: 30 bar Spraying temperature: room temperature

This study has proved that the best deposition conditions were those obtained by imposing the lower inlet pressure. In fact, from Fig. 12, it can be seen that more material can be deposited on the polymer surface for a gas pressure of 5 bar. The reason for such behaviour is that for lower gas pressures, meaning that for lower impact energies, the ductile thermoplastic material embeds the particles without showing any significant damage. Therefore, the first layer was proved easily to form by mechanical interlocking mechanism (Fig. 12a). Unfortunately, the upcoming particles do not have the impact energy required to adhere to the first-metallic layer duo to the insufficient speed. The deposition mechanism, in fact, is a metal-to-metal interaction between the particles and surface, and the coating build-up could be difficult, if not impossible, under this condition. When the gas pressure increases to 30 bar, which means an increase in particle speed, the substrate can fail and show voids and craters leading a substantial erosion of the PC/ABS surface, as shown in Fig. 12b. Despite the particle speed being that required to create a copper coating, the relative impact energy released to the polymer substrate is so significant to create voids and cavities on the surface, leading to substrate erosion and degradation. The result is that the higher the gas pressure, the higher the erosion, resulting in poor cold spray deposition efficiency.

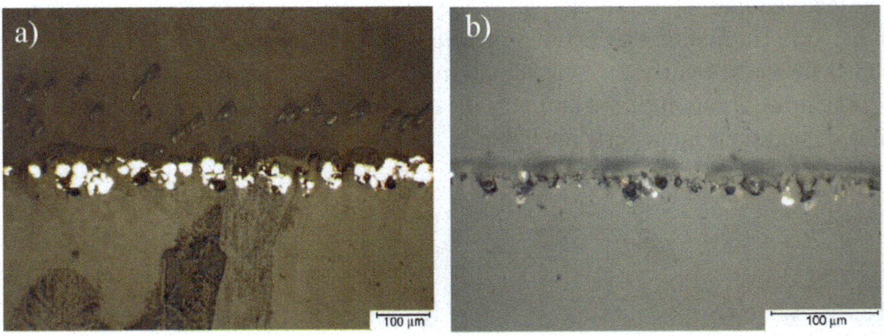

Fig. 12 Section images of the cold sprayed spherical copper particles onto PC/ABS substrates; the gas pressure was set to 5 bar (**a**) and 30 bar (**b**)

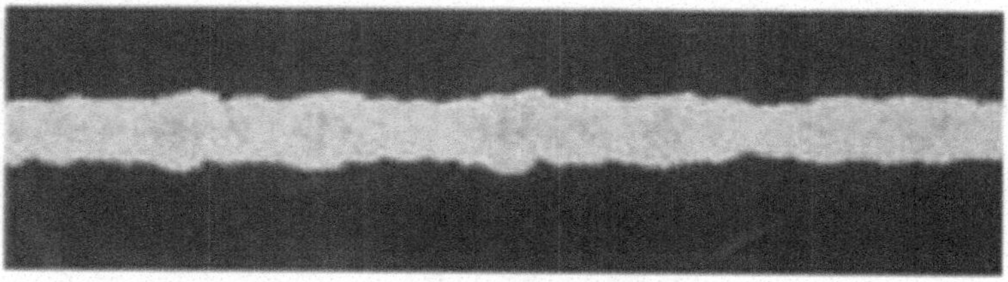

Fig. 13 Tin tracks on PC/ABS

Aluminum

The authors [10] also studied the possibility of using prismatic aluminum particles to coat pure polycarbonate by CS. This attempt proved that aluminum could not be efficiently sprayed on polymer substrate because no deposition was achieved, also by setting high values of gas pressure (30 bar in this activity). The possible explanation is that aluminum is a lightweight material characterized by a relatively low-density value and it is not capable of generating the required impact energy for bonding; aluminum coatings could be formed only if the carrier gas is heated and higher particle velocities can be reached (about 600 m/s).

Tin

Experimental trials are shown in the close-up picture in Fig. 13. Experiments were carried out by injecting the tin particles axially in the divergent part of the nozzle, after the throat section, through a low-pressure powder feeder. The reason for such choice is due to the fact that this kind of material, which is a soft thermal-sensitive material, could clog the nozzle as the particles can adhere on the internal surface of the nozzle itself, compromising the effectiveness of the CS process. Spherical Sn particles were cold sprayed following the process below:

Standoff distance SoD: 103 mm
Substrate transverse speed: 14 mm/s
Rotational speed: 200 rpm
Spraying pressure: 30 bar
Spraying temperature: room temperature.

The outcomes are shown in Fig. 14, where the tin particles can be easily identified after the chemical etching. Typical tin coatings on PC/ABS substrates have a thickness in the range 45–100 μm. It can be seen from the figure that the particles deformed significantly and changed their geometry if compared to their initial shape. The deformation mechanism allowed the particles to deposit one on top of the other, making it possible to create a dense and thicker coating. Again, a second deposition pass caused the erosion of the first layer in all the cases investigated proving difficult of coating grow-up. The reasons were explained above; however, the new researches have been carrying out on this topic in the last years.

Conclusions

The studies carried out on PC/ABS substrates proved that the sprayed particle typology affects the deposition mechanism; in particular, it was found that the copper

Fig. 14 Section image of tin particles sprayed on PC/ABS substrate

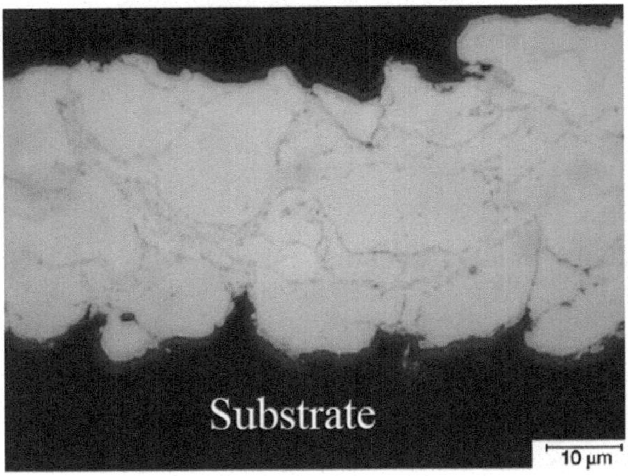

material particles, which are dense and relatively high strength powders, produce an erosion of the substrate when they impact on the surface due to the relatively high impact energy (in the order of 0.02 mJ for a single impact). With the increase of the particle diameter to micrometre size, the level of the impact energy increases, and the result is that the stress transmitted to the substrate can overcome the strength of the material, leading to the formation of voids, craters, and failures. As for aluminum coating, it was found that no effective deposition can be obtained due to the lightness of the material. In this case, deposition occurs when the spraying temperature (that is the critical velocity) is increased. Regarding tin deposition, which is a lower strength material, the critical velocity required for bonding is relatively lower than the above-mentioned metals, resulting in a decrease of the impact energy transmitted to the substrate; that means that the deposition can be effectively obtained. Unfortunately, the mechanical properties of this material are very poor, so it could be used as an interlayer material to initialize the coating formation on polymer substrates.

Polystyrene—Polyamide-6—Polypropylene

Lupoi and O'Neill [10] reported the results obtained by spraying spherical tin on Polystyrene, Polyamide-6, and Polypropylene.

Tin

On these polymers, the spherical tin powder was sprayed successfully following the previous process. The results are shown in Figs. 15 and 16.

Fig. 15 Tin tracks on polystyrene (**a**), polyamide-6 (**b**) and polypropylene (**c**)

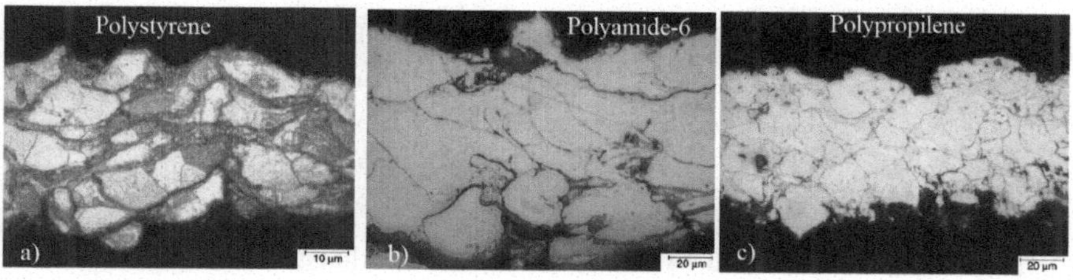

Fig. 16 Optical microscope observations of the cross-sections of tin particles sprayed on polystyrene (**a**), polyamide-6 (**b**) and polypropylene (**c**)

Conclusions
As previously stated, tin is a deformable material, and deposition occurs quickly on a wide range of thermoplastics substrates.

PVC
Ganesan et al. [12] made a bimetallic coating by using dendritic copper as well as spherical copper particles as interlayer on PVC.

Bimetallic coating: dendritic and spherical copper
A simple sketch of the interaction between a dendritic copper particle and a PVC substrate is reported in Fig. 17. It can be seen that the contact area is extremely irregular; the contact points are randomly distributed so that the particles can easily stick on the impacted surface. Unfortunately, the bonding force existing between the particle and the polymer substrate is not strong enough to face the impact energy of the upcoming particles due to the dendritic shape of copper powder. The result is that the shot peening effect produces an erosion of the metallic layer; therefore, the coating build-up and the particle-to-particle cohesion bonding seem to be very difficult under these working conditions.

However, as stated before, a dense and compact coating made of copper particles can be successfully built on a composite substrate with polymeric matrix by using an appropriate interlayer and by choosing the right typology of particles, in terms of shape and morphology. The spray parameters used for the thick copper coating are:

Fig. 17 Simple sketch of the interaction between a dendritic copper particle and a PVC substrate

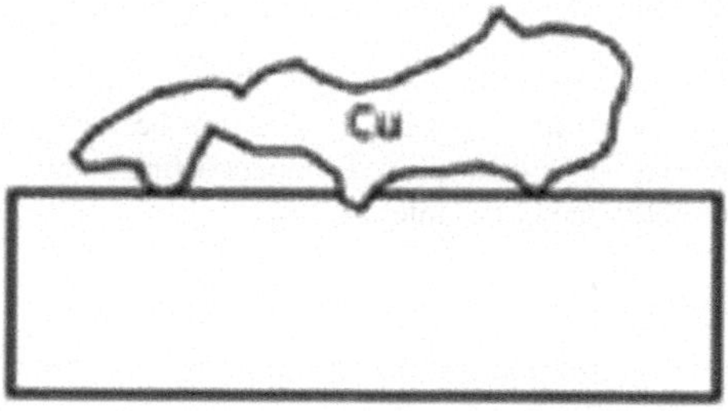

Dendritic copper	Spherical copper
Standoff distance: 30 mm	Standoff distance: 30 mm
Substrate transverse speed: 100 mm/min	Substrate transverse speed: 200 mm/min
Rotational speed: 5 rpm	Rotational speed: 5 rpm

The procedure developed by the authors foresees that at the beginning of the deposition process, the first layer of the coating is built by using the spherical copper particles so that they can deeply penetrate the PVC substrate and bond with the surrounding surface. After that, the top coating can be created by spraying dendritic copper particles with irregular and random shape, as shown by the sketch reported in Fig. 18a. In fact, dendritic particles can be mostly accelerated by the carrier gas due to the higher value of drag coefficient for dendritic and irregular particle shape, ensuring a stronger adhesion strength and a metallic bonding. Despite the higher velocity of dendritic particles, the total impact energy-reduced upon impact with the substrate because of many contact points as well as their large and irregular surface area. The result is less erosion and enhanced coating growing and build-up. Moreover, as shown in Fig. 18b, the several contact points of dendritic particles with the spherical ones which adhered on the substrate, tend to grow more significant by the effect of the upcoming particles, resulting in metallic bonding.

Conclusions
It was found that with the aid of the tin interlayer, copper particles can be successfully cold sprayed on PVC substrate to form a dense and compact metallic coating. This mechanism is confirmed by the FIB image in Fig. 19 showing that the particles with dendritic morphology make a shaped contact with the surrounding spherical particles. The contact interaction between dendritic and spherical copper particles is indicated in Fig. 19 by a dotted line. Finally, a thicker and more compact coating can

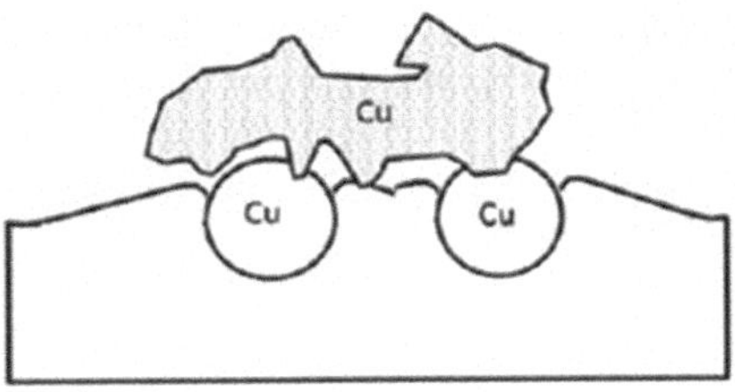

(a)

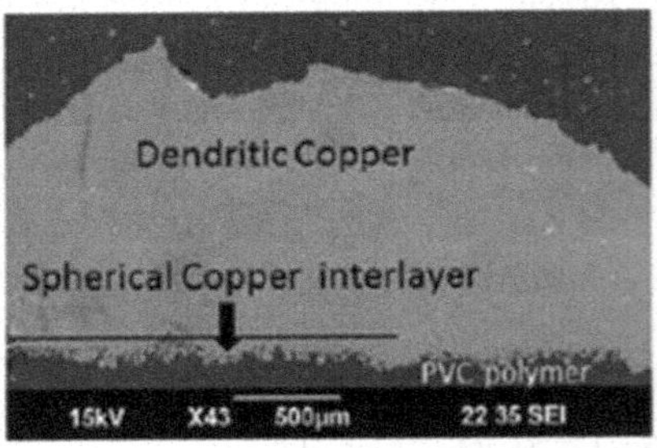

(b)

Fig. 18 Simple sketch of interaction between the dendritic particle and spherical one used as interlayer (**a**); cold sprayed copper coating obtained by using both dendritic particles and spherical particles, the last as interlayer (**b**)

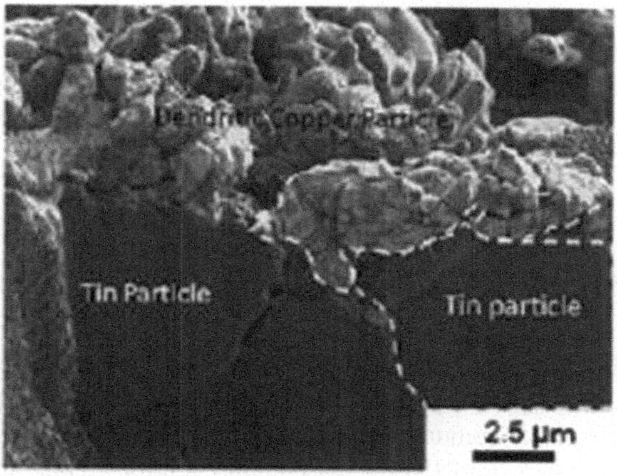

Fig. 19 FIB image showing the contact interaction between dendritic and spherical copper particles

be developed on PVC surface when the spherical particles are used as interlayer and the dendritic ones as a top coating.

Epoxy

The powders metal deposition on thermosetting substrates is more difficult because of their greater fragility. The powders impact causes the erosion of the substrate instead of its plastic deformation. Anyway, Ganesan et al. [13] made a bimetallic coating with dendritic copper and spherical tin as interlayer on epoxy substrates.

Bimetallic coating: dendritic copper and spherical tin

As mentioned in the previous paragraphs, tin particles can be easily cold sprayed on polymer substrate because of reduced mechanical properties of such material, compared to epoxy substrate, resulting in a strong deformation upon impact with the target surface. The jet formation ensures the mechanical interlocking and proper anchoring of the particles with the substrate. Again, an appropriate material used as a metallic interlayer was here proposed to form a thicker and more compact coating.

The spray parameters used for the thick copper coating are:

Dendritic copper	Spherical tin
Standoff distance: 30 mm	Standoff distance: 30 mm
Substrate transverse speed: 100 mm/min	Substrate transverse speed: 500 mm/min
Rotational speed: 5 rpm	Rotational speed: 5 rpm

A similar phenomenon that occurs in PVC was found for Sn particles cold sprayed on the epoxy substrate; in particular, due to the weakness of Sn material, the upcoming dendritic copper particles can easily stick on this first metallic layer to form a stronger bonding on the surface.

Conclusions
As the PVC case, the dendritic particles can easily adhere to the metallic layer obtained by using spherical tin particles. The result is that by using the interlayer concept, the coating built-up was proved to be more simple and effective. A summary of metal cold spray depositions on polymer substrates is reported in Appendix 1.

6.2 Spray Tests on Fibre Reinforced Polymers

From literature review, it was found that successful metallic coatings have been produced by cold spray onto various fibre reinforced composite materials; in particular, copper, aluminum, tin and also zinc particles were sprayed on glass fibre reinforced polymers (GFRPs), and carbon fibre reinforced polymers (CFRPs) with epoxy resin or PEEK matrix. The cold spray technology was used to coat hemp-PLA laminates by metal powders cold spray.

The choice of using the abovementioned metallic particles is relative to the electrical and thermal characteristics of these material typologies that make them suitable materials for the cold spray deposition on composite substrates. They also have excellent mechanical properties and reduced costs and can be effectively sprayed on metallic surfaces. In particular, copper is the most conductive, aluminum is lightweight and cold sprayable on metal substrates, and tin was proved to be the best material for cold spray deposition on polymer substrates. Finally, zinc was chosen to be mixed with tin particles for the increasing of the overall conductivity properties of the coatings.

GFRP substrates
Lupoi and O'Neill [10] reported the results obtained by spraying spherical copper particles on glass fibre composite materials. For this purpose, the cold spray facility was built at home without the heating system of the carrier gas and used in this experimentation, resulting in a room temperature spraying process.

The macrograph top view of the cold sprayed spherical copper particles is reported in Fig. 20. The coatings were obtained onto GFRP substrates by setting the pressure of the carrier gas to 5 (Fig. 20a) and 30 bar (Fig. 20b).

Copper coatings were cold sprayed following the process parameters reported below:

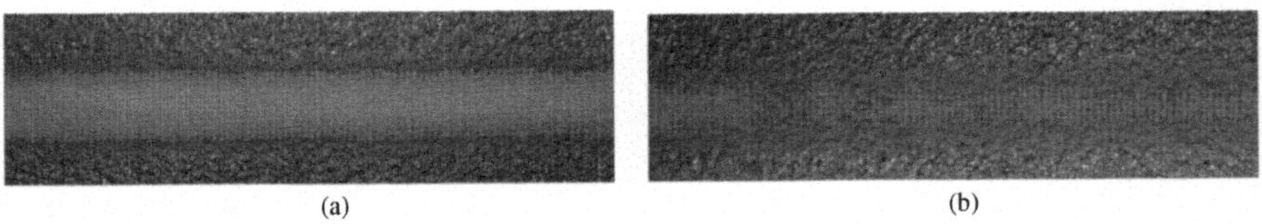

Fig. 20 Macrograph top view of the cold sprayed spherical copper particles onto GFRP substrates; the gas pressure was set to 5 bar (**a**) and 30 bar (**b**)

(a)	(b)
Standoff distance SoD: 40 mm	Standoff distance SoD: 40 mm
Substrate transverse speed: 8.3 mm/s	Substrate transverse speed: 8.3 mm/s
Rotational speed: 5 rpm	Rotational speed: 12 rpm
Spraying pressure: 5 bar	Spraying pressure: 30 bar
Spraying temperature: room temperature	Spraying temperature: room temperature

It was found that when the gas pressure was 30 bar, the deposition efficiency strongly decreased, meaning that a lesser amount of material was deposited onto substrate compared to the weaker pressure condition (5 bar). In fact, it can be seen from Fig. 20a that the coating is more compact and denser than the other case. This result is in agreement with the outcomes found by the same authors when sprayed copper particles on PC/ABS substrates, whose discussion was reported above. As it can be seen by looking Fig. 21a, for low energy impacts, a dense and compact layer of particles can be obtained; with the increase of the gas pressure to 30 bar (this value was chosen with the scope to try to develop a denser and thicker coating), an extensive erosion of the substrate was observed, as shown in Fig. 21b: in fact, the particle impact energy transmitted to the substrate is so high to lead damage and failure of the coating zone.

Conclusions

Results obtained by spraying micron-sized copper particles on GFRP substrates showed that under these working conditions the glass fibres presence within the polymeric matrix does not affect the deposition behaviour, which seemed to be strongly dependent on the inlet gas pressure. In particular, it was proved that relatively high

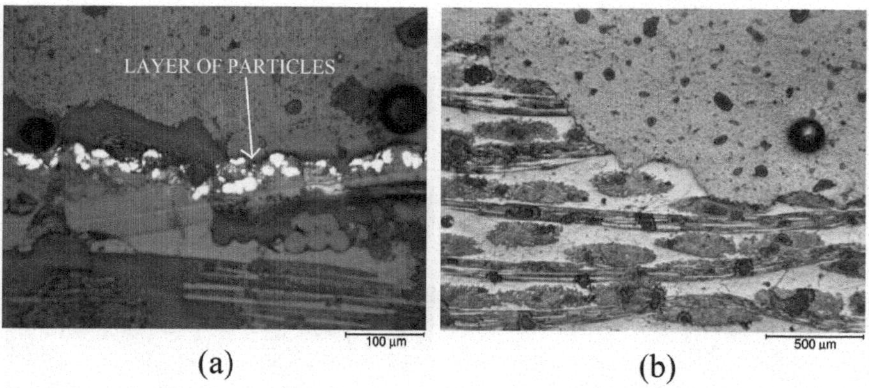

Fig. 21 Micrographs of copper tracks on GFRPs: gas pressure set to 5 bar (**a**) and gas pressure set to 30 bar (**b**)

values of gas pressure (30 bar in this activity) lead to substrate damage and erosion because of the relatively high value of contact stress generated by the single-particle impact energy.

CFRPs with PEEK matrix

In this activity, pure Al metallic coating and a double coating made of aluminum and copper particles, the former used as interlayer, were cold sprayed on carbon fibre-reinforced polymer substrate (PEEK450CA30) by Zhou et al. [1]; nitrogen was used as the particle carrier gas.

Aluminum

Micron sized spherical Al particles were cold sprayed on commercially available carbon fibre reinforced PEEK450CA30; the main CS parameters are reported below:
Standoff distance SoD: 20 mm
Substrate transverse speed: 5 mm/s
Rotational speed: 5 rpm
Spraying pressure: 12 bar
Spraying temperature: 300 °C.

The SEM image of the cross-section of the aluminum coating is shown in Fig. 22a proving the good ability of aluminum particles to be sprayed on polymer surface; in fact, the coating thickness is equal to 500 μm and the coating voids and craters are uniformly distributed and minimized. The Al coating and the PEEK450CA30 substrate seem to exhibit good bonding. No cracks and other defects were observed. The substrate interface (Fig. 22b) shows pronounced irregularities, meaning that the substrate thermal softening was induced by the heating carrier gas. Upon aluminum particles impact onto the target surface, a deformation of the substrate occurs which embeds the particle and ensures the good anchoring at the particle-substrate interface.

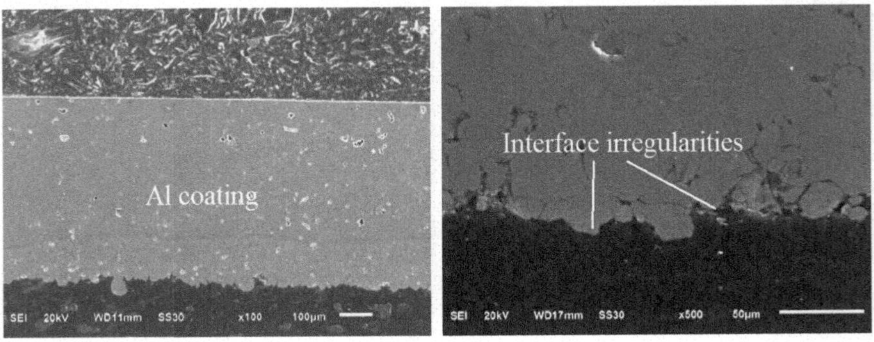

(a) (b)

Fig. 22 SEM image of the cross-section of the aluminum coating (**a**) and details at the particle-substrate interface (**b**)

Bimetallic coating: aluminum and copper
A double coating made of aluminum and copper particles, the former used as interlayer, were cold sprayed on carbon fibre-reinforced polymer substrate (PEEK450CA30). The first layer of pure aluminum was laid down by imposing the process parameters described above; the second layer consisting of spherical copper particles was developed on the Al coating previously sprayed and polished, by setting the CS parameters reported below. It is clear the concept of metallic interlayer here proposed.
Standoff distance SoD: 15 mm
Substrate transverse speed: 3.3 mm/s
Rotational speed: 0.1 rpm
Spraying pressure: 19 bar
Spraying temperature: 450 °C.

The cross-section of the double Al/Cu metallic coating is reported in Fig. 23. It can be seen that both the coating typologies seem to be quite dense and compact, without showing irregularities and significant discontinuities; also the voids percentage in Al coating was proved to reduce to 1.1%. The reason is that the shoot peening effect of the upcoming copper particles tends to improve the coating quality and morphology. By looking Fig. 23b, it is noticeable the shear instability phenomenon occurring at the Al-Cu interface: the jet mixing forming between the Al and Cu particles leads to firm adhesion strength due to the strong mechanical interlocking.

Conclusions
Aluminum coating can be successfully laid down on carbon fibre PEEK surface by means CS. The coating was proved to be a quite dense ad-free of significant defects. The carrier gas heating can have a beneficial effect on CS deposition because of the thermal softening effect on the polymer substrate that can more easily embed the impacting Al particles. The result is a better mechanical interlocking, namely a good adhesion strength. Better results can be obtained by using aluminum coating as an interlayer between the polymer surface and the copper particles. In this case,

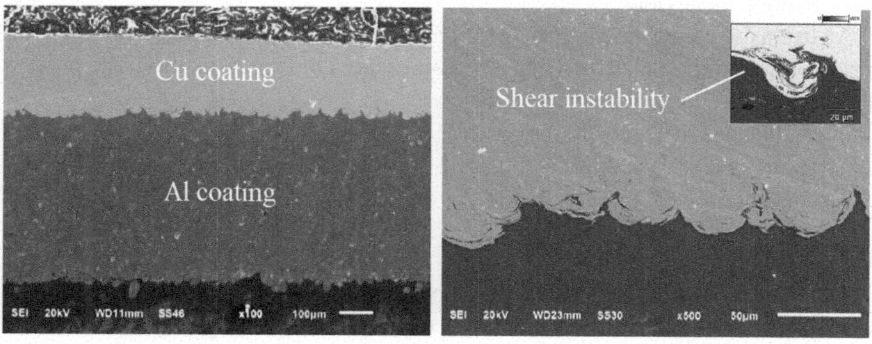

(a) (b)

Fig. 23 SEM images of Al/Cu coating (**a**) and interface (**b**) on PEEK450CA30 substrate the mechanical interlocking can be strongly enhanced by the formation of jet mixing relative to adiabatic shear instability at Al-Cu interface.

CFRPs with epoxy resin

The deposition of metallic particles on thermosetting substrates, like CFRPs, was found to be very difficult because of their greater fragility, as discussed in the previous sections of this chapter. However, some authors [4, 5, 14] investigated the deposition characteristics of cold sprayed aluminum, copper, tin and zinc particles on CFRP substrates.

Aluminum

Affi et al. [4] used a self-designed and custom-manufactured cold spray system for spraying commercially pure aluminum powders of two different sizes, 3 μm and 15 μm, on CFRP plates. The best process parameters are reported below:
Standoff distance SoD: 20 mm
Substrate transverse speed: 20 mm/s
Rotational speed: 6 rpm
Spraying pressure: 10 bar
Spraying temperature: 600 °C.

In agreement with the results found by Lupoi and O'Neill [10], for which in order to obtain the particle anchoring, the impact energy of each particle should be below a given value, large-size particles can erode the CFRP substrate due to their relatively high impact. In particular, it was proved that aluminum particles with an average size of 15 μm could erode CFRP substrates. Hence, with the scope to obtain a dense metallic coating, smaller aluminum particles with the size of 3 μm were sprayed and successfully laid down on CFRPs, decreasing the impact energy. By looking Fig. 24, the powders seem to be well attached to the matrix surface, obtaining a coating thickness of about 30 μm. Several difficulties were found when the authors proved to increase aluminum thickness. The reason is that the plastic matrix was removed by the peeling off upon particles impact the surface leaving the carbon fibres to the shoot peening effect. The fibres, in fact, cannot be coated by metallic particles due to their fragile behaviour.

An extensive study was also performed by Che et al. [14] that sprayed commercial-purity aluminum particles by using a high-pressure cold spray facility. The process parameters investigated are reported below:
Standoff distance SoD: 20–80 mm
Substrate transverse speed: 300–500 mm/s
Rotational speed: 1–5 rpm
Spraying pressure: 20–50 bar
Spraying temperature: 100–400 °C.

The results showed that, under the abovementioned CS conditions, it seems to be very difficult the coating formation as no successful coatings were observed; it was found that the substrate erosion is the main issue for developing of a continuous coating. In fact, two distinct areas can be observed characterizing the impacted surface: the zones where the particles bombardment removed the epoxy resin from

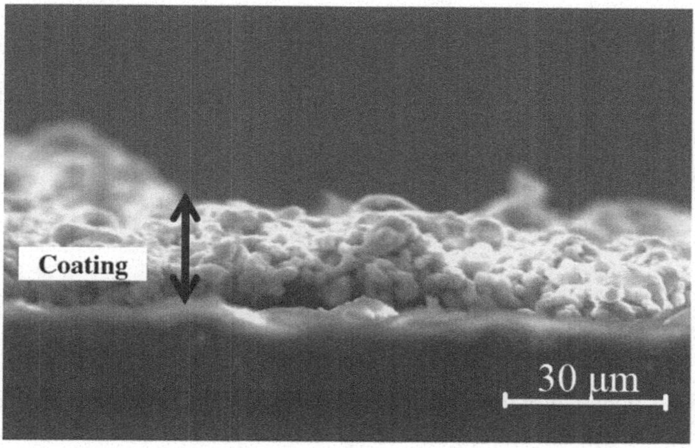

Fig. 24 Aluminum coating on CFRP by using small particles (3 μm) to reduce impact energy

the surface, and other zones of the same sample that did not damage extensively; the final effect is a discontinuous coating because the carbon fibre reinforcements can damage when exposed to the particles shoot peening (within the eroded area) and the particles cannot stick on or among the exposed reinforcement (Fig. 25).

Copper

A huge experimental campaign was carried out by Che et al. [14] who used the high-pressure cold spray facility as well as the low-pressure one to lay dawn copper particles on CFRP substrates, with the scope to analyze the influence of the gas pressure on coating quality and deposition. In particular, spherical and irregular shaped copper particles were sprayed by using both the CS systems; spherical copper particles only were sprayed by means of low pressure. The cold spray process parameters used in this experimentation are reported in Table 2.

Fig. 25 Micrograph of aluminum impacted sample

Table 2 Cold spray process parameters used to lay down copper particles

	Low-pressure cold spray	High-pressure cold spray	
	Spherical shape	Spherical shape	Irregular shape
Standoff distance SoD (mm)	18	50–200	40–100
Substrate transverse speed (mm/s)	25	50–500	300
Rotational speed (rpm)	11	1	0.5–5
Spraying pressure (bar)	3.4–13.8	20–50	20–50
Spraying temperature (°C)	425	100–800	100–700

The results obtained in this case are very similar to those observed for aluminum particles; in particular, the copper coating formation and build-up were proved to be complicated for all the tested conditions and for both spherical and irregular particle shape. As found above, there exist some zones of the impacted sample where the epoxy resin was removed, and the reinforcements are exposed to particle impingement, and other zones that embedded the sprayed powders. Differently from the previous case, it was found a stronger erosion of the substrate caused by the copper particles; the result is that the carbon fibre reinforcements are not only exposed but also fractured by the upcoming copper particles, as clearly shown in Fig. 26.

Tin
Che et al. [14] used the high-pressure cold spray facility as well as the low-pressure one to lay dawn tin particles on CFRP substrates in order to analyze the effect of the powder typology. Tin coatings were cold sprayed following the process parameters reported in Table 3.

Good results were found for tin particles. Unlike the results obtained for aluminum and copper powder materials, in this study it was found that it is possible to lay down tin particles forming clusters on exposed carbon fibre reinforcements,

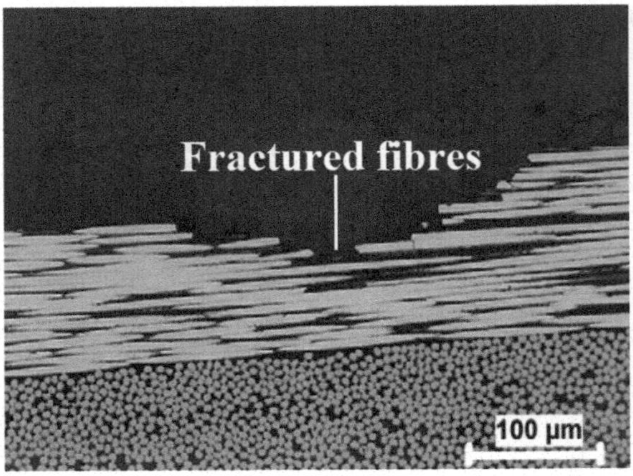

Fig. 26 Micrograph of the copper impacted sample

Table 3 Cold spray process parameters used to lay down tin particles

	Low-pressure cold spray	High-pressure cold spray
Standoff distance SoD (mm)	18	40–200
Substrate transverse speed (mm/s)	12.5–50	300–500
Rotational speed (rpm)	13	1–8
Spraying pressure (bar)	2.9–13.8	15–30
Spraying temperature (°C)	25–325	50–300

achieving a better deposition (Fig. 27). Furthermore, the impacted particles showed a huge deformation after impacting the substrate; they changed their spherical shape meaning that a good adhesion was achieved and the coating build-up was possible.

Results obtained by using the low-pressure machine seemed to reduce the level of erosion during cold spraying, resulting in a more effective deposition. The cross-section of cold sprayed Sn particles on CFRP substrates is reported in Fig. 28 for two different pressure values (4.1 and 5.5 bar) and keeping the temperature value of the carrier gas at 300 °C. By looking at this figure, the tin coating seems to be compact and quite dense in both the conditions investigated; the substrate top surface appears almost free of defects, without showing strong erosion and polymer degradation. However, a more detailed analysis of the coating-substrate interface revealed that the shot peening effect leads to the formation of irregularities and wave deformations

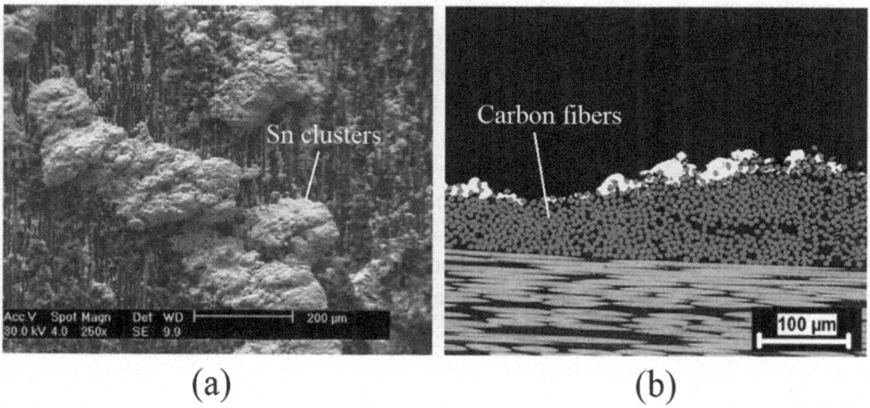

Fig. 27 Micrographs of CFRP samples after cold sprayed of Sn powder: SEM image (**a**) and cross-section optical image (**b**)

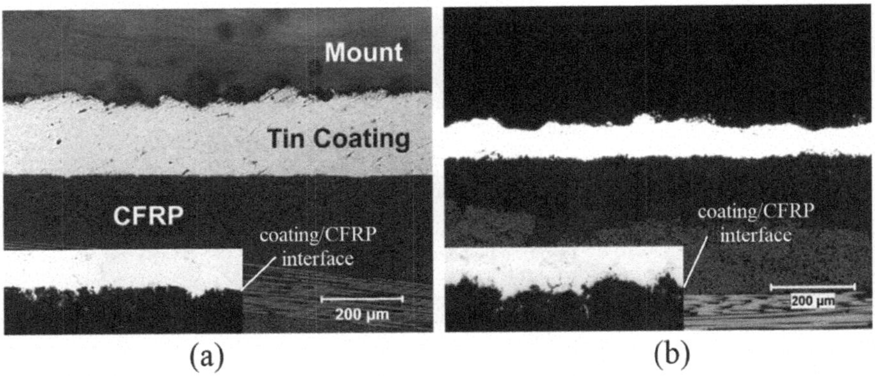

Fig. 28 Micrographs of CFRP samples after cold sprayed Sn powder at 300 °C: 4.1 bar (**a**) and 5.5 bar (**b**); a magnification of the coating/substrate interface details are also reported in the inset of the figure

indicating the coating bonding. It can be seen the presence of special tin filaments at the interface due to the mechanical interlocking taking place between the tin particles and CFRP substrate. Moreover, the relatively low melting point of Sn particles tends to partially melt upon impact the substrate if they are sprayed at a gas temperature equal to 300 °C, as proved by examination of coating morphology. The result is that the partial melting of Sn particles gives beneficial effects on coating formation and build-up due to the stronger adhesion experiencing at the coating interface.

Bimetallic coatings: tin/zinc and tin/copper

In contrast to aluminum and copper depositions on CFRPs, tin particles seem to be better cold sprayable on polymeric surfaces due to their mechanical and thermal properties, as detailed in the previous section. Unfortunately, the results found in the literature proved that the deposition efficiency of pure Sn particles on CFRPs is relatively low. Therefore, with the scope to enhance the effectiveness of the process making it more economically useful, the researchers are developing innovative solutions that could have beneficial effects on the cold spray deposition efficiency. In this respect, Che et al. [5] mixed tin particles with other metals (zinc and copper) and sprayed the mixtures on CFRPs. The powder mixture compositions were prepared in agreement with the following weight percentages: (i) 10 wt% of Zn-Sn balance (10Zn), (ii) 10 wt% of Cu-Sn balance (10Cu), (iii) 30 wt% of Cu-Sn balance (30Cu), (iv) 50 wt% of Cu-Sn balance (50Cu). The details of the CS process parameters set to spray the mixtures are reported in Table 4.

Table 4 Details of CS process parameters for powder mixtures

Powder	Standoff distance SoD (mm)	Substrate transverse speed (mm/s)	Rotational speed (rpm)	Spraying pressure (bar)	Spraying temperature (°C)
10Zn	18	25	13	4.1–5.5	280–300
10Cu	18	25	10	4.1–5.5–6.9	280–300
30Cu	18	25	12	4.1–5.5–6.9	300–350
50Cu	18	25	11	4.1–5.5–6.9	300–350

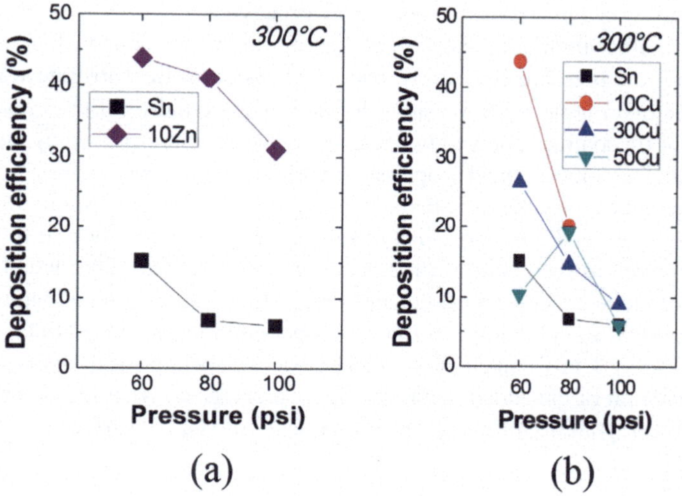

Fig. 29 The deposition efficiency of 10Zn (**a**) and three Sn-Cu mixed powders (**b**) compared to single-component tin powder

The results in Fig. 29 showed that for all the examined conditions, the cold spray DE of the powder mixtures is much higher than the relative pure tin powder. This trend was found for Zn mixture as well as for the copper ones; moreover, DE tends to decrease with the increase of the gas pressure in agreement with the trend found for the pure tin single-component. The reason is that the higher the pressure, the higher the impact particle velocity is leading an erosion of the substrate. From Fig. 29a, it can be seen that for the mixture 10Zn, DE seems to be about six times larger than that obtained for pure tin; similarly, the cold spray of 10Cu, 30Cu and 50Cu mixed powders was proved to give an increase of DE compared to pure tin particles (Fig. 29b). From the figure, it can be seen that when copper particles are added to the tin powder, a higher DE can be obtained; however, it was found that DE increment tends to decrease with the increase of copper percentage and this effect is more noticeable for lower pressure values. The reason for such behaviour is that the spray ability of the harder copper particles is very low and the higher copper percentage within the mixture tends to emphasize this critical aspect. The possible causes of the improvement of DE when both Zn and Cu particles are mixed with tin powders are to research into the tamping effect generated by the relatively harder particles, the lowering melting point from the added particles and, finally, the morphology and dendritic shape of copper powders.

Conclusions
In these studies, the feasibility of making metallic coatings on CFRP surfaces was studied, and the following results were found:

- Aluminum particles with a mean size of 3 μm can be successfully laid down on CFRP substrates by means the cold spray process;
- No continuous aluminum and copper coating can be developed when using both high and low-pressure cold spray systems. Erosion of the substrate was proved to be the main issue for the coating formation and build-up. Better results were found for tin coatings sprayed by means of a low-pressure cold spray facility; the relatively lower mechanical properties in addition to the lower melting point of the material make it suitable to be laid down on CFRPs. In fact, by examining both the microstructure and the morphology of the coating, a partial melt of tin particles was observed as the result of the impact process. The partially melted particles can better adhere to the substrate, and successful coatings can be obtained.
- Finally, it was found that when zinc or copper particles are added to tin powders, the cold spray deposition efficiency can be extensively improved; it was also found an upper limit of the added metal powder percentage above it DE reduces due to the poor spray ability of the added component onto CFRP surfaces.

Hemp-PLA laminates
In the last decades, the increasing attention of researchers toward sustainability led to the growing use of natural fibres and polymers derived from natural resources. In this respect, Astarita et al. [15] and Perna et al. [16] studied the deposition of pure metals and metal matrix composite coatings, respectively, on hemp-PLA (polylactic acid) laminate through the cold spray technique.

Pure aluminum
The authors [15] made the depositions by spraying Al particles (particle mean size equal to 40 μm) through a low-pressure cold spray machine. The best spray process parameters found in this experimentation are reported below:
Standoff distance SoD: 70 mm
Substrate transverse speed: 3 mm/s
Rotational speed: 5 rpm
Spraying pressure: 6 bar
Spraying temperature: 600 °C.

The authors found that when the momentum of the sprayed particle is too low, the particles can rebound from the substrate producing coating unsuccessfully (Fig. 30a). On the other hand, for very high values of particle momentum, severe damages on the substrate take place, as shown in Fig. 30b. That means that aluminum coating on the laminate substrate can be successfully obtained only if the cold spray process parameters are correctly chosen. The results of the effective deposition are shown in Fig. 30c, without visible damages of the substrate; in this respect, the authors suggest of using non destroying evaluations to verify that the set process parameters did not induce any kind of damage within the substrate.

By looking Fig. 31, it is possible to appreciate the effectiveness of the deposition; the coating thickness seems to reach about 100 μm; furthermore, the coating appears dense and compact without any sign of damage and deterioration.

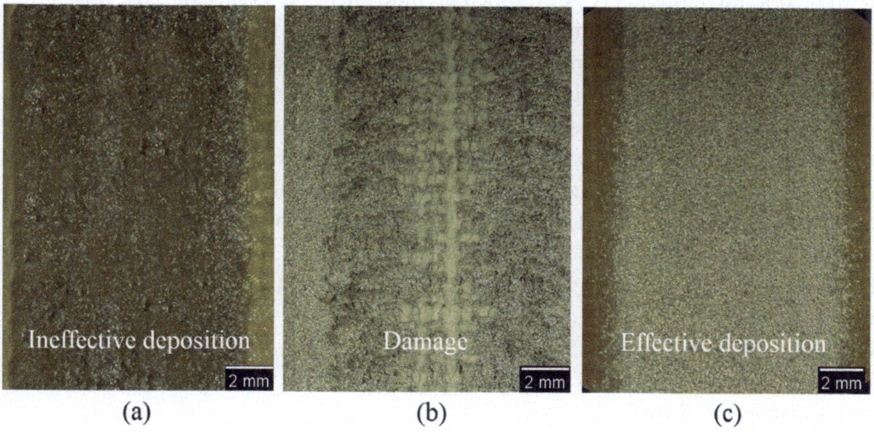

Fig. 30 Unsuccessfully deposition (**a**), substrate damage and deterioration (**b**) and effective deposition (**c**) for the aluminum coated the hemp-PLA composite surface

Fig. 31 Micrograph of Al coating on the hemp-PLA substrate

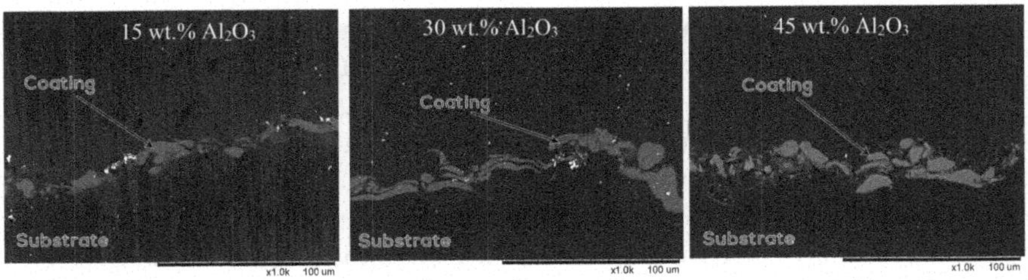

(a) (b) (c)

Fig. 32 SEM micrographs of MMC coatings varying Al_2O_3 content: **a** 15 wt%, **b** 30 wt%, **c** 45 wt%

Metal matrix composite coatings: Al and Al_2O_3

In this work, the authors [16] developed an innovative idea regarding the possibility to coat the surface of bio-compatible substrates, made of thermoplastic PLA matrix hemp fibre reinforced, by using a metal matrix composite powder mixture (MMC). In particular, aluminum particles (Al) were mixed with alumina powders (Al_2O_3), both with micron-sized diameters; the mixtures were obtained by varying the weight percentage of Al_2O_3 within Al powders. By using this approach, three different powder batches were analyzed: 15, 30 and 45 wt% of alumina particles. The aim of this activity was to point out the effects of the number of alumina particles on the substrate surface properties. A low-pressure cold spray facility was used to make the MMC coating. The CS process parameters used to lay down the mixtures are reported below:

Standoff distance SoD: 50 mm

Substrate transverse speed: 3 mm/s

Rotational speed: 5 rpm

Spraying pressure: 7 bar

Spraying temperature: 100 °C.

By looking Fig. 32, it can be observed that the coating particles are well embedded by the polymer surface, with the surrounding matrix that ensures the coating adhesion with the substrate. The particle penetration was estimated to be about 10 μm for each mixture typology, meaning that there is no significant influence of the powder composition on the particle penetration capabilities. However, it was found that the coating seems to be denser and more compact with the increase of alumina percentage; the reason is that the shot peening effect of the upcoming particles is more prominent for increased alumina mixtures due to more intense bombardment of the harder particles getting the coating denser and more continuous. Furthermore, when the harder alumina particles impact on the target surface, they produce micro-asperities on the top surface that can promote adhesion and bonding of the subsequent Al particles. In fact, the contact area between the particles and the substrate can be widened, and the coating quality can be improved.

By using this approach, the tribological properties of the composite materials can be improved due to the beneficial effect of the ceramic particles within the MMC aluminum-based coating; it was also proved that this effect tends to be more intense with the increase of alumina content. Unfortunately, there exists an upper limit for alumina content (30 wt% in this activity) above which the abovementioned beneficial effect can reduce affecting the coating quality and performance, as alumina starts to break away from the coating causing an overall worsening of the proprieties.

Conclusions

Results obtained by spraying micron-sized Al particles on hemp-PLA laminates showed that it is possible to find four different deposition cases, depending on the cold pray process parameters chosen. In particular: (i) unsuccessfully deposition, (ii) damage of the substrate, (iii) effective deposition with internal damage and, finally, (iv) effective deposition without any kind of internal failure of the substrate. Good depositions and quite dense and compact coatings can be obtained when the process parameters are correctly chosen. Moreover, when alumina particles (ceramic powders) are added within the powder mixture (up to a concentration of 30 wt%.), the surface performance of the polymeric substrate can be improved. The reason is that the relatively harder particles can cause a roughening of the top surface (improving the painting capacities of the system) and can produce micro-asperities on the top surface due to the shoot peening effect, increasing the adhesion of the subsequent layers. A summary of metal cold spray depositions on fibre reinforced polymer substrates is reported in Appendix 2.

Appendix 1: Cold Spray Depositions on Polymers

Composite substrate	CS powders	CS process parameters				Literature results		
Polymer	Type	Gas temperature (°C)	Gas pressure (bar)	SoD (mm)	Transverse speed (mm/s)	Authors	DE (%)	Best case
PC/ABS	Copper	Room	0.5-3	40	8.3–16.6	[10]		3 Mpa Room
	Tin	Room	3	103	14.0	[10]		3 Mpa Room
Polypropylene	Copper	150–350	2.54	30	0,5	[17]		2.54 Mpa 350 °C
	Tin	Room	3	103	14.0	[10]		3 Mpa Room
Polycarbonate	Copper	150–350	2.54	30	0.5	[17]		2.54 Mpa 350 °C
Polystyrene	Tin	Room	3	103	14.0	[10]		3 Mpa Room
Polyamide 66	Aluminum	150–250	1.5–3	40	0.2	[18]	6–7	2 MPa 200 °C

(continued)

(continued)

Composite substrate	CS powders	CS process parameters				Literature results		
Polymer	Type	Gas temperature (°C)	Gas pressure (bar)	SoD (mm)	Transverse speed (mm/s)	Authors	DE (%)	Best case
Polyamide 66	Tin	Room	3	103	14.0	[10]		3 Mpa Room
PTFE	Copper	150–350	2.54	30	0.5	[17]		2.54 Mpa 350 °C
Polyurethane	Copper	150–350	2.54	30	0.5	[17]		2.54 Mpa 350 °C
ABS	Copper	150–350	2.54	30	0.5	[17]	4–12	2.54 Mpa 350 °C
	Tin	200	0.5–1.4	18	0.3–0.5	[19]	5	1.4 Mpa 200 °C
HDPE	Copper	425	0.7–1.4	40	25.0	[19]		1 Mpa 425 °C
PVC	Copper	150–350	2.54	30	0.5	[17]	4–6	2.54 Mpa 350 °C
PEEK	Copper	100–400	1–3	30	0.5	[12]		1 Mpa 200 °C
	Tin	200	0.5–1.4	18	0.3–0.5	[19]	1–80	1.4 Mpa 200 °C
	Copper	425	0.7–4.9	40	25.0	[19]	3	4.9 Mpa 425 °C
PEI	Iron	425		18–40	25.0	[19]	1	1.4 Mpa 425 °C
	Tin	200	0.5–1.4	18	0.3–0.5	[19]	1–76	1.2 Mpa 200 °C
	Copper	425	0.7–4.9	40	25.0	[19]	2	4.9 Mpa 425 °C

Appendix 2: Cold Spray Depositions on Fibre Reinforced Polymers

Composite substrate		CS powders	CS process parameters					Literature results		
Polymer	Reinforcement	Type	Gas typology	Gas temperature (°C)	Gas pressure (bar)	SoD (mm)	Transverse speed (mm/s)	Authors	DE (%)	Best case
Thermoset	Glass fiber	Al	Air	Room	5–30	40	8.3	[12]	//	5 bar
PEEK450 CA30	Carbon fiber	Al, Al-Cu	Nitrogen	300–450	12–19	15–20	5	[1]	//	450 °C 19 bar SoD: 15 mm Al-Cu
Airplane Grade Epoxy	Carbon fiber TR50S-6L	Al	Nitrogen	300–600	5–20	20	200	[4]		600 °C 5 bar (plasma interlayer)
Thermoset	Carbon fiber Cycom 5276–1/G30–500	Al, Cu, Sn, Sn-Zn	Nitrogen	25–700	2.9–50	18–200	12.5–500	[14]	22	300 °C 5.5 bar Sn-Zn
Thermoset	Carbon fiber Cycom 5276–1/G30–500	Cu	Nitrogen	250–425	20	40	1000	[20]	//	425 °C
Thermoset	Carbon fiber Cycom 5276–1/G30–500	Cu, Sn, Fe	Nitrogen	200–425	3–49	18–40	25	[19]	0–5	200 °C 12 bar Sn
Thermoset	Carbon fibre Cycom 5276–1/G30–500	Sn, Sn-Cu, Sn-Zn	Nitrogen	280–350	4.1–6.9	18	25	[5]	6–44	300 °C 4.1 bar 10Cu

(continued)

(continued)

Composite substrate		CS powders	CS process parameters					Literature results		
Polymer	Reinforcement	Type	Gas typology	Gas temperature (°C)	Gas pressure (bar)	SoD (mm)	Transverse speed (mm/s)	Authors	DE (%)	Best case
Polylactic acid (PLA) films (4042D)	Hemp fabric	Al	Air	150–600 100	4–8 7	10–80 50	3 3	[15, 21]	//	600–100 °C 6–7 bar SoD: 70–50 mm
Polylactic acid (PLA) films (4042D)	Hemp fabric	Al-Al_2O_3	Air	100	7	50	3	[16]	//	30 wt% Al_2O_3

References

1. Zhou, X. L., Chen, A. F., Liu, J. C., Wu, X. K., & Zhang, J. S. (2015). Preparation of metallic coatings on polymer matrix composites by cold spray. *Surface and Coatings Technology, 206*(1), 132–136.
2. Raoelison, R. N., Xie, Y., Sapanathan, T., Planche, M. P., Kromer, R., Costil, S., et al. (2018). Cold gas dynamic spray technology: A comprehensive review of processing conditions for various technological developments till to date. *Additive Manufacturing, 19,* 134–159.
3. Archambault, G., Jodoin, B., Gaydos, S., & Yandouzi, M. (2016). Metallization of carbon fibre reinforced polymer composite by cold spray and lay-up molding processes. *Surface and Coatings Technology, 300,* 78–86.
4. Affi, J., Okazaki, H., Yamada, M., & Fukumoto, M. (2011). Fabrication of aluminum coating onto CFRP substrate by cold spray. *Materials Transactions, 52*(9), 1759–1763.
5. Che, H., Chu, X., Vo, P., & Yue, S. (2017). Cold spray of mixed metal powders on carbon fibre reinforced polymers. *Surface and Coatings Technology, 329,* 232–243.
6. Assadi, H., Gärtner, F., Stoltenhoff, T., & Kreye, H. (2003). Bonding mechanism in cold gas spraying. *Acta Materialia, 51*(15), 4379–4394.
7. Grujicic, M., Zhao, C. L., Derosset, W. S., & Helfritch, D. (2004). Adiabatic shear instability based mechanism for particles/substrate bonding in the cold-gas dynamic-spray process. *Materials and Design, 25*(8), 681–688.
8. Hussain, T., McCartney, D. G., Shipway, P. H., & Zhang, D. (2009). Bonding mechanisms in cold spraying: The contributions of metallurgical and mechanical components. *Journal of Thermal Spray Technology, 18*(3), 364–379.
9. Burlacov, I., Jirkovský, J., Kavan, L., Ballhorn, R., & Heimann, R. B. (2007). Cold gas dynamic spraying (CGDS) of TiO_2 (anatase) powders onto poly(sulfone) substrates: Microstructural characterisation and photocatalytic efficiency. *Journal of Photochemistry and Photobiology A: Chemistry, 187*(2–3), 285–292.
10. Lupoi, R., & O'Neill, W. (2010). Deposition of metallic coatings on polymer surfaces using cold spray. *Surface and Coatings Technology, 205*(7), 2167–2173.
11. Pathak, S., & Saha, G. C. (2017). Sustainable development of cold spray coatings and 3D additive manufacturing components for repair/manufacturing applications: A critical review. *Coatings, 7*(8), 122–149.
12. Ganesan, A., Affi, J., Yamada, M., & Fukumoto, M. (2012). Bonding behavior studies of cold sprayed copper coating on the PVC polymer substrate. *Surface and Coatings Technology, 207,* 262–269.
13. Ganesan, A., Yamada, M., & Fukumoto, M. (2013). Cold spray coating deposition mechanism on the thermoplastic and thermosetting polymer substrates. *Journal of Thermal Spray Technology, 22*(8), 1275–1282.
14. Che, H., Vo, P., & Yue, S. (2017). Metallization of carbon fibre reinforced polymers by cold spray. *Surface and Coatings Technology, 313,* 236–247.
15. Astarita, A., Boccarusso, L., Durante, M., Viscusi, A., Sansone, R., & Carrino, L. (2018). Study of the production of a metallic coating on natural fiber composite through the cold spray technique. *Journal of Materials Engineering and Performance, 27*(2), 739–750.
16. Perna, A. S., Viscusi, A., Astarita, A., Boccarusso, L., Carrino, L., Durante, M., & Sansone, R. (2019). Manufacturing of a metal matrix composite coating on a polymer matrix composite through cold gas dynamic spray technique: Process development and tribological testing. *Journal of Materials Engineering and Performance*. (in press).
17. King, P. C., Poole, A. J., Horne, S., de Nys, R., Gulizia, S., & Jahedi, M. Z. (2013). Embedment of copper particles into polymers by cold spray. *Surface and Coatings Technology, 216,* 60–67.
18. Giraud, D., Borit, F., Guipont, V., Jeandin, M., & Malhaire, J. M. (2012). Metallization of a polymer using cold spray: Application to aluminum coating of polyamide 66. In *Proceedings of the International Thermal Spray Conference* (pp. 1–6), Houston, Texas, USA, May 21–24, 2012.

19. Che, H., Chu, X., Vo, P., & Yue, S. (2018). Metallization of various polymers by cold spray. *Journal of Thermal Spray Technology, 27*(1–2), 169–178.
20. Che, H., Vo, P., & Yue, S. (2018). Investigation of cold spray on polymers by single particle impact experiments. *Journal of Thermal Spray Technology, 28*(1–2), 135–143.
21. Boccarusso, L., Viscusi, A., Durante, M., Astarita, A., De Fazio, D., Sansone, R., et al. (2018). Deposition of aluminum coating on bio-composite laminates. In *Proceedings of the 21st International ESAFORM Conference on Material Forming* (pp. 1–6), Palermo, Italy, April 23–25, 2018.

Advanced Modeling and Simulation Tools to Address Build-Up Issues in Additive Manufacturing by Cold Spray

Michel Jeandin, Francesco Delloro, and Margaux Bunel

Abstract The objective of the proposed chapter is to discuss recent advances in modeling and simulations for specific application to additive manufacturing by cold spray. To meet the requirements for overall modeling of the process, two scales have to be considered, i.e. that of the powder particle and that of the deposit. These result in two parts in the chapter respectively, i.e. Sections 2 and 3, which follow a rather elaborated introductory section. The latter gives the background and a rapid state-of-the-art in the field of cold spray for additive manufacturing. Experimental and numerical approaches to coating build-up are compared, in particular. The two-fold core of the chapter then highlights both conventional finite element analysis and original morphological modeling of the basic mechanisms involved in cold spray coating build-up. The role of the number of particles to be involved in the simulations is discussed since this number is the key parameter for shape prediction in additive manufacturing.

Keywords Additive manufacturing · Cold spray · Modeling · Simulation · Finite elements analysis · Build-up · Morphological models

1 Introduction

Based on what most of materials science and engineering forums show, all the technological routes seem to lead to additive manufacturing. As all roads lead to Rome, additive manufacturing can, therefore, be said to be Roman and be renamed "Romanufacturing". Historically, one of these roads/routes comes from the thermal spray, Vardelle et al. [73], e.g. with plasma forming processes. Nowadays, the road from thermal spray to additive manufacturing is all the shorter because the cold spray is

M. Jeandin · F. Delloro (✉) · M. Bunel
MINES ParisTech, PSL Research University, MAT-Centre des Matériaux, CNRS, UMR 7633, BP 87, 91003 Evry, France
e-mail: francesco.delloro@mines-paristech.fr

M. Jeandin
e-mail: michel.jeandin@mines-paristech.fr

S. Pathak and G. C. Saha (eds.), *Cold Spray in the Realm of Additive Manufacturing*, Materials Forming, Machining and Tribology,
https://doi.org/10.1007/978-3-030-42756-6_5

used, Botef et al. [14]. Cold spray is very attractive because of a high deposition efficiency, which pushes this process at the forefront to compete and/or to complement now-conventional laser-based techniques such as Laser Powder Bed Fusion (LPBF) or Laser Metal Deposition (LMD). The road is also short because a thermal spray operator does additive manufacturing without knowing it. He is like the famous Molière's character Monsieur Jourdain – nothing to do with Jeandin – speaking prose without knowing it. This results from the mere fact that thermal spray is based on the deposition of material in the form of small-sized elementary components, in other words, powder particles. However, to play in the "big league" among laser-based processes and ride the wave of additive manufacturing, cold spray needs the development of advanced simulation/modeling of deposit build-up for prediction and control of the shape and properties of sprayed parts.

Historically, people have always been fascinated by stacking issues in all fields, i.e. artistic, literary, and scientific with, for example, Acimboldo, Sade, Utagawa, or Ren Hang (Fig. 1, censored for the latter in this figure). It is therefore quite normal they still fascinate today. The only difference, in the context of additive manufacturing, is that the ingenuity of the previously-mentioned creators is now replaced by numerical simulation: this simulation is coupled with an extensive characterization of the building blocks.

The objective of this chapter, the content of which is given in the abstract, is to show recent advanced development of modeling of the cold-spray build-up process for real additive manufacturing, knowing that cold spray is more and more claimed to be promising in the whirlwind of additive manufacturing [38]. The modeling/simulation route is mandatory for the control of the process to result in the achievement of specific parts, i.e. beyond mere coating application.

Fig. 1 Stacking issues over the centuries, **a** Arcimboldo in the sixteenth, **b** Sade in the eighteenth century, **c** Utagawa in the nineteenth century

2 Background and State-of-the-Art

2.1 Coating Build-up Approach to Thermal Spray

In this section, there is no intention of giving a comprehensive view of modeling of coating build-up in thermal spray. The purpose is more to show a rapid overview of the significant milestones (including the first ones) in the development of deposition models in the field, which will put into perspective those for additive manufacturing by cold spray.

When talking about modeling in thermal spray, a few decades ago, say 25 years ago and over a long time, one thought of work on interaction, between a given sprayed particle and the substrate. Interaction meant impingement and splashing, which involved rapid cooling, solidification and various physicochemical phenomena. Basically, the related studies authored by names synonymous with this type of issues, i.e. E. J. Lavernia, A. Vardelle, M. Pasandideh-Fard, and J. Mostaghimi, resulted in the understanding of splat formation though modeling with some pioneering articles [51, 61, 72]. Anecdotally, a major challenge at this epoch and for a rather long time has been to know the maximum number of splats which could be considered at the same time in a finite element analysis of deposition. At this time, the considered particle was a droplet since spraying was obtained by plasma processing, i.e. prior to the advent of cold spray. From this time, researchers have never ceased to keep forging ahead in this type of approach using more and more powerful computing facilities and related numerical method [13, 50]

In parallel with this finite element analysis of impinging phenomena, other routes were initiated, primarily stochastic modeling [29] which could involve rather conventional Monte-Carlo simulation [46] or original (in the thermal spray field) "lattice-gas" models [20]. The latter was undoubtedly the feeding sap of further development based on mathematical morphology and morphological concepts which were considered to be specifically tailored for cold spray even though works on plasma spraying did not stop (Beauvais et al. [10]). Subsequent works seemed (and continue to seem) to prove the point. When restricting to the morphological approach in line with that elaborated in Sect. 2 of this chapter, the main milestones date back to 2010 and 2014 [23, 37]. The most recent advances in the field, in addition to a comprehensive description of corresponding models, are given in the second section of this chapter, i.e. "Modeling at the scale of the powder particle". This section shows the so-called "morphological approach", which combined to a finite element analysis of particle impact, can be claimed to be the decisive step to general modeling of coating build-up by cold spray.

2.2 *Towards Additive Manufacturing by Cold Spray*

Any development in modeling needs validation from experiments. This is the reason why the major requirements for additive manufacturing had (and still have) to be met in parallel. These consist primarily in achieving a high thickness and/or a shape control/mastery for the deposit.

The first requirement for a high thickness, i.e. a high deposition rate, can result in an actual scientific and technological bolt when spraying some specific materials. These are, for example, Al-based alloys for which deposition rate had to be dramatically improved [15]. Fortunately, in the past few years, heat treating the powders was shown to be a suitable solution to remove the bolt [15, 67]. Efforts now focus on industrial feasibility of treating powders, for example using fluidized bed facilities, which, however, is no longer a bolt for the development of cold spray additive manufacturing.

The second requirement, i.e. that for shape control/mastery of the deposit, led to the development of strategies for the nozzle trajectory, possibly including advanced robotics. A significant impetus resulted from works by NRC/Boucherville-Canada up to now [55, 74, 75] as an extension of the tessellation approach to this issue in the pioneering paper by Pattison [62] one decade before. These efforts were pursued by Wu et al. [77] in particular, to name among the most recent examples. These works should be complemented by work on numerical simulation, which can be considered as the 3rd requirement for successful additive manufacturing as a fabrication process as predicted by review and prospective studies of the topic in the key year 2017 [65, 69] in a sort of preview of this same book with this chapter for the modeling part of the build-up issue.

As already said in the introductive part of the chapter, dual modeling of the build-up process, i.e. at the scale of the sprayed particle and at the scale of the sprayed bead respectively, is claimed to be the best for an extended approach to cold spray additive manufacturing. This results in the two subsequent sections in this chapter.

3 Simulation at the Scale of the Particle

The process of a cold spray coating formation consists of the iteration of elementary phenomena which are the impact of a particle onto the substrate or onto already-deposited particles. Each impact is characterized by a number of features (e.g. particle shape, size, velocity, temperature, substrate local topography, etc.) which causes a large variability inside the set of possible configurations, even if materials and process parameters are fixed. These elementary events can be considered, as a first approximation, independent. Indeed, it can be shown that, given a typical powder granulometry, feed rates and particle velocities, the probability that two impacts take place at the same time and at the same place is rather low, as shown in [27]. When looking at the time and spatial scales of these events, experimental observation techniques do not allow nowadays access to any measurement, except in few notable cases that will be introduced later in this chapter. Modeling of particle impact, thus, takes in this perspective all its sense as the only way to access important physical information on what is happening during the process. The scale of particle impacts, both spatial (1 μm and less, depending on the mesh refinement) and temporal (ns), introduces new questions regarding the material behavior: even for well-known metals, as for example Al and Cu, mechanical behavior at the small scale and at extreme strain rates is not known. Moreover, several researchers focused in the last years on the cold spray of non-metallic materials, i.e. ceramics and polymers, thus opening new questions on their impact mechanics and constitutive behavior. A lot of work has still to be done for the understanding of the process when involving those materials, which is likely to involve different elementary phenomena. In this chapter, the focus will be on metals only, since the application of the cold spray technique to this class of materials can be considered more mature, especially from an additive manufacturing point of view. Despite the maturity of some applications, further investigation is still needed to elucidate and model various phenomena at the small scale which are key to the process and not yet completely understood, as dynamic recrystallization, jetting and extreme plastic deformation in localized regions. To this aim, specific physical approaches, involving dislocation dynamics, ultra-dynamic material behavior and polycrystalline plasticity models, need to be studied.

In this chapter section, the focus will be at the scale of the particle. Most of the publications and ongoing works consider particles as made of a homogeneous material, thus neglecting finer scale properties and features, among which the grain structure, dislocation dynamics, twins, shear bands, etc. Three main subjects will be addressed here: powder characterization, impact simulations and morphological modeling (iterative deposit build-up models, i.e. particle by particle).

3.1 Powder Characterization

Powders are the feedstock material of the cold spray process and, thus, their properties largely influence the resulting coating. Their production route is fundamental in determining morphological and metallurgical characteristics which, in turn, heavily affect the deposition efficiency and the mechanical properties of the coating. Powders can be characterized in different aspects: metallurgical, mechanical, surface cleanliness and morphological. Nevertheless, most of these characteristics are not independent, because they are all related to the manufacturing process and the thermal history seen by the material. The division made in this chapter, thus, is somehow artificial, but still useful to understand and classify the role that each powder characteristic can assume with respect to the cold spray process.

3.1.1 Metallurgy

The microstructure of powder particles is the result of material composition and of the thermal history, thus reflecting their production route. When dealing with pure metal powders, the main descriptors of the metallurgical state are grain size and shape distributions, dislocation density and internal porosity. In general, particle microstructures can range from dendritic to finely cellular for an atomized powder, as a function of the cooling rate during the manufacturing process. The picture is far more complex for alloy powders, where, in addition to the features already discussed for pure metals, the presence of numerous phases in the form of inclusions and precipitates contribute to the particular microstructure of the material (Fig. 2). An exhaustive discussion of powder metallurgy and its role in the cold spray process is not in the scope of the present work and is already the topic of several research articles. A general review of these issues will be left to future publications. One may say, as a general remark, that alloy powders are extremely sensitive to the production route and to the experienced thermal history, which is the key factor influencing the microstructure. Thus, a great variety of microstructures and, as a consequence, mechanical properties, can be obtained for the same material composition, as a result of manufacturing process parameters and, eventually, post-production thermal treatments [67].

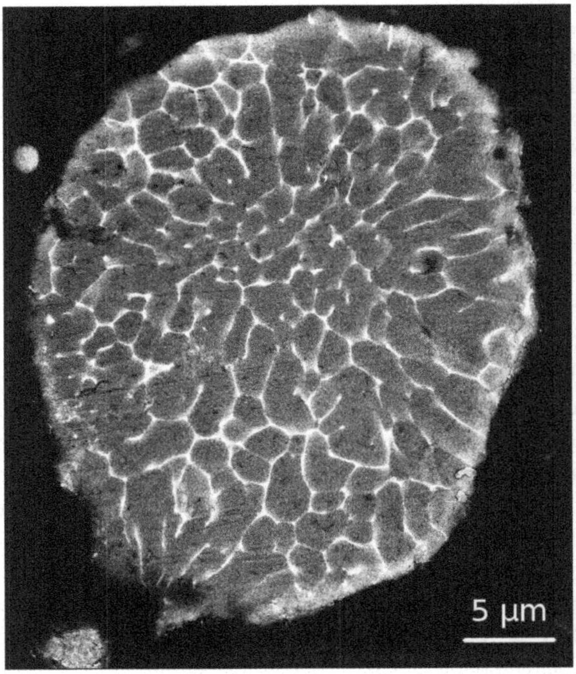

Fig. 2 SEM BSE cross-section image of an Al 2024 particle, produced by nitrogen gas atomization; the grain boundaries are revealed by the higher concentration of alloying elements

3.1.2 Mechanics

Particle mechanical properties come into play at the impact with the substrate or with already deposited particles and are intimately related to the metallurgical state of the material. As a consequence, the interplay between microstructure and mechanical properties is even more complex during the deformation because both characteristics tend to evolve. When colliding, a given particle can undergo extreme plastic deformation, concentrating in the region close to the newly formed interface. In particular, strain and heating rates can respectively reach 10^9 s^{-1} and 10^9 K s^{-1} typically. In these conditions, various microstructure changing mechanisms can occur, i.e. adiabatic shearing instabilities, dynamic recrystallization, melting at interfaces, twinning, dislocation rearrangement and amorphization. For further details, the reader may refer to various comprehensive descriptions, such as those by Moridi et al. [56, 57], Jeandin et al. [40], Cinca et al. [19]. Due to the richness and variety of these elementary phenomena, the modeling of particle mechanics at impact is a very challenging task. The approach generally adopted in the vast majority of the simulations is to neglect the complex microstructure and its evolutions. In this approximation, features at the grain scale or below are not explicitly described and the particle is considered as made of a homogeneous material. In this framework, material behavior can be described by a variety of mechanical models, which will be briefly addressed later on, in Sect. 3.2.1.

Independently of the choice of the material model, a common problem is the origin of the parameter set to be used in cold spray impact simulations. When testing material behavior in the high strain rate regimes, the usual technique is the Split Pressure Hopkinson Bar test (SPHB) (Fig. 3).

One may refer to the voluminous existing literature for an exhaustive description of the technique, for example [28]. Different variations of the test allow for the characterization of the material in compression, tension and torsion. Moreover, the effect of the temperature can also be taken into account in special testing configurations where the sample can be heated. The material model parameters are then identified

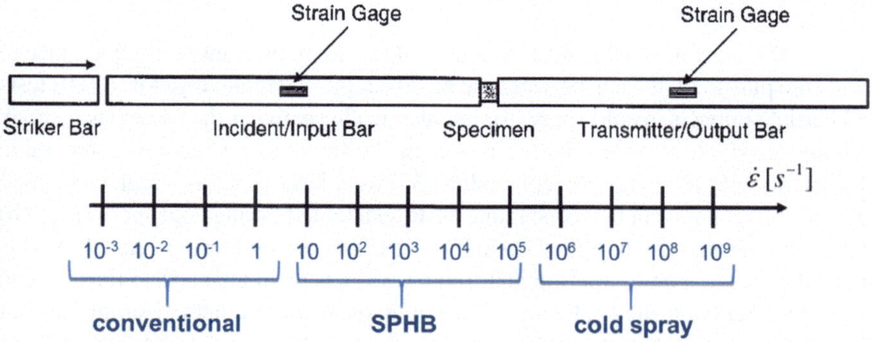

Fig. 3 Schematic illustration of the Split Pressure Hopkinson Bar (SPHB) test, after [28] and comparison of strain rate domains of conventional mechanical tests, SPHB and cold spray

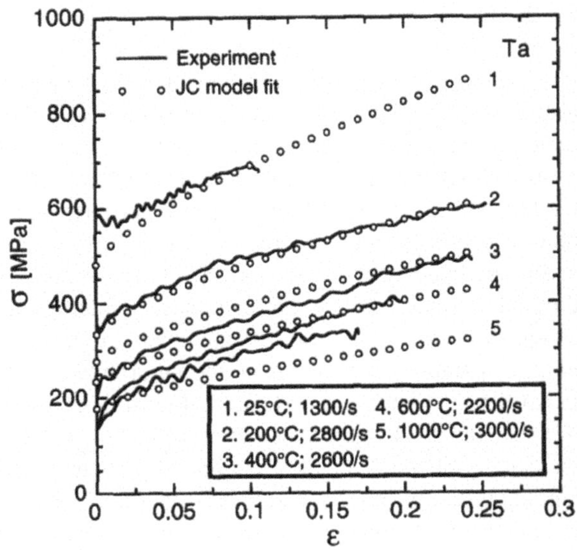

Fig. 4 SPHB test results for Ta at different temperatures and strain rates (curves), while dots represent the numerical simulations with fitted parameters after [18]

by the comparison of simulation results and experimental data obtained through the use of strain gages and post-mortem analysis of the deformed specimen (Fig. 3). In the SPHB test, deformation rates up to 10^5 s^{-1} can be reached. When comparing to typical impact conditions of a particle in cold spray, two critical aspects, namely the size of the sample and the maximum strain rate, can be identified. In one hand, the strain rates experienced by the particles are 3 to 4 orders of magnitude higher (Fig. 4).

On the other hand, the size of the samples is macroscopic for SPHB, consisting usually in cylinders with diameter and height of about 10 mm, while it is 3 orders of magnitude smaller for typical cold spray particles. In addition, the particular metallurgical state of atomized particles, exposed to fast cooling rates, is hardly reproducible in a macroscopic sized sample, as those used in SPHB. The material tested is therefore inherently different, even if the chemical composition might be the same.

From the analysis of what presented here, it is evident that new techniques have to come into play to fill the gap between the macroscopic and relatively slow SPHB tests and particle impacts in cold spray. An interesting alternative is the use of laser shock techniques, which are able to induce mechanical solicitations in regimes comparable to cold spray [9] (Fig. 5a). Recent studies [33] used laser shock to accelerate single particles to velocities in the same range as those attained during cold spray (Fig. 5b) in experiments that can be called LASHPOL ("LAser Shock Powder Launch"). This opens the way to a new experimental framework, where single particle impacts can be produced at controlled velocities. The comparison with numerical simulations of single particle impacts is then expected to result in a reliable set of material behavior parameters, in a procedure known as "reverse modeling". Having chosen an experimentally accessible goal function (e.g. the particle flattening ratio), the procedure

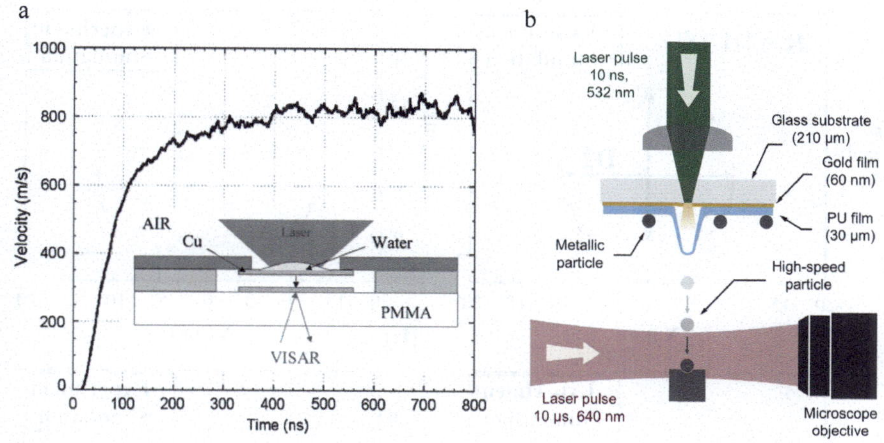

Fig. 5 Laser shock flier experiments and cold spray, **a** high velocity impact of a thin Cu foil onto Al, reproducing the same metallurgical transformations at the interface as observed in cold spray, after [9], **b** high velocity impact of a single particle onto a hard substrate, after [33]

consists in fine tuning the parameters until a good match between experimental and simulation results is obtained (Fig. 6). A recent work moved the first steps in this direction [17].

The identification of material behavior is a complex task and the coupling of different techniques can be beneficial. In particular, a study [7] focused on the quasi-static compression of single particles. A nano-indenter with a flat head was used to this purpose and produced force-displacement data sets which, in conjunction with post-mortem particle shape characterizations and in comparison with the process simulation, allowed to fit the parameters of a strain rate independent constitutive law (Fig. 7). The coupling of static and dynamic techniques will certainly give a richer material characterization than using either one or the other.

3.1.3 Surface Cleanliness

A common problem that every cold spray operator or researcher has surely been confronted to at least one time is to spray an aged powder. Even if properly stored, powders that performed well when fresh tend to lose their ability to adhere to the substrate and to form a coating. This fact suggests that particle surface plays a very important role in cold spray. Different contaminants, inclusions and/or external phases such as oxides, nitrides, etc. can be present at the surface of powder particles (Fig. 8), especially if not properly stored. In particular, external phases are rather easily formed on reactive and oxygen-sensitive materials such as Cu, Ti or Ta [24, 40].

Powder oxidation is a complex phenomenon and its description is behind the aim of this chapter. To remind some of the fundamental processes altering particle surface one may refer to adsorption, dissolution, diffusion and oxide formation.

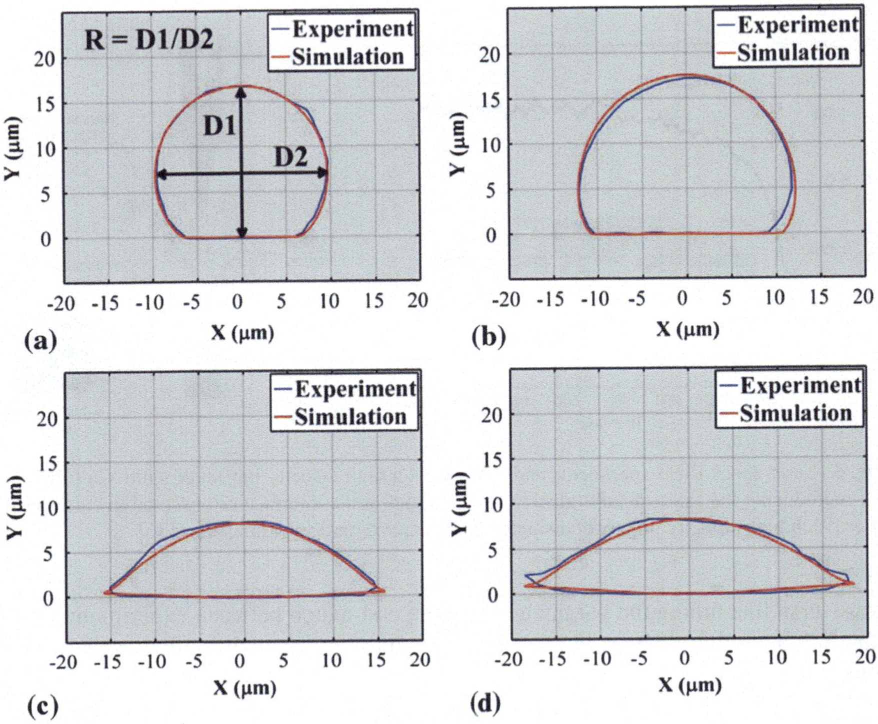

Fig. 6 Comparison of experimentally observed and simulated splat shapes, enabling the tuning of material parameters for a bi-linear Johnson-Cook model, derived from [17]

The kinetic of these phenomena is mainly controlled by their activation energy, temperature and oxygen concentration. Oxidation process starts from the surface but can affect also particle internal microstructure. With aging, the surface oxide layer increases in thickness and oxygen diffusion towards particle interior can take place, thus modifying the mechanical properties of powders via those mechanisms.

In many works on cold spray, adhesion is generally considered as depending on materials and particle velocity: a particle can adhere when its velocity is in an appropriate range, i.e. the so called deposition window [68]. In other words, particle has to flight faster than the critical velocity and slower that the erosion one (Fig. 9). Nevertheless, when looking closer, other characteristics than velocity and material are important with respect to adhesion and surface cleanliness is certainly among these. Naturally, every particle in a powder possesses its oxide layer at the surface, which can be formed during the production process or later, when particles experience contact with oxygen in the atmosphere. Few materials make an exception to this rule, probably gold being one of these.

The oxide layer is hardly avoidable and it is not an insurmountable obstacle. When not too thick, it can be easily broken at impact assuring intimate contact

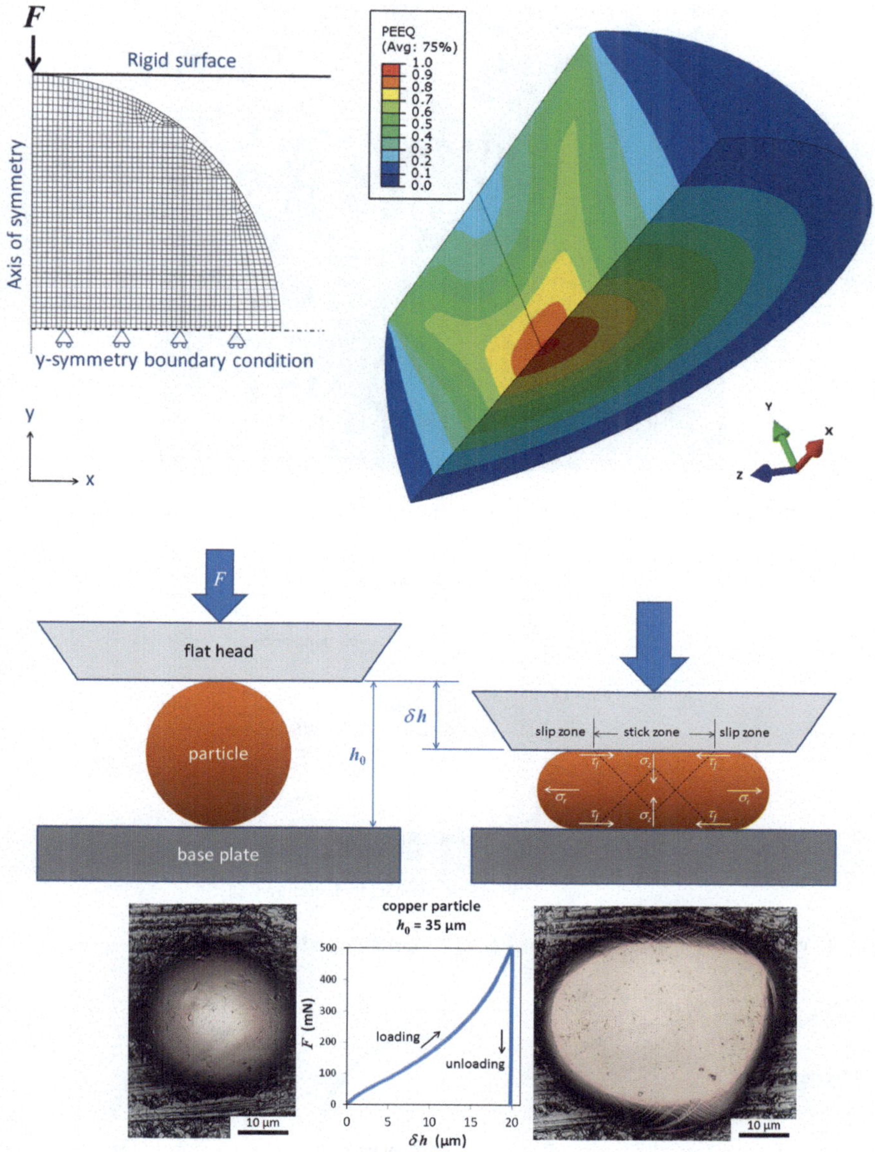

Fig. 7 Schematic illustration of a single particle compression test, after [7]

between the freshly generated metallic surfaces. This is the so-called "oxide break deposition model" which, in combination with the observed material elimination at impact through the jetting phenomenon [27–34], offers a convincing explanation for the experimentally observed metallurgical bonding [35, 48]. Figure 10 shows such a mechanism: during the plastic deformation of the particle and the substrate, the brittle oxide layers locally brake, especially in the periphery where the shear strain is

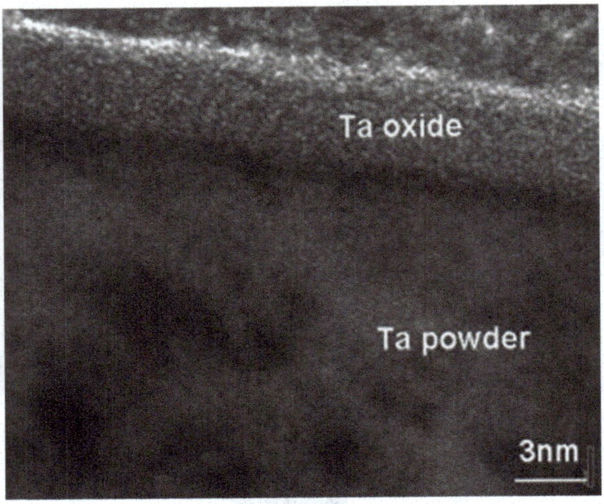

Fig. 8 HR (high resolution) TEM image of a commercial feedstock of Ta powder, after [40]

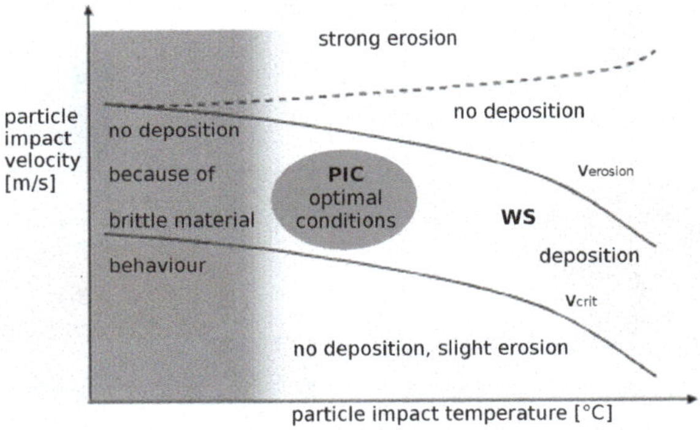

Fig. 9 Window of sprayability as a function of temperature and velocity, after [68]

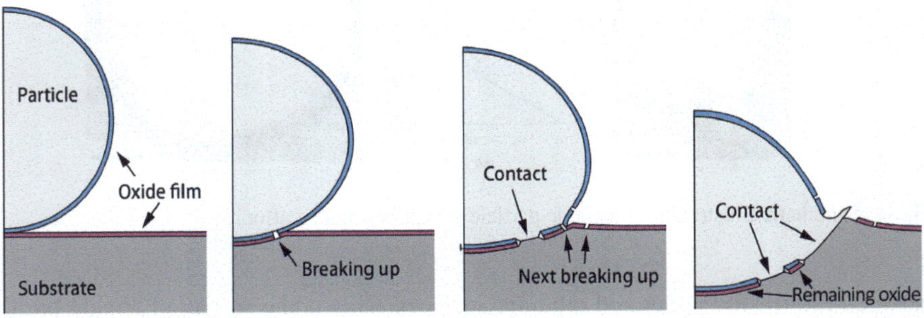

Fig. 10 Schematic view of the "oxide break deposition model", after [36]

higher. Instead, at the initial point of collision (south pole of the particle), where stress is purely compressive, they were shown [35, 36] to remain intact, thus degrading the local adhesion in that particular zone.

3.1.4 Morphology

The first use of the term "morphology" is attributed to Goethe, in relation to comparative anatomy. We can use it in a similar sense when comparing the external features (i.e. shape and size) of any object and, in particular, of powder particles. When looking at the appearance of different powder samples, one can notice that the morphological features are related to the manufacturing process. For example, water atomization tends to produce irregularly shaped particles and gas atomization more spherical ones, the degree of sphericity being related to the reactivity of the atomizing gas. The most spherical particles are obtained using nitrogen and controlling the surrounding atmosphere, while employing air results in more irregular shapes. Morphological characteristics of atomized powders seem, thus, to be related to the reaction kinetics of superficial oxide formation during the rapid cool down. Other production routes can give more complex morphologies. For example, the most intricate structures are obtained by the agglomeration of smaller particles, which can be made by different materials and contain porosity (Fig. 11a); dendritic or coral shaped particles can be produced by electrolytic methods [5] (Fig. 11b); fused and crashed powders show a very "angular" morphology (Fig. 11c). In these cases, classical observations techniques, as SEM images of the loose powder or cross sections of particles embedded in resin, fail to capture the essential geometrical characteristics, mainly because of their limitation to 2D. In order to give a correct representation of the shape of complex objects, a 3D approach is needed.

In [22], a 3D observation method was developed using X-ray microtomography (XMT) (Fig. 12). XMT is nowadays a well-developed technique, extensively used in many scientific fields and, in particular, in thermal spray as, for example, in [1–3, 31,

Fig. 11 Exotic particle shapes, **a** XMT image of Ag-SnO_2 powder, Courtesy of Y. Zeralli, MINES ParisTech, 2013, **b** SEM cross section of a coral-like particle, after [52] and **c** SEM top view of the loose tantalum powder Amperit 151.065 (HC Stark, Munich, Germany), produced by the fuse and crash method, after [22]

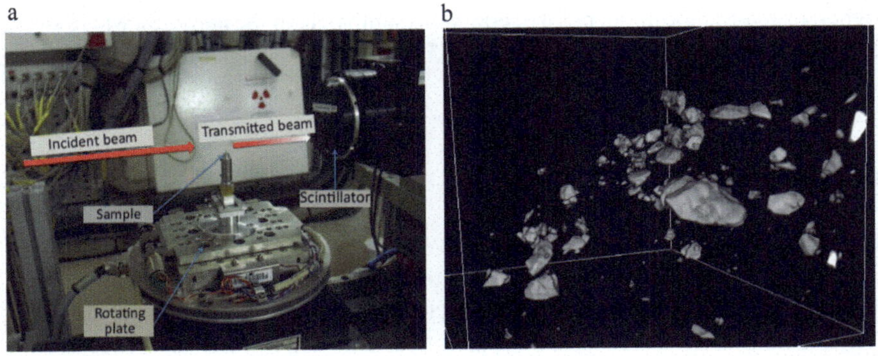

Fig. 12 Application of X-ray micro tomography to powder characterization, **a** experimental set up at ESRF; **b** 3D image after the segmentation showing a volume containing some Ta particles (taken from the powder batch shown in Fig. 11c)

39, 66]. X-rays of a given energy, capable of penetrating the sample through all its thickness, illuminate the material to be observed. The transmitted beam is captured by an imaging device (a scintillator followed by a CCD camera with high resolution, which can be below the μm), resulting in a 2D projection of the volume of the object. The sample is then rotated by a small angle and another image is taken. The iteration of these steps results in a collection of thousands of projections that can be combined in a proper 3D image of the object, through specific algorithms in a process called reconstruction. In the 3D image, the gray level of each voxel (i.e. the 3D analogue of a pixel) is proportional to the local X-ray absorption of the sample. The powder, to assure some distance between particles, was dispersed in a resin and a cylindrical sample suitable for the observation was created. In the reconstructed image of such a sample, metallic particles appeared brighter than the embedding resin. The 3D image needs to be further treated for assigning each voxel either to the powder or to the resin, a process called "segmentation". At the end of the analysis, a particle library is obtained, made up by all the particles contained in the imaged sample, which could sum up to several thousands.

For certain production routes and, in particular, those giving irregular morphologies, particles can come with a variety of shapes within the same powder batch. A classification method is thus needed for a relevant description of the powder morphology. This was developed, in 3D, in [22], where all particles imaged by XMT were regrouped in a finite set of shape classes (Fig. 13). To achieve this, quantitative shape measurements were introduced, based on the choice of suitable morphological indicators (Table 1), namely MIs. More details on their description and on the classification procedure, which is quite complex, can be found in [22, 31, 60].

To summarize, first an independent set of MIs was created checking the correlations and excluding MIs that were not independent. Then, dimension of the MIs space was reduced applying the Principal Component Analysis (PCA), a technique that consists in linearly combining the MIs and identifying the smallest set of these combinations that assure a good description of the variability of the original data set.

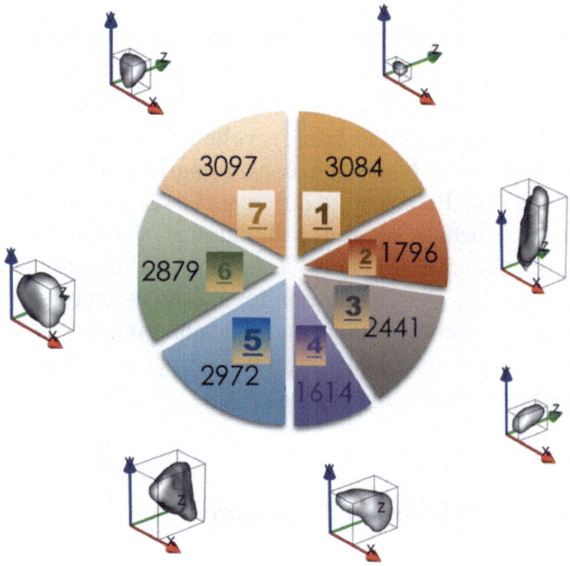

Fig. 13 Particle classification based on morphological criteria, the numbers indicate the number of particles in each class and they sum up to more than 18,000 analyzed objects, after [22]

Table 1 Size and shape measures. Underlined parameters are vectors (more than one component) and PAI stands for Principal Axis of Inertia, after [22]

Symbol	Definition
Basic measures (not used directly)	
V	Volume
r_{eq}	Radius of the equivalent sphere: $(3\ V/4\pi)^{1/3}$
r_m	Mean radius (mean distance of the surface from the barycenter)
Size	
r_1	$(r_m + r_{eq})/2$
<u>Bb</u>	Bounding box (along PAI) sides
S	Surface
Shape	
r_1	$(r_m - r_{eq})/2$
Sph	Sphericity: volume fraction contained in the equivalent sphere
S/V	Surface over volume ratio
<u>Λ</u>	Normalized PAI: $\lambda_i = PAI_i/(\Sigma_j\ PAI_j)$
<u>nbb</u>	Normalized <u>Bb</u>: $nbb_i = Bb_i/(\Sigma_j\ Bb_j)$
imbr	Mean imbrication, as defined in [22]

More details are given, for example, in [43]. Finally, the attribution of each object to a predefined shape class was obtained by the K-means cluster analysis technique, an unsupervised classification method, explained for example in [26]. Every particle was assigned to one of the 7 shape classes (Fig. 13). It appeared that certain classes were predominant in volume, meaning that the size distribution was non-homogeneous between classes. It must be noted that particles with a similar shape could actually belong to different classes. In fact, at least in this case, the clustering was a structure imposed on the data and not naturally present from the beginning. The shape varied continuously between particles and the classification was a somehow artificial process, still useful for a clearer understanding of shape data. This implied that two particles belonging to neighboring classes and lying close to the common boundary (on opposite sides) were expected to have very similar shapes and there is no contradiction in this.

3.2 Modeling of the Impact

When considering the individual events that, repeated millions of times, constitute the cold spray process (i.e. the impact of a particle), the temporal scale is in the order of 100 ns and the spatial in the range 20–50 μm. If we focus on sub-particle phenomena, investigating for example the deformation localized near the interface, the jetting, oxide layer brake, etc., the scales are even smaller. A part from recent laser based techniques allowing the observation of a deforming particle, as already discussed in Sect. 3.1.2, no direct technique is nowadays capable of giving information on the impact process. For this reason, modeling is the only way to access to the physical quantities of interest, such as temperature, stress, plastic deformation, etc. during the impact. The following part will firstly give a review of some constitutive laws describing the material behavior in the dynamic regime. Then, the most important frameworks for the simulation of impacts will be presented: finite element analysis (FEA) and non-FEA methods. Finally, more specific issues related to impact modeling will be addressed, as the effect of the oxide layer and the influence of substrate surface roughness.

3.2.1 Material Models

A number of constitutive laws for material plastic behavior in the high strain rate regime can be found in literature, such as for example Johnson-Cook [8, 25, 41], Preston-Tonks-Wallace [21, 59, 63], Khan-Huang-Liang [17, 44, 45] and Zerilli and Armstrong [6, 79]. The reader may refer to [18, 64] for a comparison of some of them. The interest in such kind of models did not certainly arise with the cold spray technology, because military related studies (ballistic) addressed this field since many decades. The fact that some, if not most, of these results is defense sensitive can probably explain the difficulty to find extensive and reliable material data for these models.

Due to the extremely dynamical nature of the impact, the use of the Mie-Gruneisen equation of state (EOS) is a common feature. This model, in fact, describes the material state in the high pressure domain and the propagation of shock waves. It can be written as follows, in the so-called Hugoniot formulation:

$$p - p_H = \Gamma \rho (E - E_H) \tag{1}$$

where p is the pressure, ρ the density, E the internal energy, $\Gamma = \Gamma_0 \rho_0 / \rho$, $\eta = 1 - \rho_0/\rho$, $p_H = \rho_0 c_0^2 \eta / (1 - s\eta)^2$, $E_H = p_H \eta / 2\rho_0$. There are 3 independent parameters for each material in the EOS: η, s and c_0. Further details on the Mie-Grüneisen EOS can be found in [4].

In the following, we will briefly describe the Johnson-Cook model [41] which is the most widely used for its simplicity. This model is an empirically-based representation of the yield stress and its main advantage is that, due to its popularity, material parameters are available in literature for a large number of materials. The yield stress is given by the following equation:

$$\sigma_{JC} = \left(A + B\varepsilon^n\right)\left(1 + C \ln \frac{\dot{\varepsilon}_0}{\dot{\varepsilon}}\right)\left[1 - \left(\frac{T - T_0}{T_m - T_0}\right)^m\right] \tag{2}$$

where A, B, C, n and m are material parameters, ε is the strain, $\dot{\varepsilon}$ is the strain rate, $\dot{\varepsilon}_0$ a reference strain rate, T_m the melting temperature of the material and T_0 a reference temperature. The three factors in the equation take into account respectively the strain hardening, the strain rate hardening and the thermal softening.

3.2.2 FEA Methods

Finite element analysis (FEA) is the most extensively used framework in the simulation of cold spray particle impacts, e.g. among many others [30, 48], and, one could say, in solid mechanics in general. FEA is developed since many decades and comes in a variety of different formulations, each one more or less adapted to address a given problem, ranging from simple elastic and static cases to large deformations, fracture propagation, thermo-mechanical or other multi-physical problems, etc. Many books treating FEA were published and one may refer to them for a thorough explanation of the technique in all its variations. For our purposes, it will be sufficient to mention that FEA is based on the discretization of space in cells, a process called meshing. Inside each cell, the fields (e.g. stress, strain, etc.) are described by some parametric function (typically polynomials), namely the elements.

Two main general formulations can be distinguished, the Lagrangian and the Eulerian. In the first one, the mesh is attached to the material and deforms with it, allowing interfaces to be easily tracked. In the second, the mesh is fixed and the material can flow through it. Eulerian elements may be partially or completely void, so that interfaces in general do not coincide to element boundaries. For the accuracy of the discretization, it is important for the mesh to be regular, i.e. the cells should have a homogeneous aspect ratio in all directions. In particle impact simulations, the local deformation can be very high, so that the elements in the Lagrangian formulation can be extremely distorted. As a result, the accuracy of the discretization is lost and the simulation could fail due to an excessive distortion of the mesh. A possible solution for this problem is the use of a hybrid formulation, namely ALE (Adaptive Lagrangian Eulerian), which consists in an adaptive meshing tool, performing a regularization of the distorted mesh by displacing its nodes and remapping the fields to the new mesh. This technique proved to be effective in reducing element distortion, but still was not sufficiently robust to easily perform impact simulations (see, for example, Delloro et al. [22]). An important issue linked with the remeshing, was the loss of the correspondence between mesh nodes and material points: the displacement fields were lost in the remeshing steps. This problem could be overcome with the Abaqus® tool called "tracer particles", which allows following the position of a certain number of material points during the simulation and, thus, retrieving the displacement fields by interpolation.

An alternative technique to ALE is the Combined Eulerian Lagrangian (CEL), also available in Abaqus®. In this framework, the interaction between pure Lagrangian and pure Eulerian domains is implemented. This approach is well suited for complex fluid-structure interaction and large deformations, although it can suffer from the interface tracking issues typical of the Eulerian formulation, especially in multi-particle simulations when the substrate is Lagrangian and the particles Eulerian [76, 78].

3.2.3 Other Non-FEA Methods

Smoothed particle hydrodynamics is a mesh-free discretization technique where the space is divided in non-connected particles, representing interacting mass points. Moreover, they are used for interpolating fields, based on data from neighboring particles, scaled by a weighting function, namely the kernel. Classical conservation equations are then imposed and interaction between particles is set to reproduce plastic material behavior. SPH method, due to its meshless characteristic, is well suited to the large deformations experienced in cold spray (Fig. 14), as shown for example in [49, 53, 54], but suffers from a more expansive computational time compared to Lagrangian FEA.

Another interesting approach, which is relatively new in comparison to all other techniques presented here, is the Discrete Element Method (DEM), which allows modeling simultaneous flow dynamics and growth and microstructure evolution of deposited matter [58]. This framework provides the ability to explicitly track all inter-particle and particle-target collisions and hence provide an alternative to traditionally used methods.

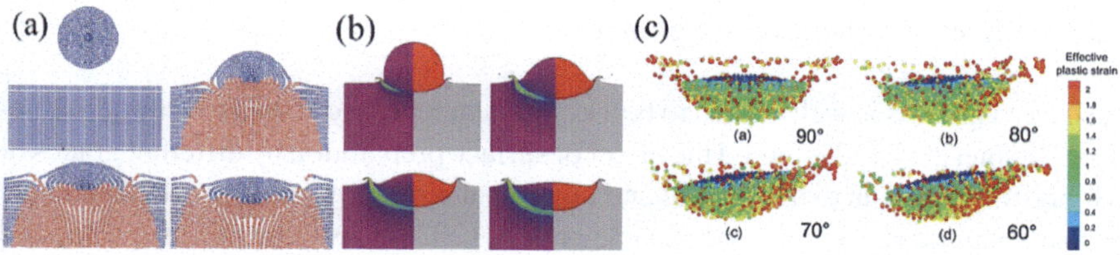

Fig. 14 Impact simulations using SPH. **a**, **b** Show a comparison between respectively SPH and Eulerian techniques for a Cu particle impacting onto a steel substrate at 700 m s^{-1}, after [53]; **c** SPH method to study the effect of the angle of impact, after [49]

3.2.4 Simulation of Surface Oxide Layers

Few studies dealt with the simulation of oxide layers. In [47] simulations of an Al particle impacting onto an Al substrate in presence of oxide layers, both on the substrate surface and around the particle, were performed using LS-DYNA finite element software, at different velocities (Fig. 15). Three kinds of simulations were performed: without the oxide layer (alumina), with a thin one (0.4 μm) and with a thicker one (0.8 μm). A brittle behavior was given to alumina by the Johnson–Holmquist plasticity damage model [42]. It was found that the region close to the interface experienced intensive plastic deformation, with its maximum on the sides, far from the initial contact point. The jetting phenomenon could also be simulated, especially for impact velocities higher than 400 m s^{-1}. The alumina layer showed a considerable effect not only at the interface, but in the whole deformability of the particle: oxide break absorbed part of the initial kinetic energy, which therefore was no longer available for the plastic deformation of the metal. A thicker layer thus implies higher particle velocity to have the same material deformation, so that the critical velocity is shifted to higher values.

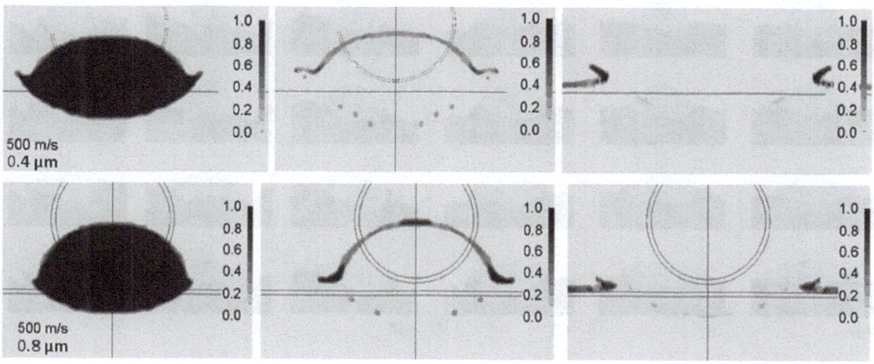

Fig. 15 Simulation results for an Al particle impacting onto an Al substrate at 500 m s^{-1}, with two different oxide layer thickness, respectively 0.4 and 0.8 μm, after [47]

3.2.5 Effect of Substrate Roughness

Surface roughness is well known to be a key parameter for cold sprayed coating adhesion. Blochet et al. [12] studied the effect of surface preparation by different grades of grit blasting, from an experimental and numerical point of view. In particular, single impact simulations of a pure Al particle onto AA2024-T3 substrate were performed (Fig. 16). A very fine roughness, made up by small peaks and valleys, vanished due to high kinetic impacts. For coarse roughness, the impact onto a valley or a peak gave different results but in both cases the particle could find itself well anchored into the substrate, showing a large contact area. The most prominent surface defects could not be smoothed by particles impact, so that the residual interface roughness was correlated to the initial substrate treatment. The simulations, thus, confirmed the importance of the substrate surface state and helped in explaining the mechanisms favoring particle adhesion.

As a general conclusion, independently of the material couple and of particle velocity, it can be said that the important characteristic determining the outcome of particle-substrate interaction is the relative size of the particle with respect to the characteristic roughness size of the substrate. Three main scenarios can thus be identified (Fig. 17).

1. Macro-roughness (roughness bigger than the particle): equivalent to impact with an angle (the local angle near the contact point).
2. Meso-roughness (roughness at the same scale as the particle): the deformed state can be approximated by the sum of the initial roughness and the deformation induced by the impact on a flat surface (linear combination).

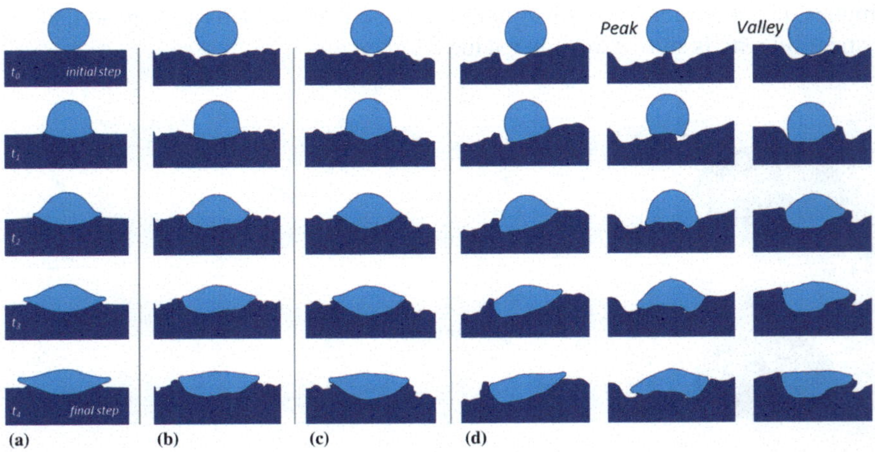

Fig. 16 Effect of the AA2024-T3 substrate roughness at the impact of a pure Al particle (diameter 35 μm, velocity), **a** without surface treatment, **b** fine grit-blasted, **c** medium grit blasted and **d** coarse grit-blasted; $t_0 = 0$, $t_1 = 50$ ns, $t_2 = 100$ ns, $t_3 = 150$ ns, $t_4 = 200$ ns, after [12]

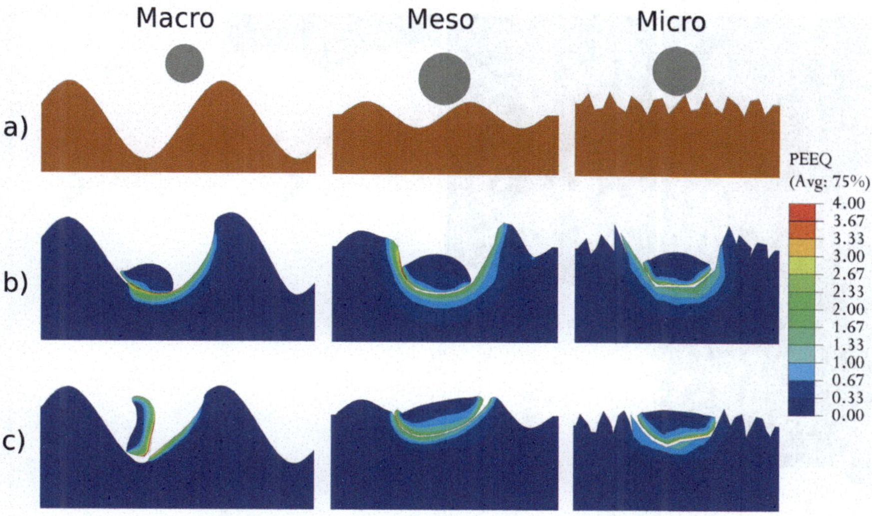

Fig. 17 Effect of substrate roughness at the impact of a Ta particle (diameter 20 μm, velocity 500 m s^{-1}). **a** Initial state, **b** after the impact onto a Cu substrate, **c** after the impact onto a Ta substrate; the color scale represents the plastic equivalent strain (PEEQ)

3. Micro-roughness (roughness smaller than the particle): the surface features are simply erased by the impact. As a consequence, deformation is concentrated in the small prominences. When comparing to a flat surface, the small-scale roughness dumps the deformation and penetration of the particle.

3.3 *Morphological Modeling*

Single particle impact models give rich information on small scale phenomena but, when it comes to the simulation of the build-up of a whole coating, computational costs in terms of CPU time and memory become prohibitive due to the high number of particles involved. Another approach is possible, where coating formation is modeled by adding particles one by one in an iterative mode, miming the building up of the coating. These models are somehow more empirical and phenomenological than single particle impacts because the physical laws are not explicitly considered here. In literature there are very few studies on this kind of modeling, being outnumbered by more classical FEA impact simulations.

In [71] a purely phenomenological 2D model was set up, combining observations of deformed particle shapes and deformation mechanisms. The simulation domain consisted in a regular mesh where particles were added one by one. These were already deformed spheres, thus consisting in ellipses with experimentally measured shape ratios and velocities. When a particle crashed onto the substrate or the forming coating, pixels coming into contact were displaced in a number proportional to the

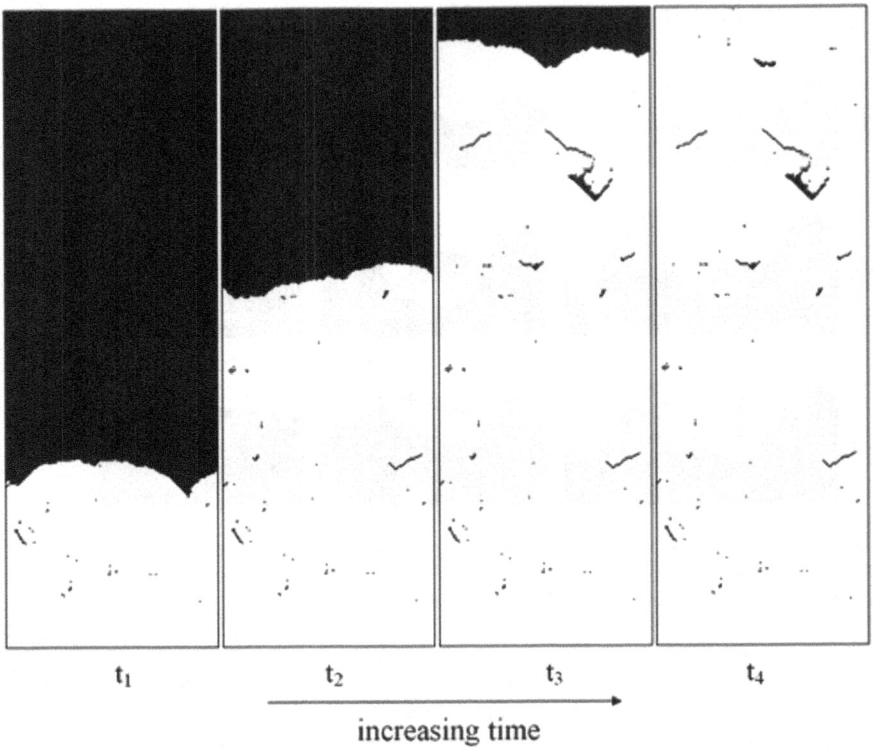

Fig. 18 Iterative coating build up model, for the simulation of porosity in 2D, after [71]

kinetic energy of the particle. After this deformation stage, the particle became part of the coating and a new incoming particle was generated (Fig. 18). This model intended to reproduce porosity formation mechanisms and was tuned to this aim.

A different approach, again miming the coating formation process in 2D by successive depositions of single powder particles, was developed in [22]. Here, data from finite element simulations of single disc-like particle with different sizes impacting with different velocities were combined into a build-up model capable to follow the evolution of the microstructure (splat boundaries). Firstly, a data set consisting of 2D discretized displacement fields was extracted from finite element simulations, constituting the "impact library". To this aim, the already mentioned tracer particles technique was necessary because remeshing techniques were used. For each simulation, an interpolation of the displacement field was obtained using Delaunay triangulation on the set of displaced tracer particles (Fig. 19) and could be used as a basis for the 2D coating build up model.

To simplify (more details can be found in the original paper), the coating build-up model consists in the iteration of the following steps. First, a particle impact is chosen from the "impact library", thus selecting a particular displacement field. Then, the substrate, with the already deposited particles, is deformed as dictated

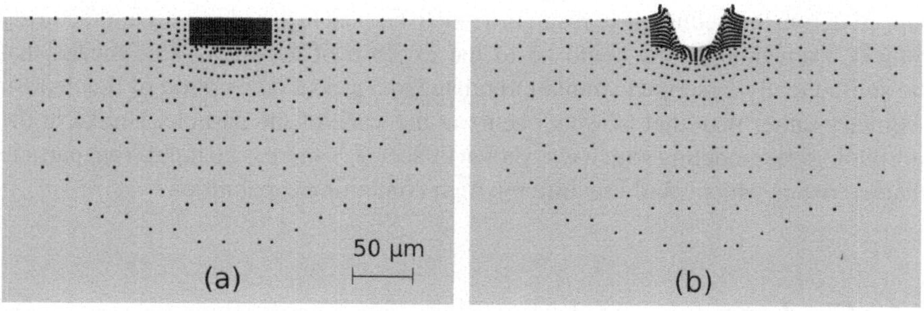

Fig. 19 Positions of tracer particles, before (**a**) and after (**b**) particle impact, after [22]. Only the substrate is shown here, in gray

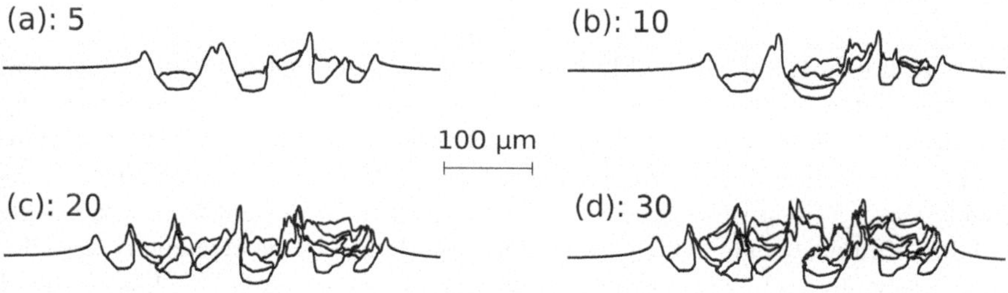

Fig. 20 Results of the coating build up simulation, after 5 (**a**), 10 (**b**), 20 (**c**) and 30 (**d**) iterations, after [22]

by the interpolated deformation field. Finally, the impinging particle is added. This procedure can then be repeated, in an iterative way, up to a large number of particles (Fig. 20). This modeling approach showed to be powerful and promising for the simulation of the cold spray process. Nevertheless, many features have yet to be introduced in the model, as for example a full 3D approach and the effects of the substrate roughness.

4 Modeling at the Scale of the Deposit

A major drawback in the finite element (FE) analysis of the coating build-up rests on calculation times. For this reason, to scale up from the particle dimension to that of the deposit needs to move on to simulation methods other than those based on finite elements. Section 2 in this chapter showed the use of advanced statistical methods to complement previous a FE analysis. This ideal dual two-scale approach to build-up still remains to be completed despite recent significant progress. Besides this dual approach, recent FE-free direct simulation at the deposit scale was developed, as considered in this third part of the chapter. The first target to be met in this development was computing time reduction. Better flexibility and efficiency are expected from this reduction to result in a tool to be used in the additive manufacturing context, even though this could be to the expense of accuracy. This means that the subsequently-described simulation relates to a global description of the deposit geometry rather than that of what occurs at the scale of the particle. Unlike in the morphological modeling which was shown in Sect. 3.3, there is no input from particle characteristics which would require too high computing capabilities.

4.1 Fundamentals and Assumptions for Simulation of Deposition Build-Up

4.1.1 General

The core principles of this build-up model are to claim that the spray angle does govern the build-up process. This is quite in keeping with the additive manufacturing process in which the role of the shape of the deposit is prominent for both the (build-up) process itself and its result. The term "spray angle" is to be understood in its broadest and most diverse sense. This includes that of conventional spray angle and that of impact angle, which relate to the nozzle and to the elementary particles respectively. As could be guessed, the deposition efficiency is behind this notion of angle through the implicit involvement of the particle rebound phenomenon.

In practice, the simulation from the model consists in building-up of successive beads (i.e. one-pass deposits), the profile of which was determined initially. The model/simulation was developed in a 2D first then extended in a 3D (or more exactly, a pseudo-3D) version. A great part of the discussion in this section will be based on 2D results to alleviate the approach. The 3-dimensional part will be considered mainly at the validation stage. Moreover, the model was tested using Al 2024 (Aluminum 2024 Alloy) powders. These powders had been heat treated prior to spraying to result in a high deposition efficiency to be suitable for additive manufacturing as discussed in Sect. 2.2. All conditions for heat treatment and for subsequent cold spray are given in (Bunel et al. 2016). Unless otherwise specified, anything which follows will refer to these experimental conditions. To give an idea of the experimental cold spray parameters which were used, one may however mention the main cold spray parameters, i.e. a powder low rate of 15 g min^{-1}, a stand-off distance of 30 mm, a pressure of 5 MPa, and a temperature of 450 °C.

These considerations led to the establishing of the two basic elements of the model, i.e. the deposition efficiency versus spray angle curve and the one-pass deposit (cross-sectional) profile. It is thus understood easily that the model therefore involves a scale higher than that involved in conventional FE analysis (Sect. 2), i.e. thousands of particles rather than given elementary particles.

The overall simulation route therefore results in 3 steps (Fig. 21).

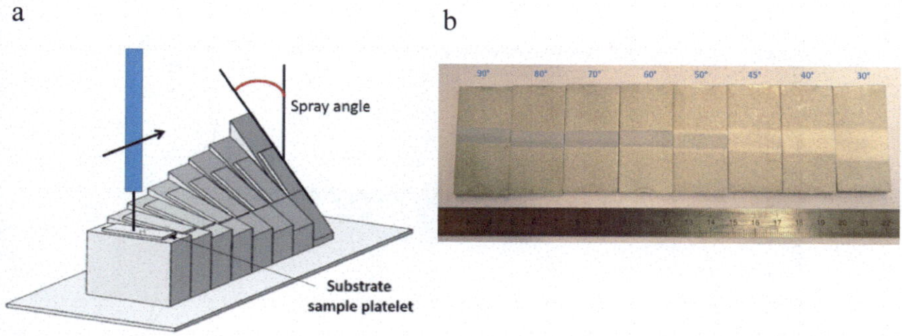

Fig. 21 Cold spray at various spray angles, **a** schematic illustration of the specific experimental set-up, **b** optical top view of the central sprayed beads onto the platelets (of 2.5 mm thick) as a function of spray angle

The first step is a spraying stage made of spraying tests to achieve beads (made of single pass deposition) for given spray angles. Incidentally, with a spray angle as defined in Fig. 21, perpendicular spraying corresponds to a spray angle of 90°.

The second step is the characterization of as-sprayed beads, which leads to the two master curves of the model, i.e. the so-called "nominal" bead profile (Fig. 22a), i.e. for a normal incidence of the nozzle, and the deposition efficiency versus spray angle curve (Fig. 22b).

The third step is that for application of the developed computing code, which results in a typical calculated profile for a deposit which can be achieved onto a large area from multi-passing of the nozzle (Fig. 23). In Fig. 23, this profile is given for perpendicular spraying (i.e. for a spray angle of 90°). The corresponding spraying parameters (those relevant to the code) are given when required.

The parameters relevant to the model and simulation are the kinetic parameters linked to the displacement of the cold spray nozzle, i.e.:

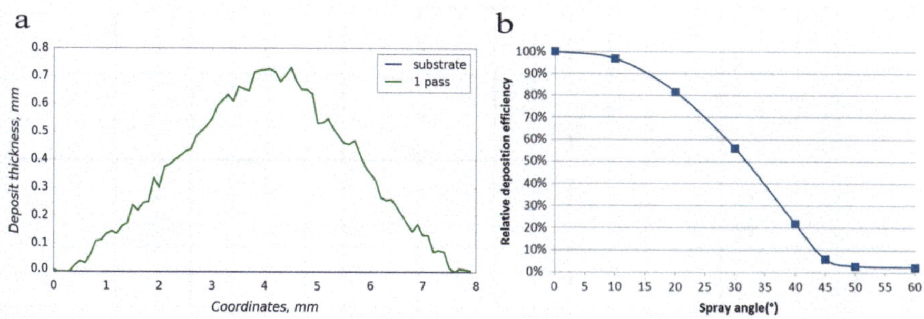

Fig. 22 Input curves for simulation, **a** nominal bead profile, i.e. deposit thickness versus (x, y) coordinates, **b** deposition efficiency (D.E.) versus spray angle

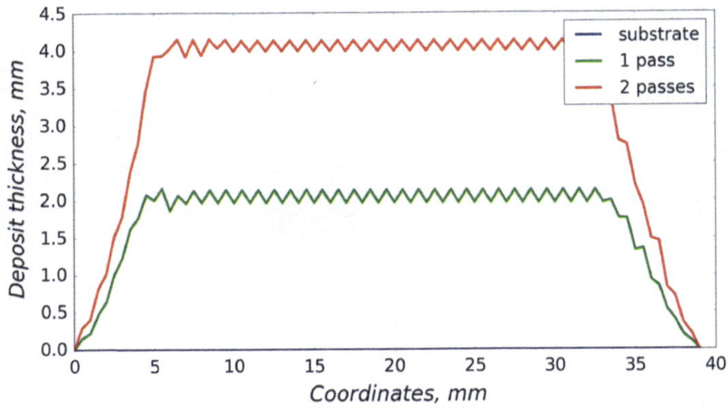

Fig. 23 Example of simulated profiles, for 1 and 2 passes

- The *traverse speed*, i.e. nozzle passing velocity (relative to the substrate).
- The *number of passes*, knowing that a pass is defined as the whole trajectory of the nozzle which leads to an elementary layer, e.g. the whole slotted track in a given color (read or blue) in Fig. 24. The blue and red tracks in Fig. 24 were consecutive and therefore corresponded to even and odd passes respectively, for example. The starting point of a given pass is the same as that of the previous pass in the same series, i.e. odd or even.
- The *step size*, i.e. the width (arrowed in green in Fig. 24) of an elementary up and down in the slotted nozzle track.
- The *spatial shift,* between 2 successive passes, taken as half of the step size, for example, in the subsequent discussion of the model.
- *Possible direction reversal* for nozzle passing after every pass. If not, at the end of a given pass, the robot track moves outside of the area to be coated and goes back to the starting zone of the substrate for subsequent passing.

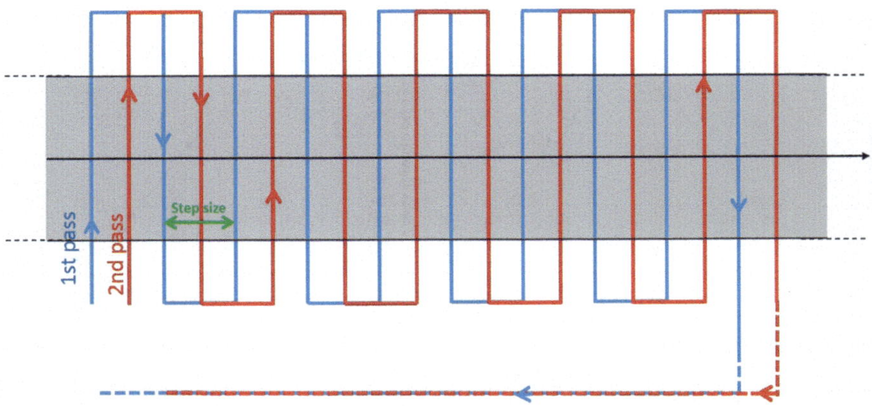

Fig. 24 Schematic top view of nozzle passing showing the key parameters for simulation

The result of the two-dimensional simulation must be compared to observation from a transverse cross-section along the longitudinal axis (black line) of the area (in gray) to be coated (Fig. 24). Every 2D profile after a given pass can be achieved and visualized (e.g. for 2 passes in Fig. 23). The model leads to simulation of the coating thickness and roughness and their evolution as a function of the number of passes. It starts from real build-up profiles and can use robot trajectory parameters to result in the prediction of the final shape of the deposit. The simulation process is therefore opposite to what can be encountered in robot programming. When programming a robot, one starts from the targeted geometry of the part and the smoothed bead profile to infer the robot trajectory [16].

The subsequent sub-sections show how the two key parameters of the model, i.e. bead profile and impact angle, are involved to result in simulation inputs. Their processing conditions and influence will be discussed.

4.1.2 Bead Profile

The nominal bead profile consists of a building block, the characteristics of which, primarily the height and the width, depend on the cold spray processing conditions and therefore integrate them. A given profile therefore corresponds to a given set to spraying parameters. Among these parameters, the stand-off distance has only little influence provided that its variation remains rather limited.

A given profile is discretized prior to be used as an input to the model. The profile is thus made of a series of n height bars at given coordinates (bright vertical lines in Fig. 25). The height of the *n*th bar is h_n and the distance between 2 successive bars is the discretization interval. The calculation step, consequently the computing time and precision of the model, depends on the discretization interval.

By definition, the value of the calculation step cannot be lower than that of the discretization interval. In contrast, the reverse is acceptable. For example, for a discretization interval of 20 μm, the calculation step will be taken as a multiple of 20 μm, e.g. 20, 40, 60, … μm. The calculation step must not be too low for two reasons. The

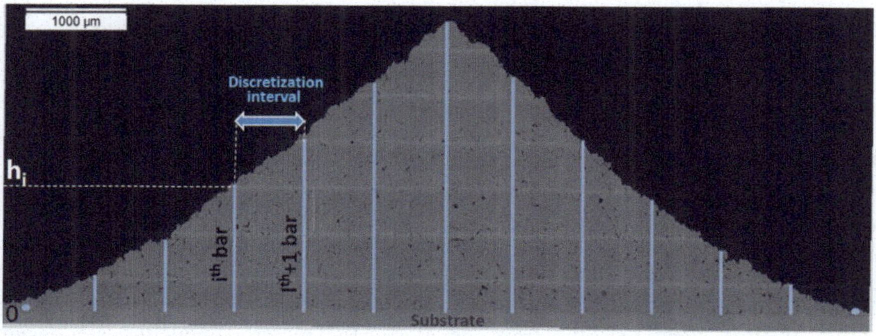

Fig. 25 Discretization of a given experimental deposit profile (2D cross-section)

first reason results from the fact that the lower the calculation step, the more precise the model is but to the expense of the computing time, the latter being proportional to the calculation step. The second reason relates to the powder particle size, which governs the roughness of the bead profile, due to a random particle build-up process. Choosing a calculation step rather higher than the mean particle size can limit the influence of this same roughness. For example, when using a powder with a Dv (50) of 20 μm, it is recommended not to use a calculation step lower than 100 μm.

4.1.3 Impact Angle

Unlike the conventional spray angle as previously defined in Fig. 21, the impact angle is a local parameter. It corresponds to the actual angle with which the sprayed particle collides the substrate. This angle therefore depends on the profile outline at a given build-up time and refers to the local slope at the point of impact in the profile outline. At the first pass for the achievement of the first bead, i.e. starting from a flat substrate, the impact angle is the same as the spraying angle. However, as from the first overlap when moving back the nozzle after the first bead, the impact angle evolves due to changes in the substrate geometry.

The spray angle has a great influence on deposition efficiency (Fig. 22b) through the variation of the degree of rebounding [11]. Deposition efficiency is all the lower as the spray angle decreases. The impact angle shows the same trend. Consequently, the impact angle has to be calculated prior to every nozzle pass to result in the proper height increment for deposit. The calculation of the impact angle at the *n*th discretization coordinate involves that of the difference between the height of the deposit at this same coordinate, i.e. h_n, and, on one hand, the height of the previous coordinate, i.e. h_{n-1}, and, on the other hand, the height of the subsequent coordinate, i.e. h_{n+1}. The signs of the two resulting height differences, namely d_{n-1} and d_{n+1}, depend on that of the corresponding slopes.

$$d_{n-1} = h_{n-1} - h_n \tag{3}$$

$$d_{n+1} = h_n - h_{n+1} \tag{4}$$

Three cases are to be envisaged according to the respective signs of the height differences, which correspond to (Fig. 26):

- A mere slope when d_{n-1} and d_{n+1} are of the same sign
- A valley when d_{n-1} is positive whereas d_{n+1} is negative
- A peak when d_{n-1} is negative whereas d_{n+1} is positive.

The overall difference must involve offsetting, i.e. positive offsets for valleys and negative offsets for peaks, but nothing for conventional slopes. The implementation of this offsetting system resulted from considering that material deposition was easier in a valley in which the angle deceased gradually rather than on a peak.

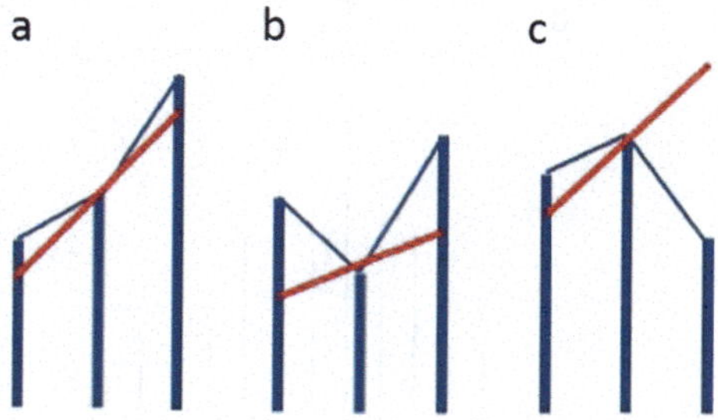

Fig. 26 Schematic illustration of the calculation process for the impact angle in the 3 possible geometries

Consequently,

- For slopes and valleys, the overall difference takes the following value of:

$$d = |d_{n-1} + d_{n+1}|$$

- For peaks it takes:

$$d = |d_{n-1} - d_{n+1}|$$

The impact angle can thus be calculated using the mere following expression:

$$\alpha = 90 - \arctan\left(\frac{d}{x_{n+1} - x_{n-1}}\right) \tag{5}$$

where x_n is the coordinate of the *n*th bar along the x-axis in the 2D bead transverse cross-section (Fig. 25).

Once the impact angle is calculated at a given point, the height of the deposit to be added takes into account the deposition efficiency curve (Fig. 22b) using the rule of proportionality. For near-normal angles, the height to be added will remain nearly the same as previously. In contrast, for rather low angles, the height can decrease significantly. The effect of the involvement of the impact angle in the model is significant, as shown on the example of deposition using conventional cold spray conditions for Al 2024 with a step size of 1 mm (Fig. 27), starting from a typical bead profile such as that in Fig. 25.

The involvement of the impact angle as determined using the previously-described rules is therefore crucial at every calculation step in the model.

4.2 *Influence of Basic Model Parameters*

The basic parameters of the models were defined in Sect. 3.1.1. To be relevant, all discussions for comparison between the influences of the parameters of the model

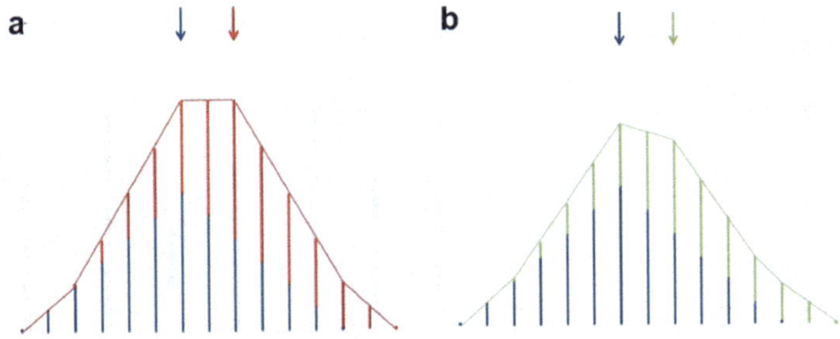

Fig. 27 Role of the impact angle on profile calculation from a starting profile in blue, **a** with no involvement of the impact angle for calculation of the subsequent profile (in red), **b** with involvement of the impact angle for calculation of the subsequent profile (in green)

will be conducted for a given spraying time, i.e. in iso-spraying time conditions in which the quantity of sprayed powder is the same.

4.2.1 Traverse Speed

Changing the traverse speed results in modifying the bead profile dimensions, i.e. bead width and bead height in a 2D cross-section. Since the width does depend on pressure, temperature and stand-off distance but not on the traverse speed, the height only has to be modified as a function of this latter parameter [70]. From experimental results showing that traverse speed has no effect on the (absolute) deposition efficiency for beads (Fig. 28), provided that the measurement errors are neglected, one can therefore state that the deposited material mass, i.e. bead height, is inversely proportional to the traverse speed. To set the definitions, one may note that, in contrast with the absolute deposition efficiency, the relative deposition efficiency compares to that obtained when spraying perpendicularly (spray angle of 90°), taken as 100%.

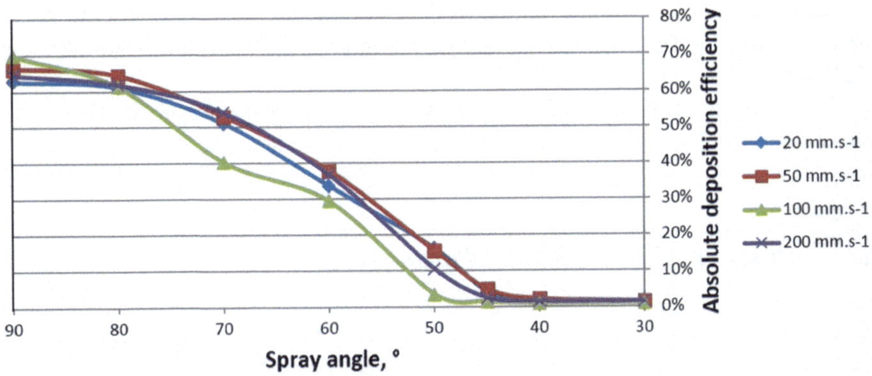

Fig. 28 Deposition efficiency versus spray angle curves for different traverse speeds

However, in iso-time conditions, the traverse speed shows a significant influence on deposit thickness, (Fig. 29). Moreover, the latter reaches a limit value after a certain traverse speed increase, e.g. approaching a thickness of 4.9 mm when above 50 mm s^{-1} typically. This can be explained easily by the fact that the thickness of

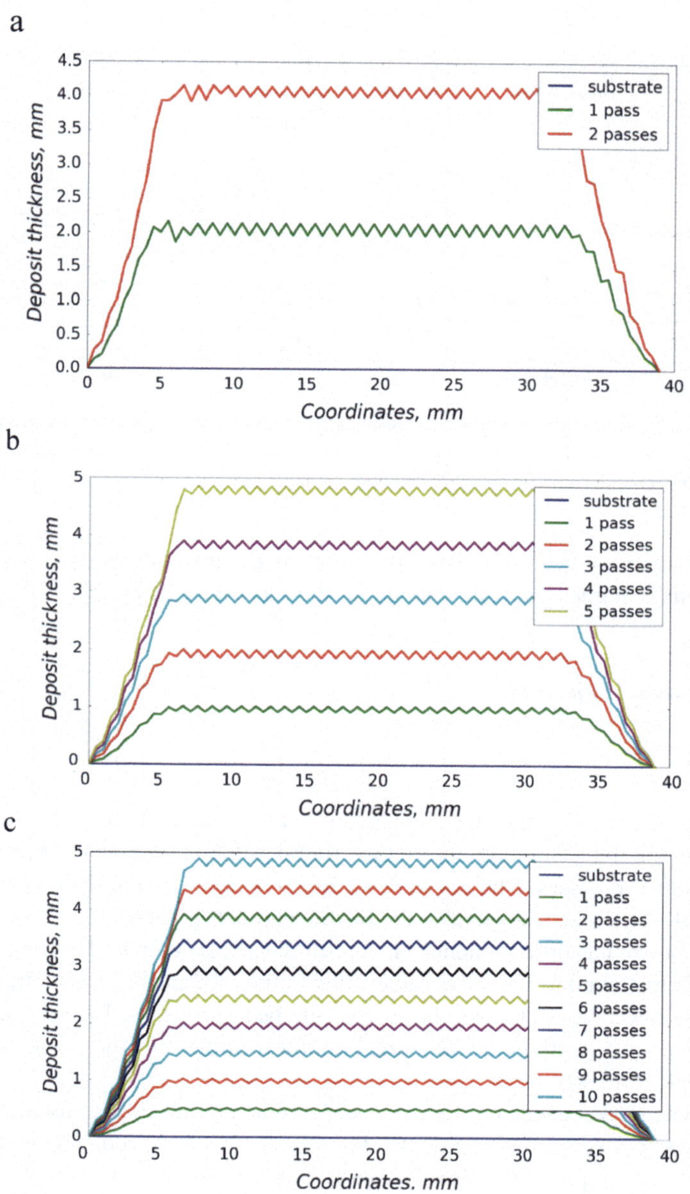

Fig. 29 Two-dimensional simulation of a deposit using a given step size of 1 mm, **a** 2 passes at a traverse speed of 20 mm s^{-1}, **b** 5 passes at 50 mm s^{-1}, **c** 10 passes at 100 mm s^{-1}

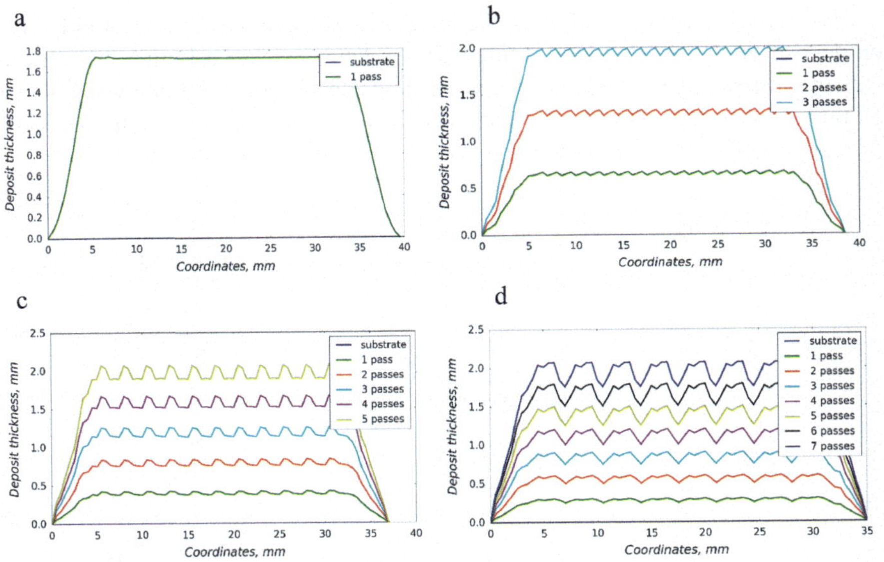

Fig. 30 Simulated profiles for a given traverse speed of 50 mm s^{-1} with, **a** a step size of 0.5 mm in 1 pass, **b** a step size of 1.5 mm in 3 passes, **c** a step size of 2.5 mm in 5 passes, **d** a step size of 3 mm in 6 passes

a given elementary bead pass is all the lower as the traverse speed is high, which corresponds to higher local spray angle for the successive beads (Fig. 29).

4.2.2 Nozzle Trajectory

Step Size

As already stated, any relevant comparison between deposits must be made for a given spraying time. Simulations are therefore developed for a given traverse speed with a number of passes to be selected as a function of the spatial shift. For example, doubling the step size leads to also double the number of passes.

The step size has little influence on deposit thickness compared to traverse speed unless it is restricted to very low values, due to the influence of the particle impact angle. The lower the step size, the steeper the bead profile is. In contrast, surface roughness does depend on the step size, both for the peak-to-valley ratio and for the profile itself (Fig. 30).

Unlike what can be seen for traverse speed, no clear trend emerges for the influence of the step size. Every case is particular. For a given traverse speed there is an optimal step size. For example, in the already-taken conditions for deposition of Al 2024 (see figures and previous sections), the optimal step size is 1.5 mm for a traverse speed of 50 mm s^{-1} though it is 1.0 mm for a speed of 20 mm s^{-1}.

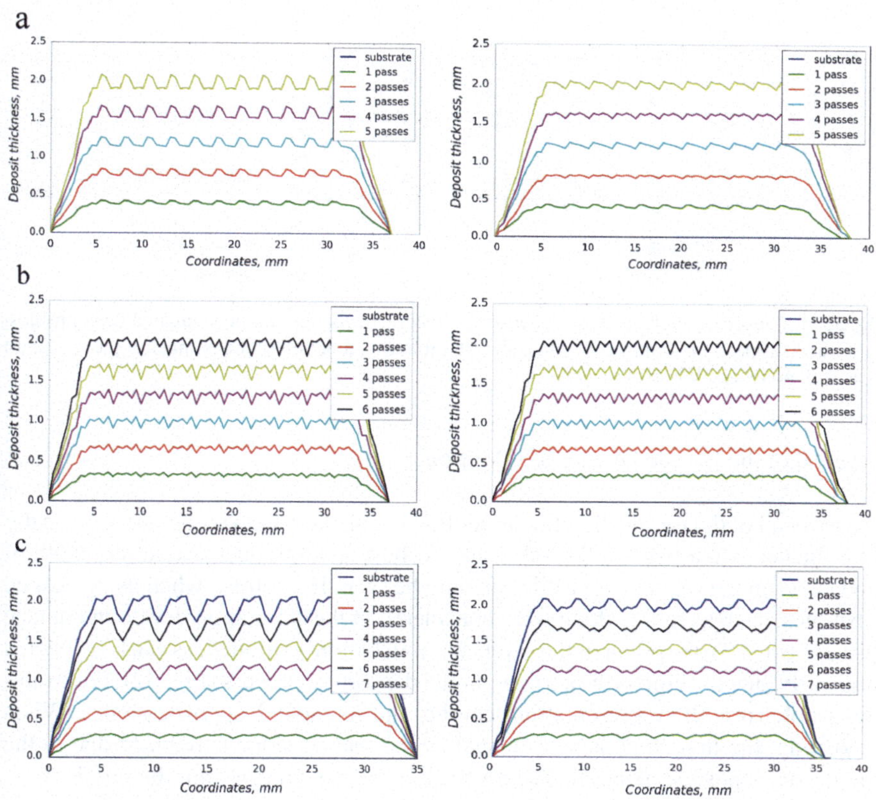

Fig. 31 Influence of a spatial shift of half the step size on the simulated bead profiles, at a traverse speed of 50 mm s^{-1}, for a step size of **a** 2.5 mm, **b** 3 mm, **c** 3.5 mm. Left diagrams, without shifting: Right diagrams, with shifting

Spatial Shift

The spatial shift between 2 successive passes, as defined in Fig. 24, can show an influence on surface roughness as done by the step size. These two parameters, as could be expected, are not independent. There is no systematic correlation. One may say that a spatial shift of half of the step size, typically, is beneficial for roughness, but the extent of improvement for this given spatial shift depends on the value of the step size. In the example of Fig. 31, for given thickness and traverse speed, the improvement is of a lesser extent for the intermediate value of the step size, i.e. 3 mm, compared to that obtained for 2.5 and 3.5 mm. Moreover, there is no improvement at all for a low value of step size such as 1 mm, insofar as the simulated curves with and without shifting are superimposable (no figure shown in this article). The improvement is not therefore systematic, the explanation of which is not yet elucidated, even though one may assume this might depend on simulation accuracy.

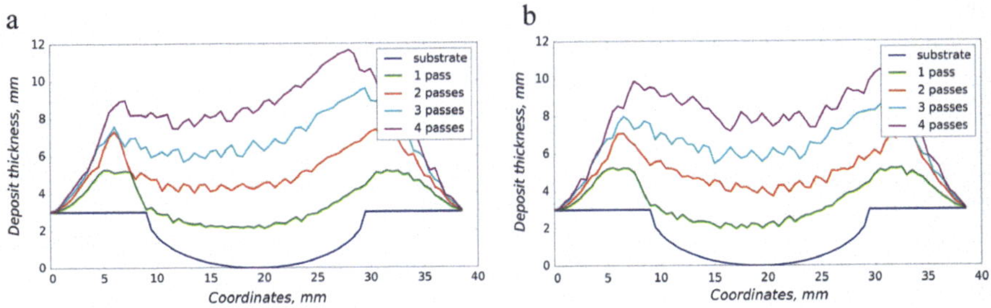

Fig. 32 Influence of the reversal of the nozzle direction of motion on the simulated bead profiles, for a traverse speed of 20 mm s^{-1}, a step size of 1 mm as a function of the number of passes (up to 4), **a** with no reversal, **b** with reversal

Reversal of the Nozzle Direction of Motion

The reversal of the nozzle direction of motion i.e. in alternate passing, has practically no influence when using a flat substrate. At best, a slight decrease in the depth of the valleys in the roughness profile can only be seen. In contrast, when using shaped substrates, such as those which were considered in the simulation development (see further Sect. 3.3), the effect of the reversal is much more exhibited, as can be expected intuitively, due to some symmetrizing in the deposition process (Fig. 32). For example, a hole (in a 2D representation) can be said to show 2 opposite slopes. When depositing, the first slope is accentuated as the second slope is reduced due to the previously-deposited material. The reversal in the motion can limit the problem.

4.3 Improvement of the Model

4.3.1 Multi-profile Simulation

To better reproduce the randomness of the surface of the sprayed deposit, a major improvement in the simulation can result from the use of a series of bead profiles, rather that one profile only as the first model input (see Sect. 3.1.1). The previous sections discussed results starting from 1 bead profile to better exhibit the influence of a given parameter or another.

When using a single typical bead profile, the final roughness of the deposit resulted from repeated processing of the same bead profile in the simulation. As can be expected, this leads to a sort of uniformity in the roughness distribution, which can be vanished when involving several beads (all the more efficient that they are numerous, 10–15 typically) (Fig. 33). These bead profiles come from cross-sections of a bead achieved in given spraying conditions.

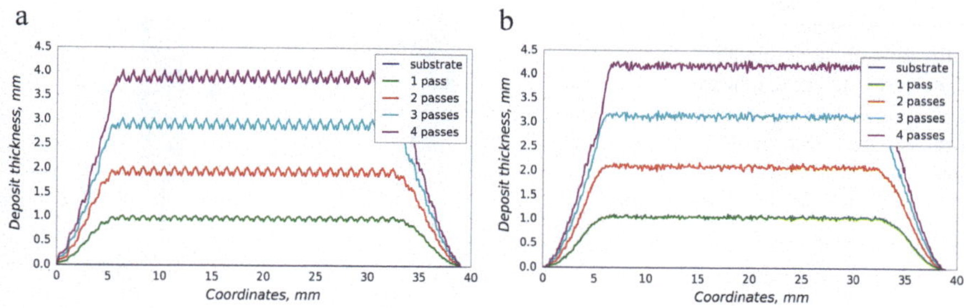

Fig. 33 Results from simulation using similar parameters but starting with, **a** 1 bead profile input, **b** several bead profiles

4.3.2 Extension to 3D Simulation

A 3D, or more exactly a pseudo-3D version of the build-up model was developed. The resulting surfaces are obtained from mere juxtaposition of the 2D profiles. However, the price to pay is that the impact angle in the direction parallel to that of the nozzle motion, i.e. the y axis, cannot be involved. This is not the case for the x axis, i.e. perpendicularly to the nozzle direction of motion. Consequently, there cannot be any slope in the y direction (Fig. 34).

In the y direction, changing the spraying direction also changes the spraying distance, which makes them hard to be discriminated. Moreover, depending on the nozzle direction of motion along this same axis, the effect of the spraying angle can be either promoted or reduced (Fig. 35).

This ascertained that the spraying angle was not involved deliberately in the development of this pseudo-3D simulation.

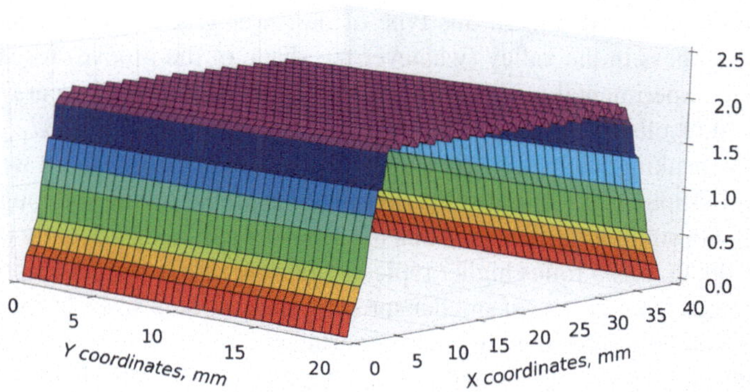

Fig. 34 Three-dimensional simulation of a 4-pass deposit for a traverse speed of 50 mm s^{-1} and a step size of 2 mm

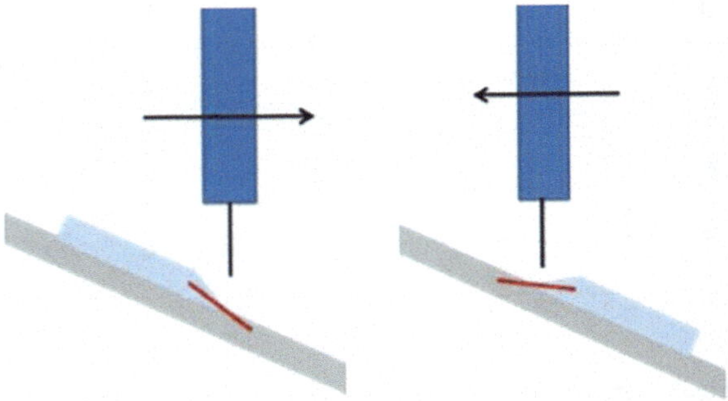

Fig. 35 Schematic illustration of the influence of the spraying angle (shown by the red line) as a function of the nozzle direction of motion

4.4 Testing and Validation of the Model

Results are discussed using both 2D and 3D simulations. However, for testing and validating the fundamentals of the model/simulation, 2D is enough.

Validation mainly rests on testing the model when applied to shaped substrates. Three typical shapes, i.e. V, U and round shapes were tested.

As a general result, simulation can be claimed to be in keeping with experiments. For reasons of brevity, the subsequent discussion is given in the example of the V-shaped groove only. The consistency between experimental tests and simulation is ascertained by typical features such as dissymmetry of the deposit profile (Fig. 36).

Moreover, the lowest point at the bottom of the groove shows the same coordinates and the various slopes almost the same angles in both profiles. The sole but rather significant discrepancy between the two profiles relates to bumps which formed in the simulated left slope of the V (Fig. 36). The presence of these bumps is assumed to result from small errors which could cumulate after every nozzle pass.

Simulation is quite satisfactory, even for rather small-sized grooves, e.g. of 2 mm in width for 1 mm in depth. In this type of narrower grooves, most noticeable is a higher roughness in the valley (whatever the shape of the groove) for simulation compared to experimental results. Otherwise, simulated results compare well with experimental results which vary in the so-called "gray range" in Fig. 37.

Roughly speaking, the results are similar for the other shapes which were tested. For the semi-elliptical in particular, the experimental global profile and roughness are in keeping with simulation, provided that the groove width is significantly larger than the groove depth, e.g. 5 times higher typically. For the smallest slots, improvements can be expected from the use of smaller spraying spot and step sizes. More generally, the comparison between simulation and experimental results can be problematic for shapes with rather steep slopes and low impact angles. This pleads in favor of the development of miniaturization of spraying systems to be used in cold spray additive manufacturing specifically.

Based on 2D results, at this stage of development the build-up model used in simulation can already be deemed robust. Further three-dimensional approach ascertains this. In this pseudo-3D approach, the basis of which was described in Sect. 3.3,

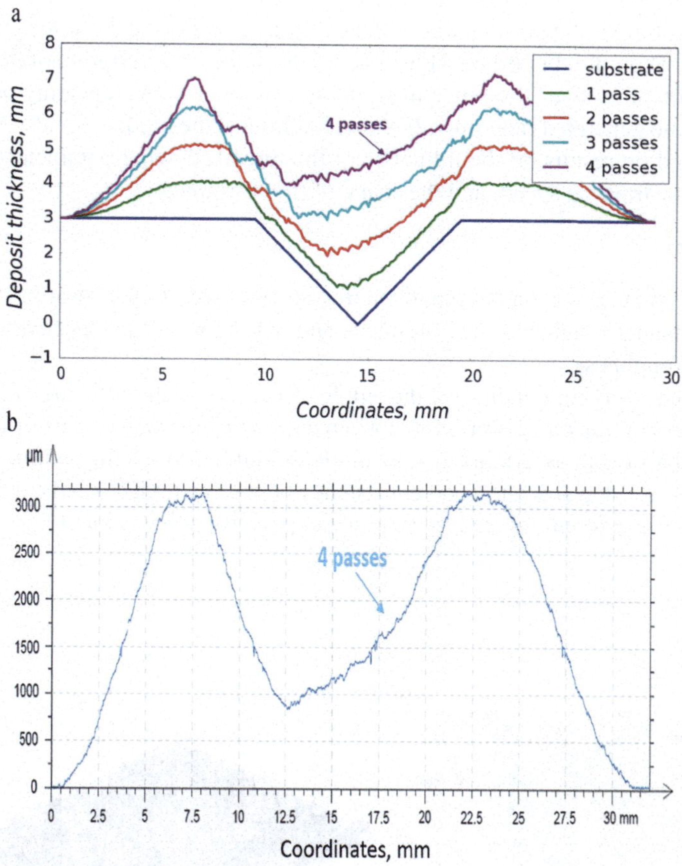

Fig. 36 Bead profile for a traverse speed of 50 mm s^{-1}, a step size of 1 mm, and 4 passes, onto a given V-grooved substrate, **a** simulated, **b** experimental

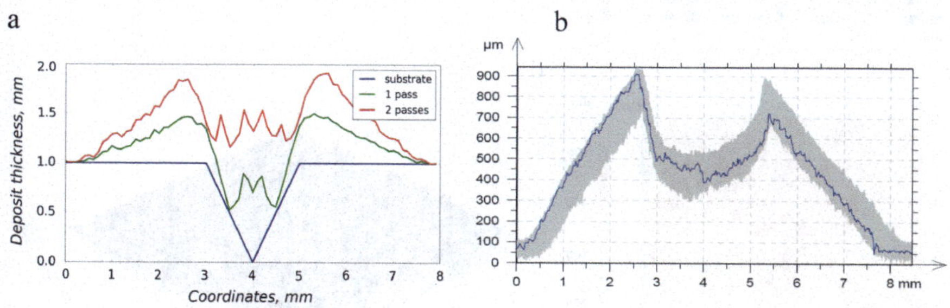

Fig. 37 Bead profiles for a traverse speed of 20 mm s^{-1}, a step size of 1 mm, and 2 passes, onto a given V-grooved substrate, **a** simulated, **b** experimental

numerous different 2D profiles were selected from a profile database for given bead conditions. Their number is taken high enough for relevant description of the deposit as a function of its size. The repeating number and order in the repeating process in the model are generated randomly. In these conditions, the model was also validated in 3D, based on results on the influence of the three following parameters, i.e. the step size, the traverse speed, and the shape of the substrate.

- Step size

For better visualization and comparison, results are given in the form of flat views for experimental roughness measurements and in the form of perspective views for simulated roughness.

For given spraying conditions, the ripple effect due to the influence of the step size in the robot trajectory is promoted when decreasing the step size from 2 to 1 mm (Fig. 38). This result ascertains that the model is valid, through the exhibiting of the influence of the step size and good consistency between simulated and experimental results regarding roughness ranges (width and peak-to-valley values).

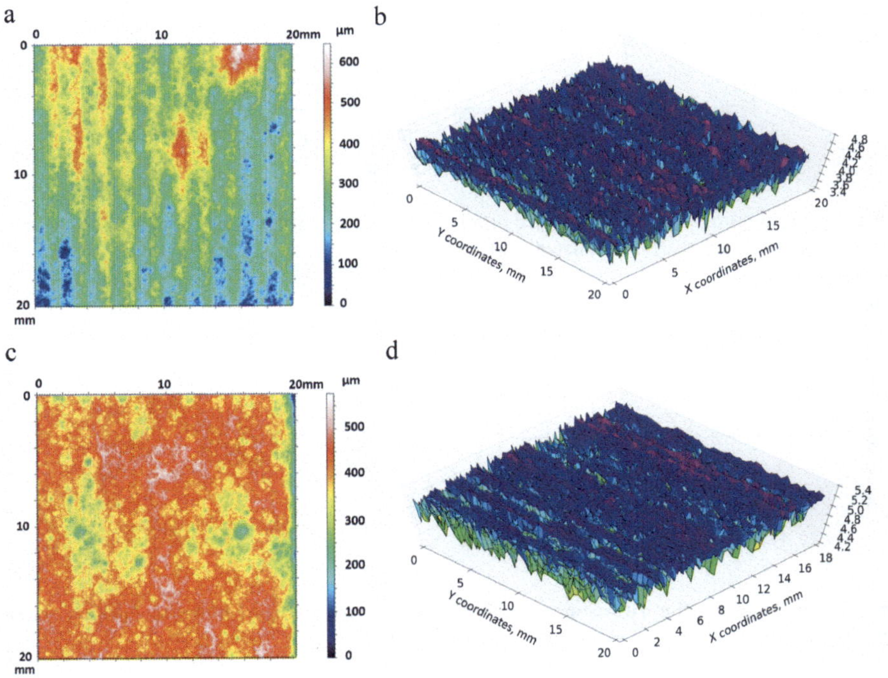

Fig. 38 Surface roughness of a $20 \times 20\ mm^2$ area for a traverse speed of $20\ mm\ s^{-1}$, **a** 2 passes with a step size of 1 mm, **b** corresponding simulation, **c** 4 passes with a step size of 2 mm, **d** corresponding simulation

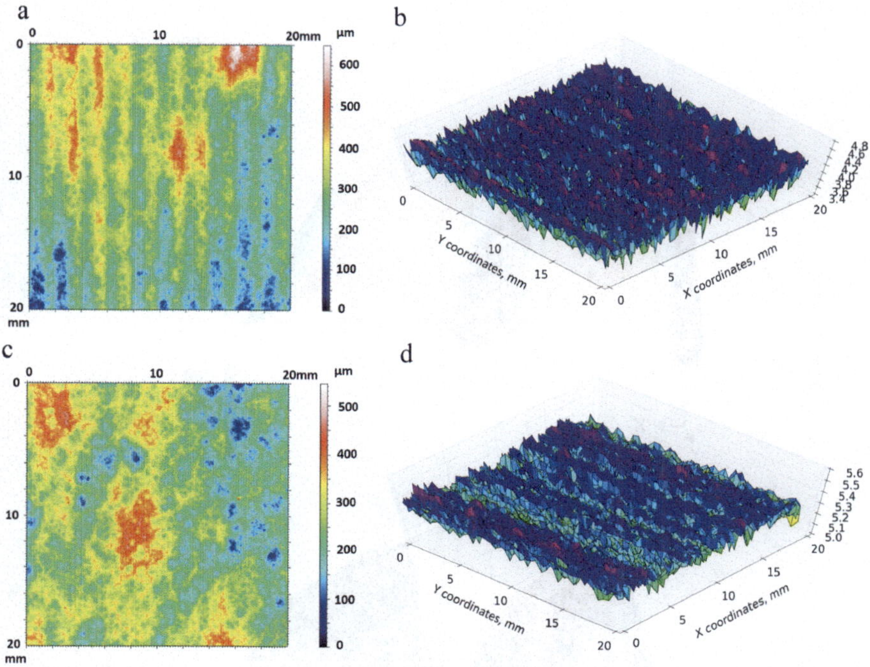

Fig. 39 Surface roughness of a 20 × 20 mm^2 area for a step size of 1 mm, **a** 2 passes at a traverse speed of 20 mm s^{-1}, **b** corresponding simulation, **c** 5 passes at a traverse speed of 50 mm s^{-1}, **d** corresponding simulation

- Traverse speed

The conclusions on the influence of the traverse speed are similar to those for that of the step size, which can justify to merely copy/paste a good part of the previous section, that is, the ripple effect due to the influence of the traverse speed is promoted when decreasing the step size from 50 to 20 mm s^{-1} (Fig. 39). This result ascertains that the model is valid through the exhibiting of the influence of the traverse speed with good consistency between simulated and experimental results regarding roughness ranges (width and peak-to-valley values).

- Shape of the substrate

Three-dimensional profilometer data compare fairly well with simulation when applied to various shaped substrates, provided these are not too complex, i.e. within the already-mentioned limits. Otherwise, this should result in a case-by-case assessment (e.g. for a deep slot). Figure 40, even though different z scales do not make the comparison very easy, ascertains that the model is valid through the exhibiting of a good consistency between simulated and experimental results.

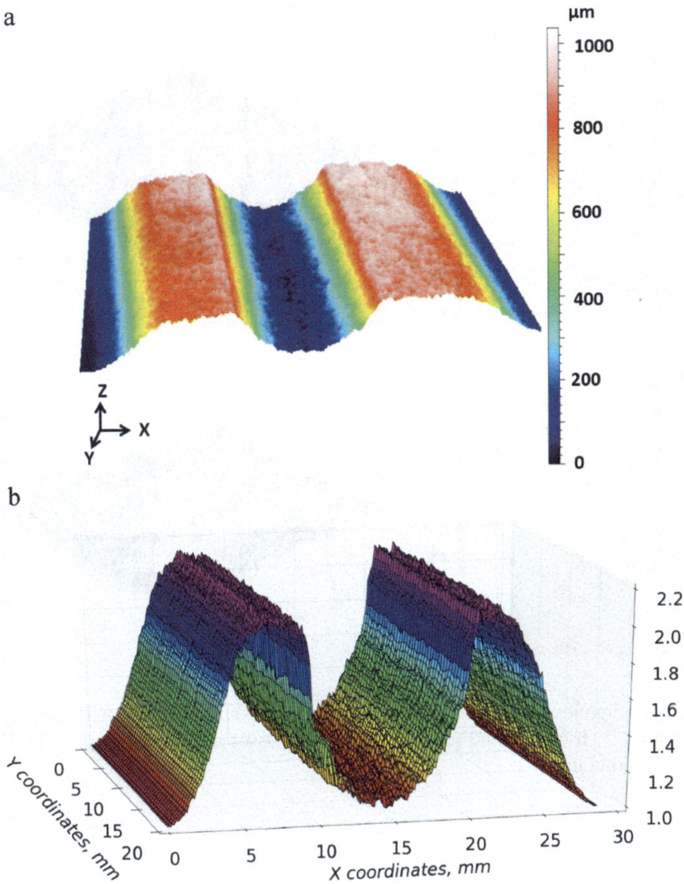

Fig. 40 Surface mapping of a deposit onto a U-shaped groove (elliptical, of 1 mm in depth and 10 mm in width), for 1 pass with a step size of 1 mm, **a** experimental, **b** simulated

4.5 Applications

Modeling at the scale of the deposit as shown in Sect. 3 of this chapter was shown to be suitable for predictive simulation of the shape and roughness of this same deposit. At the current stage of development, this type of simulation can be used for going faster, i.e. limiting the number of preliminary spraying experiments, in the establishing of relevant kinetic parameters and spraying conditions for a given application, i.e. given shape and thickness. A major asset of this type of simulation rests on the short computing time, i.e. a few minutes typically for mm-sized parts. For example, the simulated surface in Fig. 39b needs a computing time of no more than 6 s to be obtained. The development of this class of simulation can therefore be considered as a significant step on the road towards additive manufacturing.

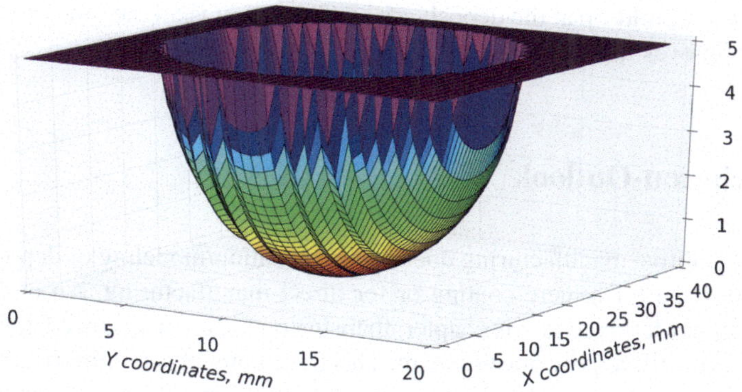

Fig. 41 Simulated surface of a hole prior to filling

A typical example of application is, for restoration/repair, that of filling a semi-ellipsoidal hole of 5 mm in depth, 30 mm in width along the x-axis and 15 mm in width along the y-axis (Fig. 41).

When fixing nominal spraying conditions including a traverse speed of 50 mm s^{-1} and a step size of 1 mm, simulation shows that 5 nozzle passes are not enough to fill the hole entirely (Fig. 42a, in which part of the deposit surface remains below the 5 mm z-plane) and that 6 passes are needed (Fig. 42b).

This build-up model needs no more than one minute to operate and leads to the simulation results shown in Fig. 42. A major asset rests on its capability in estimating the amount of powder to be used for a given application, since all the spraying parameters are known as inputs in the model. This results in an estimation of the spraying time and, consequently, of the amount of powder knowing the powder flow rate. In the previous example (Fig. 41), the filling of the considered hole needs about 16.1 g. Some discrepancy can however exist between estimation and actual powder consumption due to speed variation for the robot when changing direction. Moreover, some dead times such as that for the robot to reach the starting point for spraying are not involved in the estimation. Another consequence of any direction change for the robot is to reduce the reliability of the simulation, especially for deposit thickness prediction. This results from speed reduction for the robot, which leads to

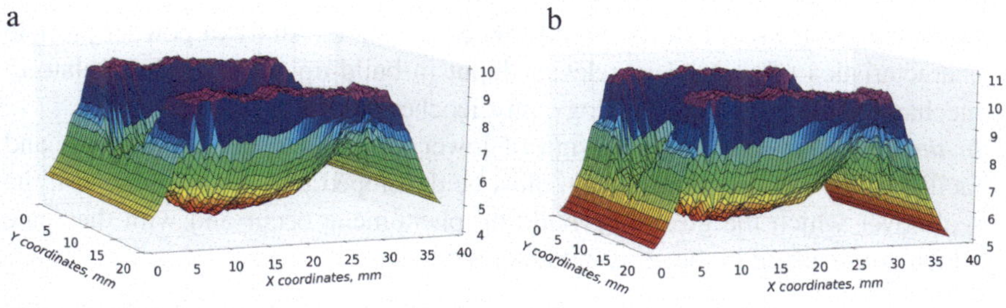

Fig. 42 Simulated surface of the hole shown in Fig. 41, **a** after 5 passes, **b** after 6 passes

some excessive build-up at the deposit edges. This effect is all the more pronounced as the robot moves fast.

5 Conclusion-Outlook

Cold spray additive manufacturing does need simulation/modeling of deposit build-up, whether it be for a mere coating or for direct manufacturing. An overview of modeling tools was given in this chapter, focusing on the most advanced approaches, which however still require development. This development is required due to extension of the range of materials which can be cold sprayed now (i.e. beyond conventional ductile powders), and to expectations from cold spray to achieve more and more complex shapes and higher properties. It is useless to say that all this must be done in the best possible cost effective conditions, i.e. with the best efficiency (powder consumption and productivity).

To be efficient for additive manufacturing using cold spray, these advanced tools for modeling of build-up issues, as shown in this chapter, have to be considered at two scales, i.e. that of the powder particle and that of the sprayed bead. Discussing build-up issues and related modeling/simulation tools resulted in the two main parts of the chapter. The philosophy behind the use of modeling in cold spray additive manufacturing is to claim that, depending on the targeted application and requirements, the user would have to use tools working either at the particle scale or at the bead scale. Moreover, as can be guessed, the best might be to combine both approaches, knowing the large variety of issues to be addressed in such a versatile process. This combination is all the more prominent because hybrid processes can also be envisaged for additive manufacturing, e.g. cold spray coupled with laser processing.

If there is just one thing to bear in mind in the development of these advanced models, this would be the crucial and often original contribution of the morphological approach to the sprayed materials (at both scales), which can involve the use of advanced tools from mathematical morphology.

In parallel with the development of modeling and simulation tools as shown in this chapter, one may infer from the content of this same chapter that action is required in two areas:

- *In the materials area*: Efforts should concentrate on the study of powder particle characteristics to be used as relevant input in build-up models. They relate to mechanical properties, morphology, physicochemistry, and metallurgy.
- *In the processing area*: The design of lower size cold spray components and facilities, not to say miniaturizing, must be developed to be in keeping with the scale over which the governing build-up phenomena occur and with the scale which are involved in the corresponding models.

Acknowledgements The authors would like to thank warmly Gilles Surdon and Kevin Roche from Dassault-Aviation/Mérignac, Erick Meillot from CEA/Le Ripault, and Pascal Lemeille from Mallard SA/Saint-Antoine-la-Forêt, all of them for technical assistance, fruitful discussions and financial support.

References

1. Ahmadian, S., Browning, A., & Jordan, E. H. (2015). Three-dimensional X-ray micro-computed tomography of cracks in a furnace cycled air plasma sprayed thermal barrier coating. *Scripta Materialia, 97,* 3–16.
2. Amsellem, O., Borit, F., Jeulin, D., Guipont, V., Jeandin, M., Boller, E., et al. (2012). Three-dimensional simulation of porosity in plasma-sprayed alumina using microtomography and electrochemical impedance spectrometry for finite element modeling of properties. *Journal of Thermal Spray Technology, 21*(2), 193–201.
3. Amsellem, O., Madi, K., Borit, F., Jeulin, D., Guipont, V., Jeandin, M., et al. (2008). Two-dimensional (2D) and three-dimensional (3D) analyses of plasma-sprayed alumina microstructures for finite-element simulation of Young's modulus. *Journal Materials Science, 43*(12), 4091–4098.
4. Antoun, T., Seaman, L., Curran, D. R., Kanel, G. I., Razorenov, S. V., & Utkin, A. V. (2003). *Spall fracture*. New York, NY: Springer.
5. Armstrong, D. R., Borys, S. S., & Anderson, R. P. (1999). Method of making metals and other elements from the halid vapor of the metal. US Patent 5958106.
6. Armstrong, R., & Zerilli, F. (1994). Dislocation mechanics aspects of plastic instability and shear banding. *Mechanics of Materials, 17*(2–3), 319–327.
7. Assadi, H., Irkhin, I., Gutzmann, H., Gärtner, F., Schulze, M., Villa Vidaller, M., et al. (2015). Determination of plastic constitutive properties of microparticles through single particle compression. *Advanced Powder Technology, 26*(6), 1544–1554.
8. Bae, G., Xiong, Y., Kumar, S., Kang, K., & Lee, C. (2008). General aspects of interface bonding in kinetic sprayed coatings. *Acta Materialia, 56,* 4858–4868.
9. Barradas, S., Guipont, V., Molins, M., Jeandin, M., Arrigoni, M., Boustie, M., et al. (2007). Laser shock flier impact simulation of particle-substrate interactions in cold spray. *Journal of Thermal Spray Technology, 16,* 548.
10. Beauvais, S., & Decaux, O. (2007). Plasma sprayed biocompatible coatings on PEEK implants. In: *Proceedings of the International Thermal Spray Conference (ITSC '07)* (pp. 371–376), Beijing, China, May 14–16, 2007, ASM International.
11. Blochet, Q., Delloro, F., Borit, F., N'Guyen, F., Jeandin, M., Roche, K., & Surdon, G. (2014). Influence of spray angle on the cold spray of Al for the repair of aircraft components. In J. Jerzembeck et al. (Eds.), *Proceedings of the International Thermal Spray Conference and Exposition (ITSC'14)* (pp. 69–74), Barcelona, Spain, May 21–23, 2014. Düsseldorf: DVS Media GmbH. ISBN 978-3-87155-574-9.
12. Blochet, Q., Delloro, F., N'Guyen, F., Jeulin, D., Borit, F., & Jeandin, M. (2017). Effect of the cold-sprayed aluminum coating-substrate interface morphology on bond strength for aircraft repair application. *Journal of Thermal Spray Technology, 26*(4), 671–686.
13. Bobzin, K., Ote, M., Knoch, M. A., Alkhasli, I., & Dokhanchi, S. R. (2019). Modelling of particle impact using modified momentum source method in thermal spraying. *IOP Conference Series: Materials Science and Engineering, 480,* 012003.
14. Botef, I., & Villafuerte, J. (2015). Overview. In J. Villafuerte (Eds.), *Modern cold spray*. Cham: Springer.
15. Bunel, M., Borit, F., Delloro, F., Jeandin, M., Bacciochini, A., Lemeille, P., et al. (2017). Experimental and numerical study of the influence of powder characteristics in the cold spraying

of Al-based alloys for additive manufacturing using low-pressure, medium-pressure and high pressure cold spray facilities. In: *Proceedings of the International Thermal Spray Conference and Exposition (ITSC '17)* (pp. 714–718), Dusseldorf, Germany, 7–9 June 2017, ISBN: 978-1-5108-5822-0.
16. Cai, Z., Liang, H., Quan, S., Deng, S., Zeng, C., & Zhang, F. (2015). Computer-aided robot trajectory auto-generation strategy in thermal spraying. *Journal of Thermal Spray Technology, 24*(7), 1235–1245.
17. Chen, Q., Alizadeh, A., Xie, W., Wang, X., Champagne, V., Gouldstone, A., et al. (2018). High-strain-rate material behavior and adiabatic material instability in impact of micron-scale Al-6061 particles. *Journal of Thermal Spray Technology, 27,* 641.
18. Chen, S. R., & Gray, G. T. (1996). Constitutive behaviour of tantalum and tantalum-tungsten alloys. *Metallurgical and Materials Transactions A: Physical Metallurgy and Materials Science, 27*(10), 2994–3006.
19. Cinca, N., Rebled, J. M., Estradé, S., Peiró, F., Fernández, J., & Guilemany, J. M. (2013). Influence of the particle morphology on the cold gas spray deposition behaviour of titanium on aluminum light alloys. *Journal of Alloys and Compounds, 554,* 89–96.
20. Cochelin, E., Borit, F., Frot, G., Jeandin, M., Decker, L., Jeulin, D., et al. (1999). Oxidation and particle deposition modeling in plasma spraying of Ti-6Al-4 V/SiC fiber composites. *Journal of Thermal Spray Technology, 8,* 117.
21. Cormier, Y., Dupuis, P., Jodoin, B., & Ghaei, A. (2015). Finite element analysis and failure mode characterization of pyramidal fin arrays produced by masked cold gas dynamic spray. *Journal of Thermal Spray Technology, 24,* 1549–1565.
22. Delloro, F., Jeandin, M., Jeulin, D., Proudhon, H., Faessel, M., Bianchi, L., et al. (2017). A morphological approach to the modeling of the cold spray process. *Journal of Thermal Spray Technology, 26,* 1838–1850.
23. Delloro, F., Jeandin, M., Faessel, M., Jeulin, D., Meillot, E., & Bianchi, L. (2014). A morphological approach to the modeling of the cold spray process. In J. Jerzembeck et al. (Eds.), *Proceedings of the International Thermal Spray Conference and Exposition (ITSC '14)* (pp. 221–225), Barcelona, Spain, 21–23 May, 2014. Düsseldorf, Germany: DVS Media GmbH. ISBN 978-3-87155-574-9.
24. Descurninges, L. L., Mingault, L. T., Guipont, V., & Jeandin, M. (2011). Influence of powder particles oxidation on properties of cold sprayed tantalum. In J. Jerzembeck (Ed.), *Thermal Spray 2011: Proceedings of the International Thermal Spray Conference* (pp. 60–65), September 2011. DVS.
25. Dosta, S., Bolelli, G., Candeli, A., Lusvarghi, L., Cano, I. G., & Guilemany, J. M. (2017). Plastic deformation phenomena during cold spray impact of WC-Co particles onto metal substrates. *Acta Materialia, 124,* 173–181.
26. Everitt, B. S., Landau, S., Leese, M., & Stahl, D. (2011). *Cluster analysis* (5th ed.). Hoboken, NJ: Ed. Wiley.
27. Fernandez, R., & Jodoin, B. (2018). Cold spray aluminum-alumina cermet coatings: Effect of alumina content. *Journal of Thermal Spray Technology, 26,* 1838–1850.
28. Gama, B. A., Lopatnikov, S. L., & Gillespie, J. W. (2004). Hopkinson bar experimental technique: A critical review. *Applied Mechanics Reviews, 57*(4), 223–250.
29. Ghafouri-Azar, R., Mostaghimi, J., Chandra, S., & Charmchi, M. (2003). A stochastic model to simulate the formation of a thermal spray coating. *Journal of Thermal Spray Technology, 12*(1), 53–69.
30. Ghelici, R., Bagherifard, S., Guagliano, M., & Verani, M. (2011). Numerical simulation of cold spray coating. *Surface and Coatings Technology, 205,* 5294–5301.
31. Gillibert, L., Peyrega, C., Jeulin, D., Guipont, V., & Jeandin, M. (2012). 3D multiscale segmentation and morphological analysis of X-ray microtomography from cold-sprayed coatings. *Journal of Microscopy, 248*(2), 187–199.
32. Grujicic, M., Saylor, J. R., Beasley, D. E., DeRosset, W. S., & Helfritch, D. (2003). Computational analysis of the interfacial bonding between feed-powder particles and the substrate in the cold-gas dynamic-spray process. *Applied Surface Science, 219*(3–4), 211–227.

33. Hassani-Gangaraj, M., Veysset, D., Nelson, K. A., & Schuh, C. A. (2018). In-situ observations of single micro-particle impact bonding. *Scripta Materialia, 145,* 9–13.
34. Hassani-Gangaraj, M., Veysset, D., Champagne, V. K., Nelson, K. A., & Schuh, C. A. (2018). Adiabatic shear instability is not necessary for adhesion in cold spray. *Acta Materialia, 158*(1), 430–439.
35. Ichikawa, Y., & Ogawa, K. (2015). Effect of substrate surface oxide film thickness on deposition behavior and deposition efficiency in the cold spray process. *Journal of Thermal Spray Technology, 24*(7), 1269–1276.
36. Ichikawa, Y., Tokoro, R., Tanno, M., & Ogawa, K. (2019). Elucidation of cold-spray deposition mechanism by auger electron spectroscopic evaluation of bonding interface oxide film. *Acta Materialia, 164,* 39–49.
37. Jeandin, M. (2011). A Socratic approach to surface modification: The example of thermal spray, Plenary talk. In E. Beyer, et al. (Eds.), *Proceedings of the 24th International Conference on Surface Modification Technologies (SMT 24)* (pp. 3–20), Dresden: Valardocs.
38. Jeandin, M. (2018). Cold spray under the banner of thermal spraying the whirlwind of additive manufacturing. *Surface Engineering, 34*(5), 341–343.
39. Jeandin, M., Koivuluoto, H., & Vezzu, S. (2015). Coating properties (Chap. 4). In J. Villafuerte (Ed.), *Modern cold spray materials, process, and applications*. Berlin: Springer.
40. Jeandin, M., Rolland, G., Descurninges, L. L., et al. (2014). Which powders for cold spray? *Surface Engineering, 30,* 291–298.
41. Johnson, G. R., & Cook, W. H. (1983). A constitutive model and data for metals subjected to large strains, high strain rates, and high temperatures. In *Proceedings of the 7th International Symposium on Ballistics* (pp. 541–547).
42. Johnson, G. R., & Holmquist, T. J. (1993). An improved computational constitutive model for brittle materials. In S. C. Schmidt, J. W. Shaner, G. A. Samara, & M. Ross (Eds.), *High Pressure Science and Technology-1993, AIP Conference Proceedings* (Vol. 309, pp. 981–984). New York, NY: AIP Press.
43. Jolliffe, I. T. (2001). *Principal Component Analysis* (2nd ed.). New York: Springer.
44. Khan, A. S., & Liang, R. (1999). Behaviors of three BCC metal over a wide range of strain rates and temperatures: experiments and modeling. *International Journal of Plasticity, 15*(10), 1089–1109.
45. Khan, A. S., & Liang, R. (2000). Behaviors of three BCC metals during non-proportional multi-axial loadings: Experiments and modeling. *International Journal of Plasticity, 16*(12), 1443–1458.
46. Knotek, O., & Elsing, R. (1987). Monte Carlo simulation of the lamellar structure of thermally sprayed coatings. *Surface & Coatings Technology, 32*(1–4), 261–271.
47. Li, W., Liao, H., Li, C., Bang, H., & Coddet, C. (2007). Numerical simulation of deformation behavior of Al particles impacting on Al substrate and effect of surface oxide films on interfacial bonding in cold spraying. *Applied Surface Science, 253*(11), 5084–5091.
48. Li, W. Y., & Gao, W. (2009). Some aspects on 3D numerical modeling of high velocity impact of particles in cold spraying by explicit finite element analysis. *Applied Surface Science, 255,* 7878–7892.
49. Li, W. Y., Yin, S., & Wang, X. F. (2010). Numerical investigations of the effect of oblique impact on particle deformation in cold spraying by the SPH method. *Surface Science, 256*(12), 3725–3734.
50. Lin, E., Chen, Q., Ozdemir, O. C., & Müftü, S. (2018). Effects of interface bonding on the residual stresses in cold sprayed Al-6061: A numerical simulation. In F. Azarmi, et al. (Eds.), *Proceedings of the International Thermal Spray Conference (ITSC 2018)* (pp. 278–295), May 7–10, 2018, Orlando, Florida, USA.
51. Liu, H., Lavernia, E. J., & Rangel, R. H. (1993). Numerical simulation of impingement of molten Ti, Ni, and W droplets on a flat substrate. *Journal of Thermal Spray Technology, 2,* 369–378.
52. MacDonald, D., Fernández, R., Delloro, F., & Jodoin, B. (2017). Cold spraying of armstrong process titanium powder for additive manufacturing. *Journal of Thermal Spray Technology, 26,* 598.

53. Manap, A., Okabe, T., & Ogawa, K. (2011). Relationship between particle size and deformation in the cold spray process. *Applied Surface Science, 10,* 1145–1150.
54. Manap, A., Ogawa, K., & Okabe, T. (2012). Numerical analysis of interfacial bonding of Al-Si particle and mild steel substrate by cold spray technique using the SPH method. *Journal of Solid Mechanics and Materials Engineering, 6,* 241–250.
55. Martin, M., & Vo, P. (2019). Advanced robot path planning for cold spray additive manufacturing. In *Proceedings of the CASI AERO Conference 2019* (p. 174). Available via https://www.openconf.org/aero2019/modules/request.php?module=oc_proceedings&action=summary.php&id=174&a=Accept. Accessed June 2, 2019.
56. Moridi, A., Gangaraj, S. M. H., Vezzu, S., & Guagliano, M. (2014). Number of passes and thickness effect on mechanical characteristics of cold spray coating. *Procedia Engineering, 74,* 449–459.
57. Moridi, A., Hassani-Gangaraj, S. M., Guagliano, M., et al. (2014). Cold spray coating: review of material systems and future perspectives. *Surface Engineering, 30,* 369–395.
58. Mukherjee, D., & Zohdi, T. I. (2015). A discrete element based simulation framework to investigate particulate spray deposition processes. *Journal of Computational Physics, 290,* 298–317.
59. Nastic, A., Vijay, M., Tieu, A., Rahmati, S., & Jodoin, B. (2017). Experimental and numerical study of the influence of substrate surface preparation on adhesion mechanisms of aluminum cold spray coatings on 300 M steel substrates. *Journal of Thermal Spray Technology, 26,* 1461–1483.
60. Parra Denis, E., Barat, C., Jeulin, D., & Ducottet, C. (2008). 3D complex shape characterization by statistical analysis: application to aluminium alloys. *Materials Characterization, 59,* 338–343.
61. Pasandideh-Fard, M., & Mostaghimi, J. (1995). On the spreading and solidification of molten particles in a plasma spray process effect of thermal contact resistance. *Plasma Chemistry and Plasma Processing, 16*(S1), S83–S98.
62. Pattison, J., Celotto, S., Morgan, R., Bray, M., & O'Neill, W. (2007). Cold gas dynamic manufacturing: A non-thermal approach to freeform fabrication. *International Journal of Machine Tools and Manufacture, 47*(3–4), 627–634.
63. Preston, D. L., Tonks, D. L., & Wallace, D. C. (2003). Model of plastic deformation for extreme loading conditions. *Journal of Applied Physics, 93*(1), 211–220.
64. Rahmati, S., & Ghaei, A. (2014). The use of particle/substrate material models in simulation of cold-gas dynamic-spray process. *Journal of Thermal Spray Technology, 23*(3), 530–540.
65. Raoelison, R. N., Verdy, C., & Liao, H. (2017). Cold gas dynamic spray additive manufacturing today: Deposit possibilities, technological solutions and viable applications. *Materials and Design, 133,* 266–287.
66. Rolland, G. (2011). Laser-induced damage in cold-sprayed composite coatings. *Surface and Coatings Technology, 205,* 4915–4927.
67. Sabard, A., de Villiers Lovelock, H. L., & Hussain, T. (2018). Microstructural evolution in solution heat treatment of gas-atomized Al alloy (7075) powder for cold spray. *Journal of Thermal Spray Technology, 27*(1–2), 145–158.
68. Schmidt, T., Gärtner, F., Assadi, H., & Kreye, H. (2006). Development of a generalized parameter window for cold spray deposition. *Acta Materialia, 54*(3), 729–742.
69. Sunil, P., & Gobinda, S. C. (2017). Development of sustainable cold spray coatings and 3D additive manufacturing components for repair/manufacturing applications: A critical review. *Coatings, 7*(8), 1–27.
70. Trifa, F. I., Montavon, G., & Coddet, C. (2007). Model-based expert system for design and simulation of APS coatings. *Journal of Thermal Spray Technology, 16*(1), 128–139.
71. Trinchi, A., et al. (2011). Copper surface coatings formed by the cold spray process: Simulations based on empirical and phenomenological data. *Journal of Thermal Spray Technology, 20*(5), 986–991.
72. Vardelle, M., Vardelle, A., Leger, A. C., Fauchais, P., & Gobin, D. (1995). Influence of particle parameters at impact on splat formation and solidification in plasma spraying processes. *Journal of Thermal Spray Technology, 4*(1), 50–58.

73. Vardelle, A., Moreau, C., Akedo, J., Ashrafizadeh, H., Berndt, C. C., Berghaus, J. O., et al. (2016). The 2016 thermal spray roadmap. *Journal of Thermal Spray Technology, 25*(8):1376–1440.
74. Vo, P., & Martin, M. (2017a). Layer-by-layer buildup strategy for cold spray additive manufacturing. In J. Jerzembeck, et al. (Eds.), *Proceedings of the 2017 International Thermal Spray Conference* (pp. 714–718). Düsseldorf: DVS Media GmbH.
75. Vo, P., & Martin, M. (2017b). Layer-by-layer buildup strategy for cold spray additive manufacturing. Pdf presentation available via https://www.coldsprayteam.com/Vo_Martin_CSAT2017%20%20Vo%20and%20Martin%20vfinal.pdf. Accessed June 2, 2019.
76. Weiller, S., Debray, A., Gaslin, F., Sennour, M., Delloro, F., Jeandin, M., & Garion, C. (2019). Cold spray as an innovative process to develop leak tight aluminum coatings for ultrahigh vacuum applications in a large particle high-energy collider. In F. Azarmi, et al. (Eds.), *Proceedings of the International Thermal Spray Conference (ITSC 2019)* (pp. 831–837), May 26–29, 2019, Yokohama, Japan.
77. Wu, H., Xie, X., Deng, S., & Liao, H. (2019). A new approach to simulate coating thickness in cold spray. In F. Azarmi, et al. (Eds.), *Proceedings of the International Thermal Spray Conference* (pp. 165–171), May 26–29, 2019, Yokohama, Japan.
78. Xie, J., Nélias, D., Walter-Le Berre, H., Ogawa, K., & Ichikawa, Y. (2015). Simulation of the cold spray particle deposition process. *Journal of Tribology, 137*(4).
79. Zerilli, F. J., & Armstrong, R. W. (1987). Dislocation-mechanics-based constitutive relations for material dynamics calculations. *Journal of Applied Physics, 61*(5), 1816–1825.

Laser Assisted Cold Spray Deposition

Venkata Satish Bhattiprolu and Luke N. Brewer

Abstract This chapter will review the current state of the utilization of lasers to enhance the cold spray process both during and after deposition. While high and low-pressure cold spray has been successfully applied to a variety of metallic alloys, there is a need to increase deposition efficiency, permit the use of heavier and less expensive spray gases, and to tailor the microstructure of the sprayed material. Over the past decade, laser-assisted cold spray (LACS) has been introduced and developed to address these needs. In this process, the laser acts as a localized heat source softening the substrate and the deposited powder particles. This thermal softening in turn facilitates the deposition of hard materials and also enhances the deposition behaviour at lower powder particle velocities, allowing the use of less expensive processing gases like nitrogen. LACS has been applied in both co-axial and off-axis forms. The off-axis geometry has shown improvements in the deposition of titanium alloys, stellite, tungsten, and even oxide dispersion strengthened steel. While LACS can improve the deposition efficiency, it will also impact the microstructure of the sprayed material, either positively or negatively, depending upon the heat input used. Recent work has shown an increase in grain size and coarsening of precipitates during the LACS process. Understanding and controlling heat input during LACS are key to correctly producing the desired microstructure and mechanical properties of the deposited material. This chapter will introduce the LACS approach, present its applications to different alloy systems, discuss its advantages and disadvantages and offer some thoughts about its development in the near future.

Keywords Laser assisted cold spray · Supersonic laser deposition · Microstructure characterization · Mechanical properties

V. S. Bhattiprolu · L. N. Brewer (✉)
Department of Metallurgical and Materials Engineering, The University of Alabama, Box 870202, Tuscaloosa, AL 35487-0200, USA
e-mail: lnbrewer1@eng.ua.edu

S. Pathak and G. C. Saha (eds.), *Cold Spray in the Realm of Additive Manufacturing*, Materials Forming, Machining and Tribology,
https://doi.org/10.1007/978-3-030-42756-6_6

1 Motivation and Introduction

Cold spray (CS) is a solid-state deposition process involving severe plastic deformation of micron-sized powder particles (~40 μm) onto a metal substrate. The bonding of individual particles to the substrate is accomplished through metallurgical bonding and mechanical interlocking. The CS process has been successfully employed to produce dense coatings for a wide variety of materials as shown in Table 1. The coatings typically exhibit high hardness, compressive residual stresses, and high wear resistance [1]. In some cases, CS coatings demonstrate tensile strengths comparable to bulk materials (Table 1).

Most of the early cold spray research involved relatively "soft" materials like aluminum and copper. This emphasis was in large part due to their lower critical velocities. Assadi and Schmidt have developed the concept of critical velocity in which particles with a velocity higher than the critical velocity adhere to the substrate upon impact [23]. As shown in Eq. 1 [23], the critical velocity is directly proportional to the ultimate tensile strength of the material being sprayed. Soft powder materials such as copper, commercially pure aluminum, and nickel have a relatively low critical velocity and thus can be readily sprayed. "Hard" powder materials, such as alloyed aluminum, ferritic steels, titanium, and nickel-based alloys require higher particle velocities for successful deposition (i.e., owing to higher critical velocity). This increase in particle velocity can be achieved by higher gas pressure, higher gas temperature, or by the use of a lighter spray gas. The nozzle exit velocity for powder particles is directly proportional to the gas velocity which is in turn inversely proportional to the molecular weight of the gas being sprayed, as shown in Eq. 2.

Table 1 Representative characteristics for metallic deposits produced using high-pressure cold spray

CS Deposition property (as-sprayed)	Aluminum (Al) and aluminum alloys	Copper (Cu)	Austenitic Stainless Steel	Titanium (Ti) and titanium alloys
Porosity (area%)	0.25 (Helium) [2] 0.32 (Nitrogen) [2]	0.04 (Nitrogen) [3]	~1% (Helium) [4] ~2% (Nitrogen) [5]	~0.3% (Helium) [6] ~11% (Nitrogen) [7]
Microhardness	25–50% higher hardness than cold worked Al [1, 8]	20% higher hardness than annealed and cold rolled Cu (150 $HV_{0.3}$) [9]	157% higher than bulk SS substrate (410 $HV_{0.3}$) [10]	16% higher hardness than bulk Ti [11]
Tensile strength (MPa)	410–440 [12–14]	~140 to 300 [3, 15, 16]	~67 to 525 [4, 17, 18]	~445 [19]
Ductility (% Elongation) in as-sprayed condition	1–3.2 [12, 14, 20]	0.2–3 [3, 15, 16]	0.1–0.4 [3, 17, 18]	~0.45 to 3 [21, 22]

Table 2 Comparison of powder particle velocities and critical velocities for "soft" and "hard" powder types

Material type	Powder particle velocities (m/s)		Critical velocity of powder particles (m/s)
	Helium processing gas	Nitrogen processing gas	
CP Aluminum	1613	761	476
Ti6Al4V	1446	703	1103

These estimates were made using Eqs. 1 and 2. A powder particle diameter of 25 μm was assumed along with a particle temperature ~150 °C

Helium is the lightest gas commonly used for cold spray, and it readily accelerates powder particles to higher velocities resulting in a greater extent of plastic deformation of particles upon impact and in the dense deposited material. However, helium is expensive and as such cheaper alternatives (e.g. nitrogen) are increasingly sought for producing CS coatings. Table 2 provides typical powder particle velocities for commercially pure (CP) aluminum and Ti6Al4V using nitrogen and helium gases. It is apparent that the velocities of powder particles are higher than the critical velocity for CP aluminum even with nitrogen processing gas.

$$V_{critical} = \sqrt{\left(\frac{A\sigma}{\rho}\right) + BC_p(T_m - T)}; \tag{1}$$

where A, σ, ρ, B, C_p, T_m, T, denote a fitting constant, temperature-dependent flow stress, density, a second fitting constant, the heat capacity of the material, melting temperature, and powder particle temperature, respectively. Here the temperature-dependent flow stress of the material is assumed to be directly proportional to the ultimate tensile stress of the material [23].

$$V_g = \sqrt{\left(2\left(\frac{\gamma}{\gamma-1}\right)\left(\frac{R}{M_{gas}}\right)T_g\left(1-\left(\frac{p}{p_i}\right)^{\frac{\gamma-1}{\gamma}}\right)+V_{gi}^2\right)}: \tag{2}$$

where γ, p, R, M_{gas}, T_g, p_i, V_{gi}, denote specific heat ratio, gas pressure, the ideal gas law constant, the molecular weight of the gas, gas temperature, initial gas pressure, and initial gas velocity, respectively [24].

The main challenge with the mechanical properties of cold sprayed material is their low ductility in the as-sprayed condition (Table 1). Primarily due to the severe strain hardening of the powder particles during the CS process, the resultant coating is relatively brittle. It is however reported that post-spray annealing can recover the detrimental effects of strain hardening through static recovery and recrystallization [20]. Hall et al. sprayed commercially pure aluminum material using helium gas [20]. The resultant material had a UTS of ~140 to 195 MPa with an elongation to failure of only ~0.4 to 1% (Fig. 2) [20]. This material exhibited a very fine-grained microstructure, as is typical of a severely cold worked material (Fig. 1).

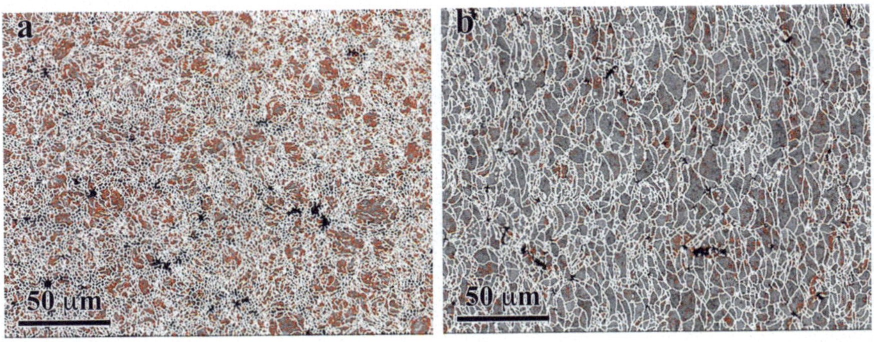

Fig. 1 Effect of annealing at 300 °C for 22 h on the microstructure of cold sprayed commercially pure aluminum. Microstructure evolved from **a** fine equiaxed grains before annealing to **b** coarse equiaxed grains after the annealing heat treatment (taken from Hall et al. [20])

After annealing at 300 °C for 22 h, the same CS coatings (Fig. 1b) possessed larger grains due to full recrystallization and subsequent grain growth. Unsurprisingly, the elongation to failure increased to 6–10% for the range of powders investigated, as shown in Fig. 2. A similar increase in the ductility and corresponding decrease in the tensile strength was observed by Coddett et al. for helium sprayed stainless steel 304 CS coatings after annealing [4]. The coatings exhibited tensile strengths of ~630 and ~425 MPa before and after annealing to 1050 °C for 4 h; the coatings exhibited ductilities of ~0 and 10% before and after annealing, respectively. Recently, however, Rokni, M. R. et al. reported an increase in both ductility and tensile strength of CS aluminum alloy Al 7075 with heat treatment [14]. These increases were attributed to the precipitation of intermetallic phases during heat treatment which subsequently hinders dislocation movement during tensile testing [14]. Accordingly, the relative increase in strength and ductility amounted to 19 and 138%, respectively. In the same work, it was reported that annealing CS Al 7075 to 412 °C for 3 h resulted in ductility of 14%, which corresponds to an increase of 338% from the as-sprayed condition [14].

In summary, the CS process has been consistently employed to deposit a range of metallic materials, but with challenges in terms of the gas required and the ductility of the as-sprayed material. The stronger the metallic material, the higher its critical velocity, necessitating higher spray gas pressures, higher spray gas temperatures, and the use of lighter spray gases. The use of helium gas has been quite successful, but it adds a high cost to the CS process. Post-spray annealing can expand the CS material selection and improve the material ductility, but these post-processing steps can quickly become financially unattractive, particularly for large parts. There is a need to balance the cost of the CS process with its mechanical properties for both soft and hard alloys. Laser-Assisted Cold Spray (LACS) is an exciting approach that builds on the cold spray process by adding in situ laser heating.

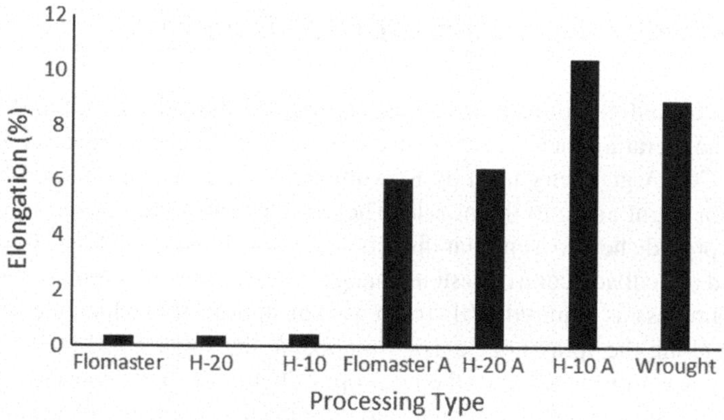

Fig. 2 Effect of annealing at 300 °C for 22 h on the ductility of CS depositions produced using Al powder procured from various suppliers (modified from Hall et al. [20]) and compared to wrought material. The letter A stands for annealed samples. The two labels Flomaster, and H, represent gas atomized aluminum powder procured from Valimet, and F. J. Brodman, respectively. The numbers 10 and 20 listed as a suffix to H indicate the average powder particle size of Valimet powder: 10 denoting ~12 μm and 20 denoting ~26 μm

2 Laser-Assisted Cold Spray (LACS) Deposition

The LACS process involves the employment of a laser along with the de Laval CS nozzle for enhancing the deformation of the substrate/deposited powder particles by softening them in situ. This process is also referred to as supersonic laser deposition in some of the related literature [25]. In addition, the in situ heating can also be used to modify the microstructure during LACS. Other work by Christoulis, D. and Sarafoglou, C. has used lasers to clean the substrate just prior to particle deposition, and while intriguing, it will not be reviewed in this chapter, but it is described in the book *Modern Cold Spray* [26]. A fair amount of work on the literature has also considered the use of post-spray laser heating. This related approach is not the focus of the current chapter, but it will be discussed in later sections.

To date, three different laser configurations have been investigated with respect to the de Laval CS nozzle: (1) Laser co-axial with the spray deposition, (2) Laser off-axis with the laser spot coincident with the centre of the spray deposition, and (3) Laser off-axis with the spray deposition, with the nozzle trailing the laser beam. The third configuration was utilized by the investigators to avoid melting of Al-12%Si powder particles if exposed directly to the laser beam [27]. Among the three configurations, the off-axis geometry with the laser and the spray nozzle coincident on the substrate has received the most attention in the literature. The experimental setup for these different configurations and the important properties of the produced coatings will be discussed in detail below.

2.1 Co-axial Laser Assistance (COLA) During CS

In the co-axial configuration, the laser beam is directed along the axis of the CS nozzle [28, 29]. The literature for COLA actually shows two different approaches. Figure 3a shows the COLA geometry used by Koivuluoto et al., in which multiple lasers are directed at a slight angle to the nozzle. These laser beams are *outside* of the spray nozzle but provide heat to a circular disc coincident with the spray [28]. Temperature sensors and a feedback control system change the laser power automatically during the LACS process as required [30]. Toom used an approach in which the laser beam is directed along the spray nozzle axis *from inside* the spray nozzle (Fig. 3b) [29]. The spray gas is introduced after the laser optics but prior to the converging section of the nozzle. The metal powder is injected after the throat into the diverging portion of the nozzle.

The coating characteristics are significantly affected by the introduction of a laser to the CS process. The coating density is shown to improve slightly with the assistance of a laser [28, 30]. In this regard, Koivuluoto et al. showed that the porosity decreased from 0.5 to 0.1% for a peak aged cold sprayed Al alloy coating with the employment of the laser [30]. Another related coating property, adhesion strength, was also reported to improve with LACS. This improvement via COLA was demonstrated for the

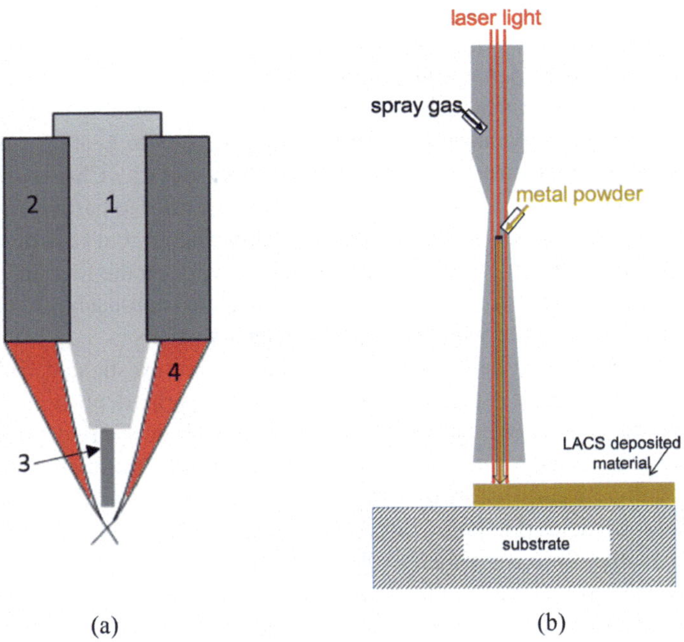

Fig. 3 Schematic of two geometries for the COLA Process: **a** (modified from Koivuluoto et al. [28]) where the numbers denote the following: 1. Spray Nozzle 2. Device for monitoring substrate temperature 3. Powder with gas and 4. Focused laser beams and **b** modified from Toom [29]

cold spray of Al, bronze alloys, Ni alloys, and Fe-Mn alloys by separate groups of investigators [29, 30]. Specifically, the adhesion strength increased from 27 to 48 MPa for cold sprayed aluminum with the employment of laser at a laser power of 1 kW [28]. Similarly, the adhesion strength increased from 27 MPa (regular CS) to 35 MPa (LACS) for Fe-Mn alloys with the internal COLA setup where the laser power was set to 200 W [29]. The laser is reported to promote substrate softening, which in turn results in better particle-particle bonding in the LACS coatings [28]. For these reasons, the cohesive strength of Cu bronze and Al alloy coatings was shown to be improved from 4 to 29 MPa and 119 to 177 MPa, respectively [30]. Koivulouto has proposed that the higher density and substrate-coating bond strength produced by the LACS process could provide enhanced barrier coatings for corrosion prevention on chemically active materials [28].

The in situ laser heating during co-axial LACS also affects the hardness of the deposited material. Due to softening of the substrate and the deposited powder particles during the LACS process, the hardness of LACS coatings is generally lower than for cold sprayed coatings [28, 30]. Again, hardness was shown to decrease from ~350–390 $HV_{0.1}$ to ~290–330 $HV_{0.1}$ for Cu bronze alloys with the application of a laser during CS. In the same study, the hardness decrease was 12–23% for COLA compared with CS of aluminum. This reduction in hardness is noteworthy as lower hardness may suggest better ductility of LACS coatings as compared to CS coatings.

2.2 *Off-Axis Laser Assistance During CS*

For most of the reported LACS literature, the laser is installed outside of and at an acute angle to the axis of the CS nozzle (Fig. 4). The laser's angle is adjusted in such a way that the focused laser spot is on the substrate and exactly below the CS nozzle, at a pre-determined standoff distance. An optical pyrometer is also often connected in parallel with the laser head. The temperature readings from the pyrometer can be used for substrate temperature control using a feedback control system, and thus the LACS experiment can be controlled according to laser power or measured surface temperature. In this configuration, the laser moves along with the nozzle during the spraying process.

One can estimate the amount of time that a particle in-flight interacts with the laser beam under the following assumptions based on typical CS conditions:

(1) the standoff distance between nozzle exit and substrate is 25 mm
(2) the laser spot diameter is 8 mm
(3) the average powder particle velocity is 800 m/s
(4) the CS nozzle and laser move at a velocity of 25 mm/s.

In this geometry, the amount of time a particle in flight is exposed to a moving laser beam is minuscule (~30 μs). Upon impact, the newly deposited particle will be irradiated for about 150 ms before it moves out of the laser spot diameter. Based on these estimates, the particle spends about 5000 times longer being irradiated after

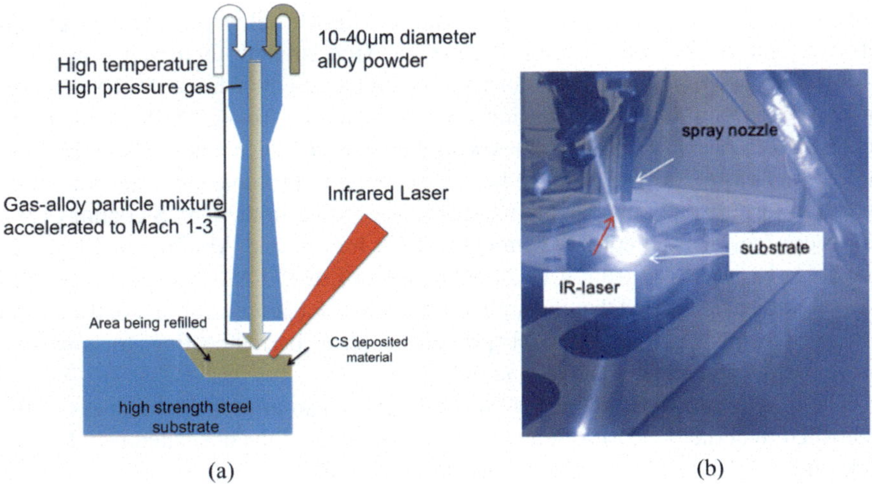

Fig. 4 The off-axis LACS process. **a** Schematic and **b** still-frame of video from LACS in operation at The University of Alabama

impact than during its flight to the substrate. This calculation strongly suggests that most of the thermal impact is from irradiation of deposited particles on the substrate, not from laser heating of the particles in flight. It should be noted that subsequent laser passes over deposited particles will continue to heat them.

The LACS process has been shown to improve the deposition efficiency of different materials. One of the earliest studies on the efficacy of the LACS process for improving the deposition behaviour of materials was reported by Bray et al. [31]. Bray et al. showed that the build rate of titanium powder increased two-fold from lower than 25 g/min to close to 45 g/min with the application of in situ laser heating [31]. Li, B. et al. reported a systematic increase in deposition efficiency with increasing surface temperature for LACS of Stellite-6 material (Fig. 5) [32]. The authors attributed this substantial increase in deposition efficiency to a decrease in the critical velocity of powder particles during impact due to an increase in powder particle temperatures (due to laser exposure). Kumala et al. made the interesting observation that deposition efficiency during LACS may be a coupled effect from both the surface temperature and the number of deposited layers (Fig. 6) [33]. For LACS of $Cu+Al_2O_3$ the impact of laser heating for only two deposited layers was minimal, while for ten deposited layers, laser heating clearly increased deposition efficiency [33]. This dependence upon layer thickness may be due to the cumulative heat input provided to the deposited material as the number of layers increases.

The LACS process has also been employed for the deposition of hard materials with good success. Several materials, including tungsten [34], Ni60 [35], Ni60-Diamond composite [36], tungsten carbide-stainless steel (W-SS) composite [37], and oxide dispersion strengthened steel (Fe-8Ni-1Zr) [38], have been successfully

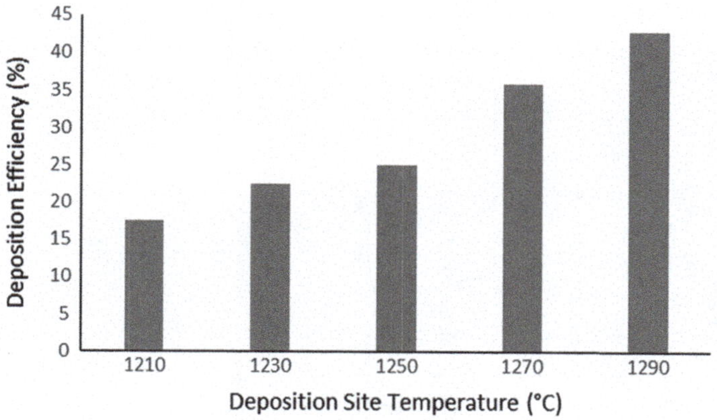

Fig. 5 DE of Stellite-6 depositions on medium carbon steel substrates at a maximum laser power of ~1.5 to 1.8 kW as a function of deposition temperature (modified from Li et al. [32])

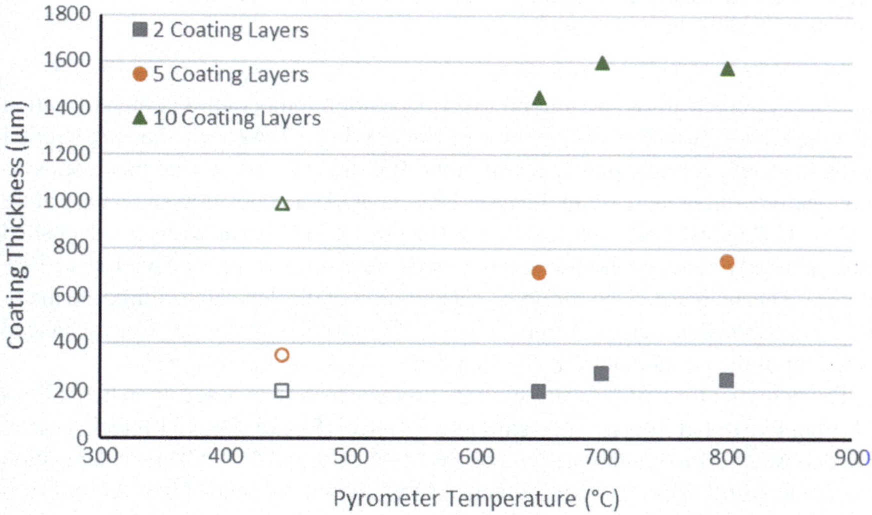

Fig. 6 Change in coating thickness as a function of both deposition temperature and a number of layers in $Cu+Al_2O_3$ cold sprayed depositions produced on low carbon steel using LACS with laser power between 1.8 and 2.4 kW. Here the open symbols represent the gas temperature for the CS depositions where the laser was not employed (i.e., regular CS) (modified from Kulmala et al. [33])

produced using the LACS process over the past five years. These results are significant, as producing these coatings using regular CS is very challenging. Li et al. [37] have demonstrated the effectiveness of LACS for challenging materials quite clearly with their recent work on WC-SS316L composite depositions. Without laser heating, CS was only able to produce coatings which retained a WC composite loading of 24% (Fig. 7). Note that the data point at a deposition temperature of 450 °C is for a regular CS process (i.e., without laser) where the temperature denotes the stagnation

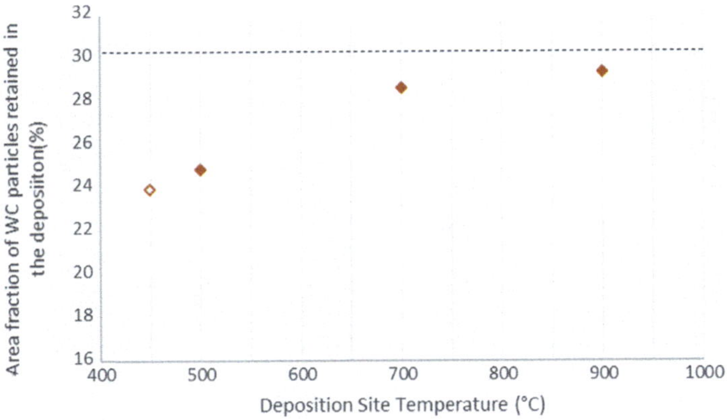

Fig. 7 Concentration of WC particles in the composite WC-SS LACS coatings produced using nitrogen gas on a SS substrate at different deposition site temperatures (modified from Li et al. [37]). Here, the deposition site temperature 450 °C represents gas temperature and this deposition was produced without a laser (i.e., regular CS). The dashed line represents the volume % of WC particles in WC-SS composite powder

gas temperature with the assumption that deposition temperature is the stagnation gas temperature (which is likely an overestimate) (Fig. 7). As seen, the area fraction of retained WC ceramic particles within the WC-SS316L composite increases systematically with an increase in the laser heat input. At a deposition site temperature of 950 °C (highest heat input), the area fraction of WC particles in the deposited material nearly matched the fraction of powder feedstock composite loading at 30% (Fig. 7). This increase in the ceramic concentration of the deposited composite material was attributed to increased embedment of WC particles, stemming from enhanced softening of the metal matrix at elevated deposition temperatures [37].

The potential for LACS to successfully produce materials based upon hard alloys is further illustrated from recent results on 4340 steel (Fig. 8). The 4340 steel material was deposited using helium processing gas (4.14 MPa and 550 °C) and a deposition site temperature of 950 °C by employing a 4 kW continuous diode laser. As observed in Fig. 8, the resultant coating was very dense with negligible (~0.1 area%) porosity and a deposition efficiency of 70%. This density is close to what is generally reported for soft materials like aluminum alloys using helium [1]. Figure 8 also shows evidence of CS tracks from multiple passes in the longitudinal direction. These tracks may be interfaces between layers of CS deposition and are typically not observed in CS coatings. Some oxidation of the steel surface between CS passes may be occurring at this elevated temperature.

LACS is also showing promise for the deposition of wear-resistant materials. Specifically, the wear resistance of LACS-produced material was compared to that from laser cladding, a widely used process for producing well-adhered [35] coatings for wear protection [39]. In general, LACS depositions were characterized by superior wear properties (i.e., lower coefficient of friction, lower volume loss, and lower depth of wear track) when compared to laser clad materials. This improvement was mainly

Fig. 8 Low magnification back scatter electron image of 4340 Steel deposited using the laser assisted cold spray process. The black dashed line represents approximate location of interface between coating (top) and the substrate (bottom)

attributed to lower heat input and to a finer microstructure in the LACS process when compared to laser cladding. In addition, LACS did not alter the phases present in the material before and after deposition.

The LACS process may also allow for the deposition of both soft and hard materials (especially titanium alloys) using nitrogen as the spray gas. As nitrogen is significantly less expensive than helium as a spray gas, the ability to still efficiently deposit high-quality material using LACS with nitrogen as the spray gas is a significant development. Evidence for producing dense LACS coatings using nitrogen gas is readily apparent from Table 3 for different material systems.

The LACS process produces characteristically different microstructures when compared to regular CS, especially when considering the prior-particle morphology. This change is clearly shown in Fig. 9, for a titanium coating produced by LACS at a deposition site temperature of 500 °C [31]. Starting with spherical titanium powder particles, the LACS process produced a coating microstructure with relatively equiaxed prior-particle morphologies, particularly at the bottom of the coating. In regular CS, the prior-particles are often much flatter (high aspect ratio) due to repeated peening by subsequent particles during the deposition. Similar results have been recently observed for LACS of 4340 steel, particularly at higher deposition temperatures. The reasons for this change in prior-particle morphology between CS and LACS is still an open question.

Table 3 Table demonstrating the laser parameters and deposition properties for various LACS depositions using nitrogen as the spray gas

	CP aluminum (Al)/Al alloys	Tungsten	Stellite-6	CP Ti/Ti6Al4V
Laser power	Al-12%Si: 2.5 kW [27]	4 kW [34]	1.5 kW [32, 40]	Ti6Al4V: 300 W to 1 kW [41] CP Ti: 650 W-1 kW [31]
Deposition surface temperature	CP Al: 450 °C [42] Al-12%Si: 200 °C [27]	Outside the range of pyrometer [34]	1210–1290 °C [43]	≤900 °C [41] CP Ti: 450–900 °C [31]
Substrate	Al-12%Si: Stainless Steel 304 [27] CP Al: Stainless Steel 304 [42]	Molybdenum [34]	Medium carbon steel [43]	Ti6Al4V: Ti6Al4V [41] CP Ti: Low carbon steel [25]
Gas	Al-12%Si: N_2 [27] CP Al: N_2 [42]	N_2 [34]	N_2 [43]	Ti6Al4V: N_2 [41] CP Ti: N_2 [31]
Porosity (area%)	Al-12%Si: 0.16 [27] CP Al: 0.01–0.04 [42]	5 [34]	NR	Ti6Al4V: 0.1 [41]
Hardness	CP Al: 46–51 $HV_{0.05}$ [42]	NR	693 $HV_{0.2}$ [43]	Ti6Al4V: Min of 331 $HV_{0.2}$ and Max of 461 $HV_{0.2}$ [41] CP Ti: 272 $HV_{0.3}$ [25]
Adhesion strength	NR	NR	NR	CP Ti: 77 MPa [25]
Tensile strength	NR	725 MPa [34]	718.3 MPa [43]	NR

Some values were not reported in the literature and thus designated by "NR"

The high temperatures during LACS can also result in an in situ, static recrystallization of the deposition microstructure, as seen from Fig. 10. This figure depicts feedstock Ti6Al4V powder characterized by a martensitic alpha grain microstructure. This microstructure evolves during LACS to a microstructure comprised of equiaxed alpha grains due to recrystallization. This substantial change in the microstructure was ascribed to high laser power and low raster speed of the CS nozzle [41]. Heat input during LACS can also result in grain growth during the deposition process (Fig. 11). In this figure, the microstructural characteristics of cold sprayed Fe-8Ni-1Zr oxide dispersion strengthened (ODS) steel depositions with and without in situ laser heating were captured using electron backscatter diffraction (EBSD) and 3D

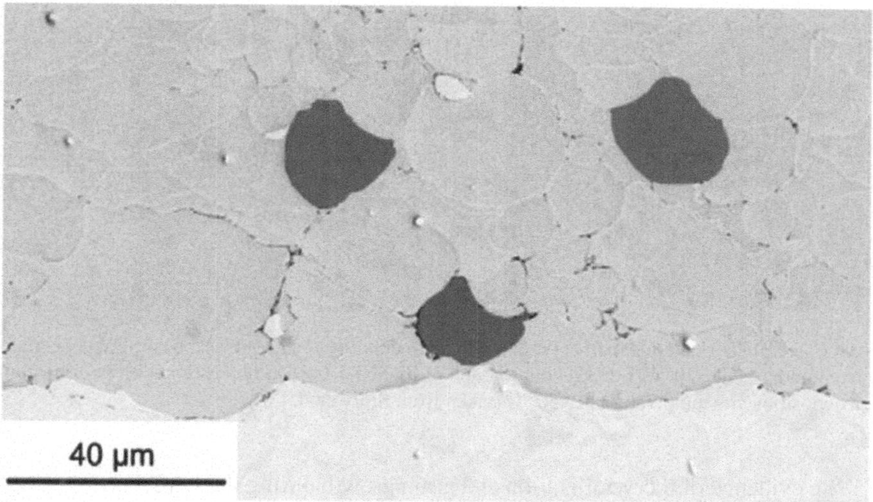

Fig. 9 Deformed particles within LACS deposition of titanium at deposition surface temperature of 500 °C (taken from Bray et al. [31]). Here, the three particles with dark contrast were shaded post-deposition to highlight the unique morphology of LACS particles

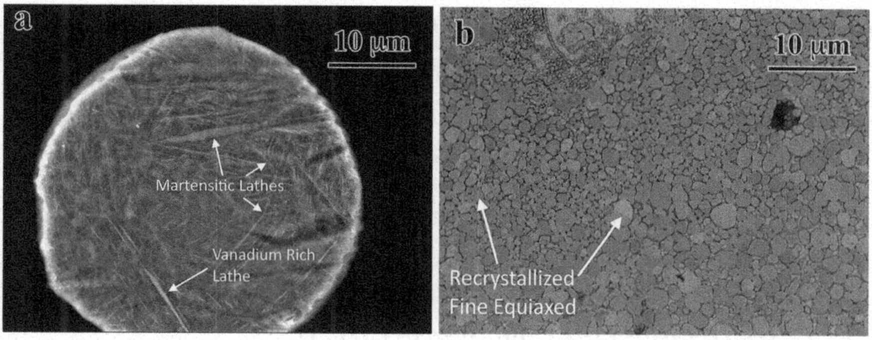

Fig. 10 **a** As-received Ti-6Al-4V powder martensitic microstructure and **b** recrystallized microstructure within Ti-6Al-4V LACS deposition at laser power of 600 W (taken from Birt et al. [41])

atom probe tomography [38]. An apparent increase in the grain size (~300%) was observed as the deposition site temperature increased from 320 °C in CS-only to 950 °C in LACS [38]. Further, the non-uniform grain growth observed at the highest LACS deposition temperature (950 °C) was attributed to abnormal grain growth due to the presence of nanoscale oxide dispersants [38]. Furthermore, a 36% reduction in microhardness was reported with the LACS process and was partially ascribed to the observed grain growth [38]. Interestingly, the nanoscale oxide particles survived the LACS process, even at 950 °C, and coarsened only slightly from an inter-particle spacing of 17 nm for CS-only to 23 nm at 950 °C [38].

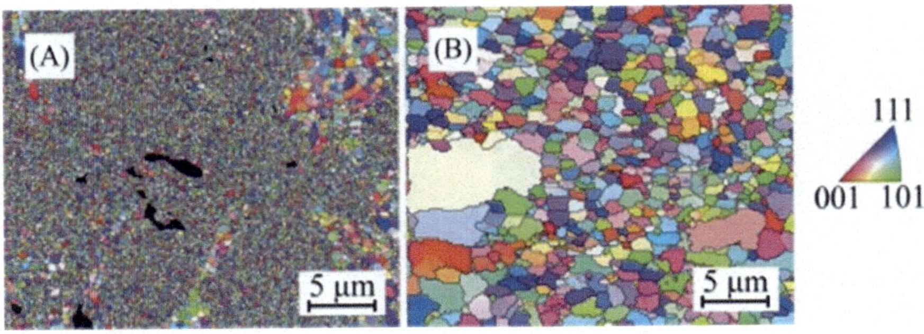

Fig. 11 **a** EBSD orientation map of Fe-8Ni-1Zr ODS deposition without laser heating at deposition site temperature of 320 °C, **b** EBSD micrograph of Fe-8Ni-1Zr ODS deposition with laser heating at a deposition site temperature of 950 °C (taken from Story et al. [38])

The evidence of recrystallization and grain growth during the LACS process suggests that an increase in ductility for as-sprayed material may be possible. Several studies of materials produced with LACS have shown recrystallized microstructures with a correlated reduction in hardness. Actual measurements of ductility for LACS deposited material are scant in the current scientific literature, however, recent results on LACS of AA7075 powders demonstrate increases in ductility [44]. As-atomized AA7075 powder was sprayed using helium gas both with and without a 1.6 kW laser (940 nm wavelength, spot size 8 mm) [44]. The cold sprayed material showed elongation to failure values of $0.7 \pm 0.07\%$ as-sprayed, while the LACS-deposited material exhibited increased elongation to failure to $2.1 \pm 0.3\%$ in the as-sprayed condition [44]. The maximum stress prior to fracture showed the opposite behaviour: CS material had maximum stress of approximately 360 MPa, while the LACS-deposited material exhibited maximum stress of approximately 320 MPa [44].

2.3 *Laser Heating Post Cold Spray (LHCS)*

Laser heat treatment can also be employed after the cold spray process. These post-deposition heat treatments have been applied as both surface re-melting and as solid-state annealing of the cold sprayed coating. [45–47]. Work by Marrocco, T. et al. involved re-melting of CS deposited CP titanium using a 2 kW CO_2 laser [45]. Similarly, work reported by Poza, P. et al. involved re-melting of Inconel 625 CS coatings using a 1.3 kW diode laser [47]. The high laser powers resulted in melting followed by pore closure due to diffusion of material to high energy surfaces. This reduction in porosity, in turn, was reported to improve the corrosion resistance of the CS coatings [45, 46] and in one case the laser re-melted material reached the properties of bulk material [45]. In addition to pore reduction, the prior-particle interfaces generally seen in CS coatings were also reported to disappear with the laser re-melting [45]. While laser post-heating has a beneficial impact on the surface of the CS coating, it could also affect the substrate in some cases. This substrate impact can be expected at a high linear heat input (either from high laser power or slow nozzle raster speed or a combination of both) [47]. Additionally, due to the large thermal gradient between the re-melted region near the surface and the region near the substrate-coating interface, tensile residual stresses may be present in the LHCS coatings. Their presence could be detrimental to the properties of the LHCS coating.

The amount of heat input during the laser re-melting process is also reported to affect the quality of the coating significantly. Whereas a low heat input may not result in a visible change, high heat input can cause cracks in the coating [48, 49]. Additionally, it is shown to stratify the CS deposition by forming three zones with three different microstructures: (1) re-melted zone (2) heat affected zone and (3) base material [49]. Among these regions, the re-melted zone is reported to exhibit a rapidly solidified microstructure with high hardness [49].

LHCS has also been recently investigated for post-cold spray solid-state annealing [50]. Four material systems were investigated; Aluminum alloy 6061 (AA 6061), Copper (Cu), SS 304, and Grade 2 Titanium (Ti). Among these materials, it was reported that only Al 6061 and Cu exhibited a reduction in hardness. This may be due to the low laser power utilized (i.e., 95 W). Interestingly, Aldwell, B. et al. also reported cracking of the SS deposits with LHCS after 10 laser passes and hypothesized that it might be due to thermal cycling [50]. This result is similar to what has been observed by a separate group of researchers for grade 2 titanium, as mentioned previously [48].

In summary, LHCS of materials has significant potential for reducing porosity, reducing the detrimental effects of strain hardening, and for improving the corrosion resistance of the CS deposition. With further work on optimization of laser parameters, it is expected that the detrimental effects of laser heating on CS materials may be mitigated (e.g., tensile residual stresses and cracks). Although, not currently reported, LHCS of certain materials (e.g., SS 304) may also result in grain boundary sensitization, especially at the high deposition site temperatures of the laser. Thus, along with mitigating residual stresses, further work on laser parameter optimization may be needed for prevention of sensitization (on susceptible materials).

3 Advantages, Disadvantages, and Directions for the Future Development of the LACS Process

The key advantage of the LACS process is its ability to provide a controlled amount of heating during the cold spray process. This laser heating can reduce the critical velocity of metallic powder particles during cold spray, ostensibly by softening the substrate and the deposited material. This reduction in critical velocity potentially allows the use of heavier, and usually, less expensive gases for LACS deposition. This benefit has already been well demonstrated for the LACS deposition of titanium [31].

In addition, "harder" metallic materials can be deposited by LACS than is possible from CS alone. This chapter has shown that a variety of difficult materials are now successfully sprayed with LACS, including Stellite, WC-stainless steel composites, ODS steel, and 4340 steel. The laser heating can also cause in situ recrystallization potentially increasing the ductility of the as-sprayed material. Finally, the amount of control from laser heating is potentially very good. The amount of laser power can be varied during the spray to achieve the desired microstructure at different locations in the deposition. It is also possible that lasers could be used for localized in situ heating in the field repair applications on structures far too large to be placed into a furnace.

The potential disadvantages of the LACS process are primarily those associated with increased thermal input. The key advantages of CS with respect to thermal spray techniques are primarily related to its low thermal input. The thermal impact of LACS is between CS and thermal spray. With excessive heat input, LACS may damage substrates under thin depositions; the thermal loading on the substrate will have to be carefully monitored and controlled, particularly in repair applications. LACS and LHCS may lead to unwanted oxide formation. While LACS may lead to increased ductility, it can also produce more softening than desired through the dissolution of precipitates, the recovery of dislocation structures, and due to excessive grain growth. In addition, LACS will produce much larger thermal gradients than CS, potentially resulting in tensile residual stresses in the deposited materials. Management of the residual thermal stresses produced by LACS will be necessary for mechanical reliability of these materials. Lastly, LACS requires the purchase and use of laser systems which add to the cost and complexity of the CS system, although the laser systems are typically less than half of the cost of the CS system itself.

LACS has shown great potential but still has several avenues which need to be explored and developed. The measurement, control, and optimization of heat input are the most important of these topics. The substrate and the newly deposited material are clearly heated by the in situ lasers, but there are other meaningful sources of heat including the heat from the CS gun itself and from subsequent passes of the laser/CS spray gun. In recent work on LACS of 4340 steel, CS conditions of 550 °C and 4.14 MPa using helium gas, resulting in a surface temperature without the laser of approximately 400 °C. Development of appropriate simulation tools will significantly aid in understanding and predicting the evolution of temperature in the substrate and the deposit during the LACS process. The raster pattern during LACS could also have more importance than in CS alone due to the potential of generating large thermal gradients and resultant residual stresses.

The interplay between thermal management and residual stresses is illustrated by recent work on LACS of 304 stainless steel (SS304) (Fig. 12). Initial LACS deposition of SS304 powder onto a SS304 substrate resulted in cracking of the deposited due to the large thermal gradients generated by the laser heating (Fig. 12a). The thermal conductivity of SS304 is particularly low and its thermal expansion coefficient is quite high, resulting in large thermal gradients and tensile residual stresses. The thermal gradients can be minimized by preheating the substrate with the laser immediately prior to LACS process. Figure 12b demonstrates the LACS SS304 coating produced

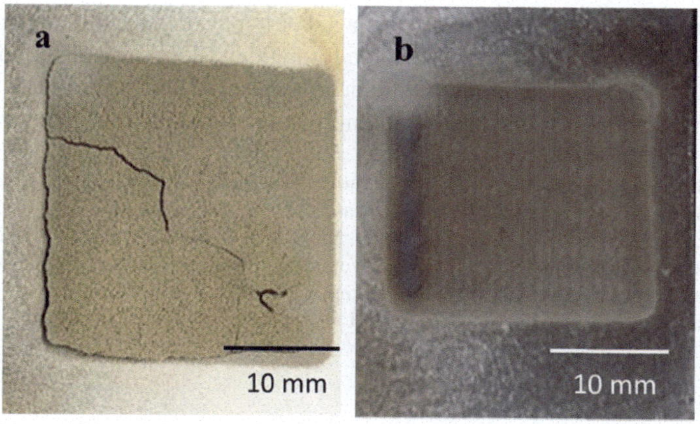

Fig. 12 SS 304 deposit produced by the LACS process **a** without laser heating the substrate prior to the process and **b** with heating the substrate prior to the process

with this laser pre-heating of the substrate. The reduced thermal gradient resulted in a successful deposition without any cracks. Because lasers can be precisely controlled, thermal cycles could be developed that can minimize residual stresses in the deposited material and substrate.

4 Conclusions

The laser-assisted cold spray (LACS) process combines the solid-state cold spray (CS) process with an in situ laser heating to lower the critical velocity required, to enhance material deposition efficiency, to evolve the microstructure during deposition, and to improve mechanical properties. LACS is performed using standard CS instruments modified with commercially available lasers, either co-axially with the spray nozzle or at an acute angle to the spray nozzle. LACS has been shown to enhance the deposition efficiency of materials that can be deposited by CS, including aluminum alloys, copper alloys, and titanium alloys. For titanium alloys, LACS allows the use of heavier, and less expensive, nitrogen gas while maintaining, and even improving the deposition efficiency as compared with CS alone. LACS has also been successfully used to deposit materials which cannot be successfully deposited by CS alone, including Stellite, high strength steels, tungsten, and even oxide dispersion strengthened alloys.

LACS has a number of advantages, disadvantages, and opportunities for future development. The key advantage of the LACS process is its ability to provide a controlled amount of heating during the cold spray process. This laser heating can reduce the critical velocity of metallic powder particles during cold spray allowing the use of heavier and less expensive gases for LACS deposition. In addition, "harder" metallic materials can be deposited by LACS than is possible from CS alone. The heat input during LACS must be carefully managed or it may damage the substrate under thin depositions and cause unwanted oxide formation. The control and prediction of heat input is a crucial topic in the future development of LACS. A related need for further development is in the control and mitigation of residual stresses.

Acknowledgements The authors are very grateful for funding to support our research into LACS from the following sources: Office of Naval Research: (Dr. J. Wolk, AMMP N000141812266, Dr. A. Rahman, SBA N00014-18-1-2519, and Mr. W. Nickerson, DURIP N00014-16-1-2576) and the U.S. Department of Energy NEUP (18-15372 Workscope FC-4.2; Contract DE-NE0008770). In addition, we would like to thank J. Colburn, W. Compton, W. Story, and D. Barton for their assistance with the measurements from the UA Cold Spray Laboratory.

References

1. Champagne, V., & Helfritch, D. (2016). The unique abilities of cold spray deposition. *International Materials Reviews, 61*(7), 437–455.
2. Sharma, M. M., Eden, T. J., & Golesich, B. T. (2015). Effect of surface preparation on the microstructure, adhesion, and tensile properties of cold-sprayed aluminum coatings on AA2024 substrates. *Journal of Thermal Spray Technology, 24*(3), 410–422.
3. Huang, R., et al. (2015). The effects of heat treatment on the mechanical properties of cold-sprayed coatings. *Surface & Coatings Technology, 261,* 278–288.
4. Coddet, P., et al. (2015). Mechanical properties of thick 304L stainless steel deposits processed by He cold spray. *Surface & Coatings Technology, 277,* 74–80.
5. Dikici, B., et al. (2016). The effect of post-heat treatment on microstructure of 316L cold-sprayed coatings and their corrosion performance. *Journal of Thermal Spray Technology, 25*(4), 704–714.
6. Bhattiprolu, V. S., et al. (2018). Influence of feedstock powder and cold spray processing parameters on microstructure and mechanical properties of Ti-6Al-4V cold spray depositions. *Surface & Coatings Technology, 335,* 1–12.
7. Birt, A., et al. (2015). Microstructural analysis of cold-sprayed Ti-6Al-4V at the micro- and nano-scale. *Journal of Thermal Spray Technology, 24*(7), 1277–1288.
8. Champagne, V. K., et al. (2010). The effect of cold spray impact velocity on deposit hardness. *Modelling and Simulation in Materials Science and Engineering, 18*(6), 065011.
9. Stoltenhoff, T., et al. (2006). Microstructures and key properties of cold-sprayed and thermally sprayed copper coatings. *Surface & Coatings Technology, 200*(16–17), 4947–4960.
10. Huang, J., et al. (2019). Influence of cold gas spray processing conditions on the properties of 316L stainless steel coatings. *Surface Engineering*, 1–8.
11. Hussain, T. (2013). Cold spraying of titanium: A review of bonding mechanisms, microstructure and properties. In *Key engineering materials*. Trans Tech Publications Ltd.
12. Li, W., et al. (2018). Solid-state additive manufacturing and repairing by cold spraying: A review. *Journal of Materials Science and Technology, 34*(3), 440–457.
13. Rokni, M., et al. (2017). Microstructure and mechanical properties of cold sprayed 6061 Al in As-sprayed and heat treated condition. *Surface & Coatings Technology, 309,* 641–650.
14. Rokni, M., et al. (2017). The effects of heat treatment on 7075 Al cold spray deposits. *Surface & Coatings Technology, 310,* 278–285.
15. Yu, B., et al. (2019). Microstructural and bulk properties evolution of cold-sprayed copper coatings after low temperature annealing. *Materialia, 7,* 100356.
16. Yin, S., et al. (2018). Microstructure and mechanical anisotropy of additively manufactured cold spray copper deposits. *Materials Science and Engineering A, 734,* 67–76.

17. Meng, X.-M., et al. (2011). Influence of annealing treatment on the microstructure and mechanical performance of cold sprayed 304 stainless steel coating. *Applied Surface Science, 258*(2), 700–704.
18. Yin, S., et al. (2019). Annealing strategies for enhancing mechanical properties of additively manufactured 316L stainless steel deposited by cold spray. *Surface & Coatings Technology, 370,* 353–361.
19. Vo, P., et al. (2013). Mechanical and microstructural characterization of cold-sprayed Ti-6Al-4V after heat treatment. *Journal of Thermal Spray Technology, 22*(6), 954–964.
20. Hall, A. C., et al. (2006). The effect of a simple annealing heat treatment on the mechanical properties of cold-sprayed aluminum. *Journal of Thermal Spray Technology, 15*(2), 233–238.
21. Chen, C., et al. (2019). Effect of hot isostatic pressing (HIP) on microstructure and mechanical properties of Ti6Al4V alloy fabricated by cold spray additive manufacturing. *Additive Manufacturing, 27,* 595–605.
22. Zhao, Z., et al. (2019). Microstructural evolutions and mechanical characteristics of Ti/steel clad plates fabricated through cold spray additive manufacturing followed by hot-rolling and annealing. *Materials & Design*, 108249.
23. Schmidt, T., et al. (2009). From particle acceleration to impact and bonding in cold spraying. *Journal of Thermal Spray Technology, 18*(5–6), 794.
24. Champagne, V. K. (2007). *The cold spray materials deposition process*. Elsevier.
25. Lupoi, R., et al. (2011). High speed titanium coatings by supersonic laser deposition. *Materials Letters, 65*(21–22), 3205–3207.
26. Villafuerte, J. (2015). *Modern cold spray: Materials, process, and applications*. Berlin: Springer.
27. Olakanmi, E., et al. (2013). Deposition mechanism and microstructure of laser-assisted cold-sprayed (LACS) Al-12 wt.% Si coatings: Effects of laser power. *JOM, 65*(6), 776–783.
28. Koivuluoto, H., et al. (2017). Structures and properties of laser-assisted cold-sprayed aluminum coatings. In *Materials science forum*. Trans Tech Publications Ltd.
29. Reddy, S. (2017). Development of Fe-Mn alloy coatings using Coaxial laser assisted cold spray process. In *Mechanical engineering* (p. 69). Dearborn: University of Michigan.
30. Koivuluoto, H., et al. (2015). A novel coaxially laser-assisted (COLA) cold spray system. In *ITSC 2015-International Thermal Spray Conference*. Long Beach: ASM International.
31. Bray, M., Cockburn, A., & O'Neill, W. (2009). The laser-assisted cold spray process and deposit characterisation. *Surface & Coatings Technology, 203*(19), 2851–2857.
32. Li, B., et al. (2018). Influence of laser irradiation on deposition characteristics of cold sprayed Stellite-6 coatings. *Optics & Laser Technology, 100,* 27–39.
33. Kulmala, M., & Vuoristo, P. (2008). Influence of process conditions in laser-assisted low-pressure cold spraying. *Surface & Coatings Technology, 202*(18), 4503–4508.
34. Jones, M., et al. (2014). Solid-state manufacturing of tungsten deposits onto molybdenum substrates with supersonic laser deposition. *Materials Letters, 134,* 295–297.
35. Yao, J., et al. (2015). Characteristics and performance of hard Ni60 alloy coating produced with supersonic laser deposition technique. *Materials and Design, 83,* 26–35.
36. Yao, J., et al. (2015). Beneficial effects of laser irradiation on the deposition process of diamond/Ni60 composite coating with cold spray. *Applied Surface Science, 330,* 300–308.
37. Li, B., et al. (2015). Microstructure and tribological performance of tungsten carbide reinforced stainless steel composite coatings by supersonic laser deposition. *Surface & Coatings Technology, 275,* 58–68.
38. Story, W. A., et al. (2018). Laser assisted cold spray of Fe-Ni-Zr oxide dispersion strengthened steel. *Materialia, 3,* 239–242.
39. Vilar, R. (1999). Laser cladding. *Journal of Laser Applications, 11*(2), 64–79.
40. Luo, F., et al. (2012). Performance comparison of Stellite 6® deposited on steel using supersonic laser deposition and laser cladding. *Surface & Coatings Technology, 212,* 119–127.
41. Birt, A. M., et al. (2017). Statistically guided development of laser-assisted cold spray for microstructural control of Ti-6Al-4V. *Metallurgical and Materials Transactions A, 48*(4), 1931–1943.

42. Olakanmi, E. (2016). Optimization of the quality characteristics of laser-assisted cold-sprayed (LACS) aluminum coatings with Taguchi design of experiments (DOE). *Materials and Manufacturing Processes, 31*(11), 1490–1499.
43. Yao, J., et al. (2016). Characteristics and bonding behavior of Stellite 6 alloy coating processed with supersonic laser deposition. *Journal of Alloys and Compounds, 661,* 526–534.
44. Story, W. A. (2018). *Processing-microstructure-property relations in high pressure cold spray of AA2024 and AA7075.* University of Alabama Libraries.
45. Marrocco, T., et al. (2011). Corrosion performance of laser posttreated cold sprayed titanium coatings. *Journal of Thermal Spray Technology, 20*(4), 909.
46. Sova, A., et al. (2013). Cold spray deposition of 316L stainless steel coatings on aluminium surface with following laser post-treatment. *Surface & Coatings Technology, 235,* 283–289.
47. Poza, P., et al. (2014). Mechanical properties of Inconel 625 cold-sprayed coatings after laser remelting. Depth sensing indentation analysis. *Surface and Coatings Technology, 243,* 51–57.
48. Carlone, P., et al. (2016). Selective laser treatment on cold-sprayed titanium coatings: Numerical modeling and experimental analysis. *Metallurgical and Materials Transactions B, 47*(6), 3310–3317.
49. Astarita, A., et al. (2015). Study of the laser remelting of a cold sprayed titanium layer. *Procedia CIRP, 33,* 452–457.
50. Aldwell, B., et al. (2018). Fundamental investigation into the effects of in-process heat treatment in cold spray. In *ITSC 2018-International Thermal Spray Conference*. Orlando: ASM International.

Development of CuAlNi Shape Memory Alloy Structures Using Cold Spray Deposition Technique with Laser Remelting

S. Shiva, L. Michaux, A. Cockburn, D. Hopkinson, I. A. Palani, C. P. Paul, and W. O'. Neill

Abstract The current chapter is about the fabricating bulk shape memory alloy (SMA) structures using solid-state deposition of nitrogen gas based cold spray deposition. Parallelly the alloying process is carried out by high power continuous wave and pulsed wave laser. The effects of laser energy density are studied in detail. It is found that the laser energy density plays a crucial role in homogeneous alloying of the cold sprayed structure. The deposition gas temperature played a crucial role in a homogeneous deposition, as the temperature is higher the powder gets blocking in the path of the D-Laval nozzle. The critical temperature level is determined to block free deposition. The laser diffusion depth and the dwell time plays a crucial role. Also, the high laser energy density ablates the surface of the sample leading to complete distortion of the sample. Hence the perfect laser energy density has opted for an efficient alloying process. Once the alloy formation is done, the fabricated samples are subjected to several characterizations in the aspect of surface morphology, crystallization, mechanical properties and shape memory effect. The scanning electron microscopy (SEM) revealed the deposition to be highly dense and high porosity free. The mechanical property reveals the fabricated alloy to have good

S. Shiva (✉)
Department of Mechanical Engineering, Indian Institute of Technology Jammu, Jammu, India
e-mail: sshivabemech@gmail.com

S. Shiva · I. A. Palani
Department of Mechanical Engineering, Indian Institute of Technology Indore, Indore, India

L. Michaux · A. Cockburn · D. Hopkinson · W. O'. Neill
Institute for Manufacturing, Department of Engineering, University of Cambridge, Cambridge, UK

I. A. Palani
Department of Metallurgy Engineering and Material Science, Indian Institute of Technology Indore, Indore, India

C. P. Paul
Laser Development Industrial Applications Division, Raja Ramanna Centre for Advanced Technology, Indore, India

HomiBhabha National Institute, BARC Training School Complex, Anushakti Nagar, Mumbai, India

S. Pathak and G. C. Saha (eds.), *Cold Spray in the Realm of Additive Manufacturing*, Materials Forming, Machining and Tribology,
https://doi.org/10.1007/978-3-030-42756-6_7

strength when alloyed with both pulsed and continuous-wave lasers. The crystalline nature of the alloys was studied using X-Ray diffraction (XRD). The crystallized nature of the alloy is the primary requirement to attain the shape memory effect in SMA. Hence this chapter will give the researchers in beginning stage a clear idea about the evolution of cold spray deposition techniques that can be used by their research works in developing novel research ideas. Also, the nature of laser surface processing techniques is discussed in detail. The idea of laser surface processing can be utilized by budding researchers to deal with any sophisticated or micro-level machining applications.

Keywords Cold spray deposition · Laser · Additive manufacturing · Shape memory alloy

1 Introduction

Additive Manufacturing (AM) has proved itself to be a promising option by overcoming several unsolved problems by other unconventional manufacturing processes, specifically in the stream of powder metallurgy. The technology is equipped with a variety of features that makes the process highly versatile, allowing the users to perform or experiment with tailored complex structure development in the stream of design and manufacture. AM was initially deployed as a process suitable only for concept modelling and rapid prototyping, but over the years the evolution of layer by layer AM gradually took over the position of directly manufacturing net-shaped metallic components almost ready to use [1–4]. Though the technology is well-established problems like porosity and lack of fusion defects are yet to address. Researchers have countered the existing issue with specific precautionary techniques with immense heed to ensure defect-free deposition. In that hunt, several spraying technologies are now deployed in developing thick films and structures. The advantage in spraying technology is the alloying happens in almost porosity free nature and with high density. But the limitation of using spraying technique to manufacture bulk component is, only standard shapes can be developed, and even that thoroughly relies on complete post-processing techniques. Also, if a tailored composition is to be used for fabrication, the process of maintaining homogeneity in composition becomes immensely tough. One more consideration to be taken is the powder particles, which are very small in nature. Hence the time taken for deposition of bulk or thick film is always high. Among the several spraying technologies, cold spray deposition has a unique advantage of maximum coating thickness efficiency, which is always of higher advantage. The segment of new generation AM processes by cold spray technology is capable of delivering complete porosity free deposition with good density without any voids in the middle. Metallic parts of tailored composition with high closely packed density, used in various fields, like thermal, aeronautical, etc. possess more significant challenges in their development are now easily addressed using cold spray based deposition technique. Layer by layer deposition technique can fabricate the entire structure from the base by stacking the tightly bonded powders to beget the desired output of good quality in the very first attempt.

In the cold spray technique, the powders are deposited and later by the assistance of heat treatment techniques, the alloy formation is successfully done. But homogeneously engineering the microstructure is not possible in the conventional technique. In that accord to control the microstructure evolution, a laser-based melting process is carried out to melt the cold spray alloys successfully. As lasers are embedded with several advantages like the control of efficient heat supply, the ability to adapt into several closed environments causing no harm, the possibility of transferring the laser beam from one station to the other without significant loss of intensity and also the superior surface finish of the products developed. Banking on the nature of continuous-wave and pulsed lasers the alloying process can be carried out in an attempt to fabricate samples with microstructures in good homogeneity. Among the two types of lasers, the continuous wave has lesser penetration depth than the pulsed lasers. But the span of a pulsed laser is very small that complete-time is not provided for the metal to melt entirely and evolve the microstructure [5–10].

The current chapter will provide an overview of the history of cold spray based deposition techniques and the role of laser in efficiently developing homogeneous shape memory alloy samples. The compilation of continuous and pulsed wave lasers in the role of melting the alloy will give the readers a vivid picture of the process going on. Laser melting process of cold spray deposited alloy, will give a new dimension in manufacturing and applications in the field of involving SMA. Also, the discussion includes the various desirable parameters and other characteristics features of the chosen three process in brief. In the end, the future prospects and recent research trends in the chosen technologies will be deliberated. This chapter will help the researchers who are at the beginning stage of their research career in an idea to pursue their research work in the field of cold spray in developing bulk net-shaped products.

2 Evolution of Spraying Techniques

The process of coating incepted in an attempt to alter the surface of components, to increase the strength and resistance of the components in the real-time application. Thermal spray techniques are deployed to counter the failures occurring on the surface of the components, eventually increasing their life. The versatile nature of the thermal spray technique provided the option of using them on any type of materials, which makes them maverick in the line of surface engineering. The thermal spray techniques improve the resistance of the components against common surface defects like wear, corrosion, and high temperature. Also, the same spraying technique can repair any surface errors on the components which attracts industries as the overhead exponentially falls down. As the evolution spraying techniques are probed, it's apparent that researchers had initially used the process of metallizing to coat low melting point temperature metals (less than 300 °C) on the substrate by heating the metal powders with an oxygen flame. Sequentially the next stage of thermal spraying was introduced with high-velocity oxygen fuel (HVOF) technique when metal wires were heated by oxygen flame, and the molten metal was atomized by compressed gas that also assisted a homogeneous coating on the surface of the substrate. Similar to HVOF technique D Gun based spraying technique got good attention in the spraying applications as for its very low porosity, strong adhesion to substrates, high cohesive strength and high hardness [11]. As an upgrading in the existing technology, in the place of oxygen flame, the electric arc was introduced. This improvised approach assisted good corrosion resistance coating of high melting temperature metals. The arc spraying process is a good choice for the on plant application due to the low cost of equipment and materials, the simplicity of operation on-site, and highest deposition rate, which is important for onsite manufacturing and repairing [12]. With similar working principle, atmospheric thermal plasma spray (APS) process the coating deposition is at a high rate, and most importantly the entire spraying can be conducted by simple equipment without any controlled atmosphere glove boxes, low-cost film deposition with short duration time can be carried out [13]. The vacuum plasma spraying (VPS) process is another type of spraying technique using electric arc under a controlled atmosphere with low pressure of inert gas, with reduced time of interaction between the plasma jet and the oxidative environment. Hence, VPS technology can be deployed in places where oxidation free deposition is the primary objective. Using VPS more controlled homogeneous coating can be achieved with less contamination [14]. In the LPPS method, the controlled atmosphere gas itself acts as a carrier gas that gushes the powder particles into the plasma jet ignited by a DC plasma torch. The gas used for the process is generally argon plus secondary gases like hydrogen, nitrogen or helium. The powder used for deposition is pre-melted and are accelerated in the plasma jet to coat them over the substrate in the form of splats, that results in the uniform coating. The entire process is carried out in a controlled argon atmosphere inside a vacuum chamber to prevent oxygen contamination. Despite all measures, the oxidation is not thoroughly inevitable as the flame used for melting is oxygen fuel. The exact adjustment of reproducible processing conditions as well as process monitoring is required to ensure fully pronounced pseudoelasticity at operating temperature. The biggest advantage of LPPS technique is to have complete control over the coatings with thickness in the range of nano to microscale and is, therefore, a promising method aiming at a compromise to keep material costs at an acceptably low level while achieving sufficient resistance against cavitation [15]. The third and latest upgrade of spraying techniques is currently using cold spray technique. In this technique, the powders are used as raw material and are deposited using solid-state deposition. The powders are compressed with preheated air and made to pass through a type of convergent and divergent nozzle in a sonic speed to get deposited on the substrate. As shown in Fig. 1, similar to cold spray technique with some mild pre-melting of powders, two more techniques are used for depositing the powders in solid-state deposition.

HVAF is mostly an assembled set up where the deposition process is not entirely reliable in the speed of gas flow. The most critical parameters in tuning the HVAF process are (1) hardware configuration such as the size of the combustion chamber,

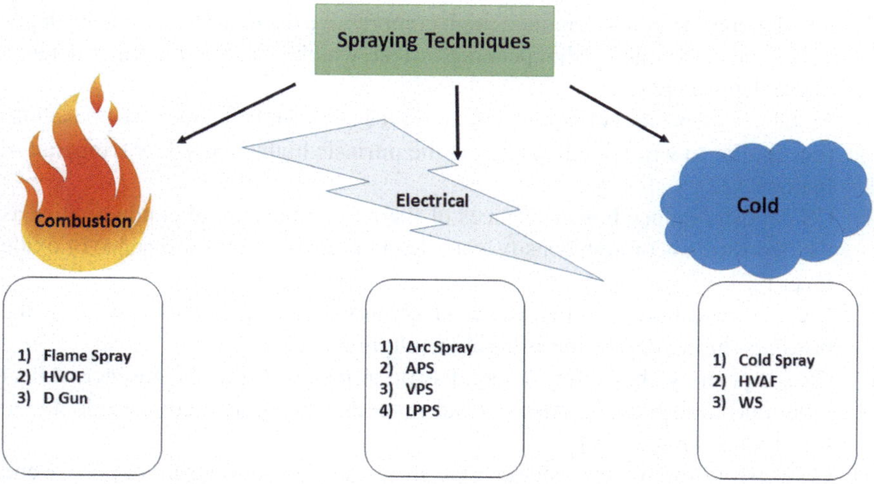

Fig. 1 Evolution of spraying techniques

nozzle, and powder injector and (2) process variables such as standoff distance and powder feed rate [16]. Regardless of the preliminary development of spray conditions specific to given powder chemistry, systematic optimization of HVAF process itself has been scarcely performed. Indeed, in order to spray coatings with required corrosion performance, systematic process parameter analysis must be applied to relate the applied parameters with the produced coating structure and properties [17]. The process-microstructure-properties-performance relationship assessment is a capable tool for process optimization and able to provide precise information on the whole process from the powder to coating performance [18]. In warm spray (WS) the temperature of a supersonic gas flow generated by the combustion of fuel and oxygen is controlled by injecting nitrogen into the mixing chamber placed between the combustion chamber and the powder feed ports. Various powder materials [19] can be deposited in the thermally softened state at high impact velocity, which allows the formation of dense coatings with limited oxidation of the particles what is extremely important in case of the anticorrosion coating.

Among the various discussed techniques, as shown in Fig. 1, cold spray technology is suited high-velocity solid-state deposition of powders. Also, there are certain advantages that keep cold spray technique ahead of the remaining spraying techniques. The advantages are as follows:

(1) CS is an apt technique to carry out the deposition of materials that are temperature-sensitive in nature such as nanocrystalline (NC) and noncrystalline materials, oxygen-sensitive materials such as copper (Cu), titanium (Ti) and aluminum (Al), and also carbide composites which are phase-sensitive materials as the deposition temperature is low [20].

(2) CS of metals, in general, induces high compressive residual stresses in the deposition due to the micro "shot peening" effect, which enhances fatigue resistance of the deposits [21].
(3) Metal CS deposits contain microstructures with high degrees of consolidation similar to wrought alloys due to the intrinsic high energy-low temperature features [22].
(4) CS deposits induce higher chances of thermal and electrical conductivities as the deposition is of high density, completely porosity free and absence of oxide layer [23].
(5) CS is a good option in the aspect of green machining as the wastage of the powders during deposition is highly negligible [24].
(6) CS is a primary choice for joining dissimilar metals due to the absence of heat input during deposition. Also, the substrate does not play a crucial role due to the absence of heat [25].
(7) CS doesn't need for any external masking as the deposition is more precise and controlled over the spraying area on the substrate. In general, the spray beam is very small and less standoff distance [26].

The CS technique can be further divided into two different types like high-pressure CS (HPCS) and low-pressure CS (LPCS). The pressure indicates the pressure of heating gas that is deployed for the deposition. The variation of high and low pressure is architected by a small design variation in the deposition nozzle. The powder injection orifice is placed at two different points, as shown in Fig. 2.

The details about the above mentioned two techniques are as discussed below.

(a) In HPCS the to be deposited powder particles are premixed with the carrier gas before entering the deposition nozzle and are injected into the nozzle that is

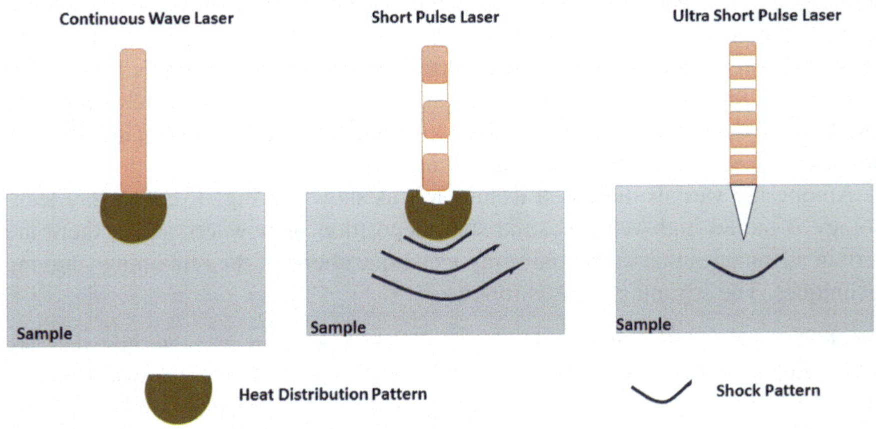

Fig. 2 Nature of heat distribution in various types of laser

Table 1 Parameters used for fabrication of CuAlNi SMA

Parameter	Unit	Values
Gas pressure	bar	15–20
Stand of distance	mm	40
Powder feed rate	g/min	30
Deposition rate	mm/min	16
Gas temperature	°C	470
Powder particles size	μm	10–30

positioned perpendicular to the substrate, preferably downstream type deposition is done. The deposition gets along the gravitational force direction is also an added advantage in the system.

(b) In LPCS the heating of carrier gas is done at the spray gun, unlike the HPCS system which preheats the gas well before entering the deposition nozzle.

(c) As the powder feed rate entirely relies on the carrier gas discharge speed, the HPCS is expected to have a higher deposition rate comparing to LPCS [27]. A detailed comparison between the two spray processes is, as quoted in Table 1.

(d) As the HPCS deals with excess pressure, the entire set up is a bit expensive compared to LPCS. Also, the LPCS has an advantage of low-cost equipment as the process compromises with low Mach speed deposition with low pressure just sufficient to supply powders to the nozzle leading to their deposition.

(e) Nozzle clogging is a significant issue to deal with in both techniques as the powder gets accumulated when exposed to heat and pressure in the sidewalls of the nozzle, hindering further homogeneous deposition. To counter this issue researchers have used premixtures that prevent powders from clogging and nozzles made of materials that prevent powder clogging on the inner surface [28].

(f) Deposition of hard particles is yet a challenge for both techniques as the particles erode the nozzle throat during the deposition process. Sequentially erosion wear in the nozzle is expected to harm the homogeneity in the deposition [29].

(g) Overall the HPCS has specific abilities better than LPCS as the process is equipped with expensive feeders and pressure managing systems that assist the HPCS to have a clear edge over LPCS by more substantial deposition efficiency [30].

3 Role of Particle Deformation in Bonding Properties

In general, the powder particles used for cold spray techniques are subjected to high strain due to the velocity generated during the high-speed deposition process [31]. The high strain-induced powders to assist porosity free deposition. The reason behind the success of cold spray in depositing hard metals and dissimilar metals is possible with the help of the powder deformation. The ability of the HPCS is to engineer the microstructure using the high speed of the powder particles during deposition. Precisely at the interface region between the coating and the substrate [32, 33]. The nature of the microstructure formed in the deposits can directly reflect on the properties like ultra-fine grain microstructure [34], recovery and recrystallization [35], cold spray precipitation [21], residual stress [36], localized amorphization [37] and phase transformation [38]. Along with the mentioned changes, also the microstructure will have a significant impact on the mechanical properties of the developed specimens in the micron to nano level. Hence before depositing the particular material a thorough knowledge about the nature of microstructure for each parameter is essential to prevent any adverse effects on the developed samples that may induce abnormality in the material property.

4 Deformation of the Powder Particles

The role of powder particle deformation has several other roles apart from the earlier discussed issues like powder substrate bonding and mechanical properties of the deposits. Also, the work hardening induced in the samples during deposition is expected to play a vital role in determining the mechanical properties of the deposition. The nature of particle deformation in a deposition is widely depending on two essential parameters of deposition temperature and particle velocity [39, 40]. The mentioned two properties are governed by three different parameters of powder characteristics, geometric effects and process parameters. The three parameters are detail discussed as follows.

4.1 Powder Characteristics

4.1.1 Particle Size

The powders opted for cold spray deposition are in general of size 20–50 μ. The particles opted are to be very small in nature as the deposition rate is maximum when the heat gets quickly dissipated from the particles after deposition. Hence with smaller powder particle, the heat dissipation is quicker. Also, as the powder particles are very smaller in nature, the porosity free deposition is also easily attainable comparative to the bigger powder particle size. Adding to more advantages the shear instability in the deposition can be halted by small powder particles as the environment is entire with higher heat radiation [41]. The heat quenching is smooth when the powder particles are smaller in nature assisting in increasing the strength of the deposition. The chances of impurities are also meagre in smaller powder particles as the surface to volume ratio is higher. Mathematically Eq. (1) is used to determine the critical powder particle size which will be efficient enough for an efficient deposition [42-Schidt].

$$D_{crit} = 36\frac{\lambda_p}{C_p \rho_p V_p} \tag{1}$$

where is the λ_p thermal conductivity, C_p is the specific heat of the particle, ρ_p is the density of particle material, and V_p is particle velocity. The properties mentioned in the equation vary for each material and hence accordingly the particle size can be determined prior to deposition. When particles used for deposition exceeds the calculated value, the heat quenching lasts long as the deformation of the powder particles is delayed, eventually leading to a delay in the powder particle bonding.

4.1.2 Powder State and Shape

The powders used for deposition are manufactured using gas atomization or cryomilling. Irrespective of the manufacturing technique, the powder particles are expected to be spherical in nature for an efficient cold spray deposition as irregular powders generate an irregularity in the deformation process that eventually reflects in the quality of the deposition. In the case of spherical powders of smaller size, the powder particle impact velocity can be achieved in cold spray deposition. Irregular shape powders generally generate a drag coefficient in the deposition that leads to irregularity in powder feed during the deposition process and the deformation time variation among the quantity of powder also leads to improper deposition [43]. The increase in powder size and irregular shape leads to poorer flowability and lower powder packing factor.

4.1.3 Surface Oxide Layer

Surface oxide layer plays a crucial role in the deposition process like oxide layer's impact reflects on deposition efficiency and particle deformation [44]. As the powder jet meets the substrate, the oxide layer on the powder surface, in general, tends to create disruption in the interface layer. Reported results claim the higher the thickness of oxide layer present in the powder surface requires higher energy to deposit the powder and lowers the rate of powder deformation [45]. The continuous deposition of the powder with surface oxide layer is deposited the mounting pressure generates cracking oxide layer which ejects out from the surrounding layer of the powder particle's perimeter. The partial ejection of the oxide layer will trap a certain amount of oxides within the deposition that reduces the bond strength between the powders [46].

4.2 Geometric Effects

The following aspects are to be addressed to satisfy the geometrical effects in the cold spray deposition.

4.2.1 Spraying Stand-Off Distance

The distance between the spraying nozzle and the substrate surface is the stand-off distance in the cold spray techniques. The higher stand-off distance decreases the deposition efficiency and deteriorates the deposition's nature. Hence the stand-off distance varies with the chosen particle size for the deposition.

4.2.2 Spraying Angle

To determine the spray angle of the deposition, a small wipe test was conducted by the researchers in the past [47]. The substrate is moved in very high-speed in front of the spraying gun. The deposition thickness is then analysed. The particular angle in which the highest deposition of powder is done. The test results reveal that when the perpendicular direction of deposition has changed the deformation of the deposited particles also change. Also, when the temperature of deposition is varied, the angular impact in the interface varies due to frictional heating leading to high instability.

4.2.3 Position in Particles Jet

The powder particles exiting from the nozzle takes a divergent flow nature before reaching the surface of the substrate. In the mentioned flow, the powder particles remain under the influence of a different velocity due to the bow shock effect [48].

4.2.4 Nozzle Geometry

The nozzle used for deposition is called de laval nozzle, which is convergent and divergent by nature. The Mach number plays a crucial role in the deposition efficiency of the powders by the carrier gas. To enhance the Mach number of the nozzle, usually, the divergent part of the nozzle is extended to achieve the required speed for the powder to get deposited. But the expansion of the nozzle is not be done randomly. The expansion plays a crucial role in the types of nozzles like under expanded, correctly expanded and over-expanded as reported by researchers in the past. Among the three types of nozzles, the correctly expanded nozzle, in general, produces no shock during deposition. The over-expanded nozzle has the ability to produce maximum outlet speed. After extensive research in the nozzle geometry, the final results concluded that the ambient air around intrudes exit of the nozzle as a result of low static and stagnation pressure. As a result, over-expansion induces unwanted shock within the nozzle that disrupts the powder flow through the nozzle.

4.3 Processing Parameters

4.3.1 Carrier Gas Type, Temperature and Pressure

The carrier gas is a vital processing parameter that assists the preheating of the powder which plays a crucial role in the bonding of the powders on the substrate. There are several types of carrier gasses like nitrogen, helium, oxygen, argon etc. Also, the deformation of powder particles on the surface of the substrate and also based upon the nature of the carrier gas been deployed for the process.

4.3.2 Substrate Hardness, Temperature, and Surface Roughness

Specific mechanical properties of the substrate material play a crucial in the deposition of the powder particles. The properties like hardness, surface roughness and surface temperature. When powder particles of soft materials are deposited on the substrate which is soft in nature, the deformation is considerably high, and when materials of highly hard nature are deposited on the hard substrate, the deformation rate is considerably low. In both cases, the material's strength plays a crucial role in the observations. In the other way around, when soft and hard the deformation rate remains highly distinct [49].

5 Laser Remelting

A laser is a tool that is used primarily for various types of surface processing techniques in modern advanced surface treatment techniques. Similar to surface processing remelting is one among the process widely used to engineer the microstructure and nature of various materials. Hence the concept of laser interaction with metal is to be analyzed in detail to extract the desired output from the process. Laser remelting is possible when the absorption of laser power is high by the material on which the laser is incident, and the laser irradiation is continuous on the material's surface. In general, laser interaction with materials can be classified into two types of resonant and non-resonant interactions. A process like localized heating and photons ionization fall under the first type of resonant type of interactions and melting and plasma generation process falls under non-resonant process. In the case of laser-based remelting, the entire process is accompanied by vaporization. It is inevitable to proceed with laser remelting without vapourization in case of metals. Latent heat of fusion plays a crucial role in the entire laser-based melting process. The required amount of heat to preserve the material in the melting point temperature depends on the nature of the material and the volume of the material to be melted.

Surface remelting is executed by passing multiple tracks of laser on the surface of the material, and subsequentially the melting of the material on the surface takes place followed by rapid solidification. This technique is highly useful in case of bulk alloying as the composition of the material can be easily varied and the process of engineering the microstructure is also quickly done. In fact multiple behaviours of the soft material are utterly unique while comparing to the conventional process. Once the surface melting process is done the solidification of the melted surface initiates when the nucleation of the solid material takes place depending upon the nature of heat flow generated by laser remelting. The liquid phase solidifies in different forms and homogeneity can be obtained in the process by entrapping heat in the material for a longer time. In case if there is undercooling mostly brittle phases are induced which is very much harmful to the materials. Also choosing the type of laser to proceed with the melting process is very important. Researchers have widely used pulsed and continuous-wave lasers. The advantages to be availed using pulsed laser is the diffusion depth is higher than continuous wave laser. Similarly the melting efficiency of the continuous wave laser is higher than the pulsed laser. The nature of pulsed and continuous-wave laser are discussed as follows.

5.1 Pulsed Wave Laser

The pulsed laser can be of two types as short pulse and ultra-short pulse lasers. The nanosecond lasers are termed as short pulse lasers whereas picosecond and femtosecond lasers come under the ultrashort laser. Widely for surface processing purposes short-pulse laser and for ablation types of work ultrashort pulse is deployed. When it comes to short-pulsed laser widely Nd:Yag nanosecond green laser is preferred for its high pulse energy and good output efficiency. As the energy of the source is stored and released in short time the generation of high intensity is possible and also the dwell time is sufficient enough to generate changes on the material in which the laser is impinged. Due to broad bandwidth, the laser pulses are very short. Among several types of pulsed lasers like a nanosecond, microsecond, picosecond and femtosecond laser, the nanosecond lasers have proved themselves much efficient when it comes to the melting of bulk materials with sufficient dwell time on the samples. The pulsed lasers apart from surface melting are widely deployed for surface processing operations like laser annealing and laser shock peening. Also in pulsed lasers an extra option of changing wavelength with various modes are possible. Hence this adjustment facility provides an extra advantage to users to achieve the requirement more precisely.

5.2 *Continuous-Wave Laser*

In continuous-wave lasers, there is no classification based on laser pulses because here the output is continuous without any break. The power of the laser depends upon the parameters like wavelength and frequency. The continuous-wave lasers are of from several sources and among them fiber optics-based continuous-wave lasers are widely used for their high power performance and cost-effective maintenance. The usage is widely deployed in all higher-end applications like cutting, drilling and other modern machining processes. Similarly they are deployed in applications where melting of powders are also done to form alloys and also in processes like laser cussing where melting of material's surface to mend them. The continuous-wave lasers in general possess very lean bandwidth comparing to pulsed lased lasers which are of broader bandwidth. In case pulsed or continuous the nature of beam chosen for various applications are TEM_{00} for their ability to transfer higher intensity from the center and distributing the intensity homogeneously in all directions comparing to other nature of beams available. The nature of the continuous and pulsed wave interaction with material is as shown in Fig. 2. Hence as discussed above the higher end applications are widely done with continuous-wave laser, in the current chapter the laser surface melting of cold sprayed powders are also to be carried out using the same type.

6 Shape Memory Alloys

Among the several smart alloys used in various scientific applications, shape memory alloys (SMA) have secured a special place for possessing a maverick property of phase transformation on the application of external load in the form of temperature. Also, the broad application of SMA is in the form of thin films in micro-electro-mechanical systems (MEMS) is used in the form of diaphragms for micropumps. The mechanism behind the functioning of SMA included process like twinning and detwinning by varying the thermal load application in the alloy where the phase transformation from austenite to martensite and vice versa takes place. The functioning of the SMA is as shown in Fig. 3. Among the various SMAs like NiTi, TiNiCu, CuAlNi is also a primary choice for applications. CuAlNi SMA is widely used in applications where vibration damping is of high priority. The Ni in the alloy has enough strength in alloy and Cu provides good ductility in the alloy. Al provides excellent damping efficiency when alloying with Ni specifically. When it comes to bulk CuAlNi SMA fabrication methods like selective laser melting (SLM) is the only method reported to be successful. Also, the report proves that using additive manufacturing technique CuAlNi can be fabricated with good ductile nature which is a unique achievement as fabricating materials of high ductile nature is not an easy achievement using Additive manufacturing technique [50-simmone].

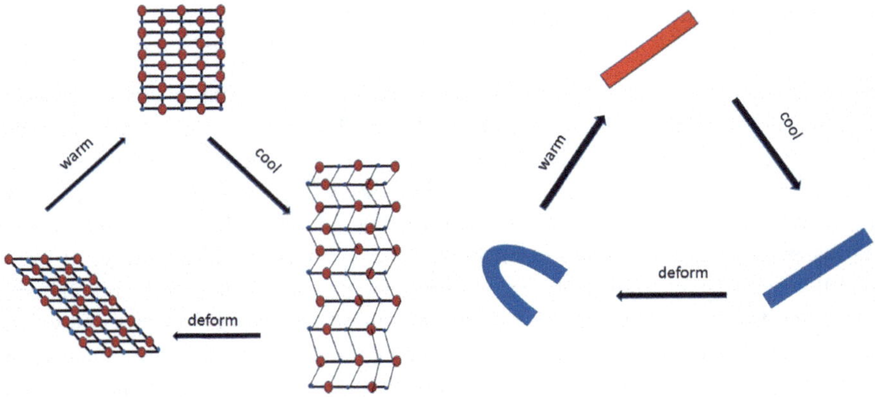

Fig. 3 Mechanism of SMA functioning

7 Experimental Approach

The powders play a crucial role in the deposition efficiency of the cold spray process. Hence the powder particles were assured to be in a size of 10–30 μm. The powder particles size were maintained within the mentioned range because the size bigger than 30 μm can lead to un uniformity in deposition and the size of powders lesser than 10 μm tends to fly away in the speed generated by the nozzle. Also, much importance was given to the shape of the powders to be in spherical shape as the deposition efficiency is high when the powders are in a spherical shape. The shape of the powders other than spherical like irregular or flakes shape powders cannot have an efficient deposition as the deformation is not expected to happen uniform throughout. Eventually, the deposition efficiency drastically decreases and also it is expected to induce porosity within the sample. The surface morphologies of the various powders used in the process are as shown in Fig. 4. The powders were mildly preheated before the deposition to get rid of the surface oxidation. Also, the powders were carefully stored in a control atmosphere to prevent any sought of reaction with open room atmosphere, which may eventually deteriorate the property of the powders.

Figure 5 shows the schematic diagram of the indigenously developed cold spray set up used for the fabrication of the CuAlNi samples. The premixed powders of the desired composition were used for the deposition process. The deposition system was designed in such a way that the powders will be fed inside the upper end of the nozzle where the preheated nitrogen air mixes get along. The nitrogen gas plays a crucial role in compressing the powders into the convergent section of the nozzle. The Mach number of De Laval-type nozzle used for the current experiment is 2.9, which assists in accelerating the pre-mixed powders to flow in the supersonic speed. As shown in Fig. 5 the powders pass through the throat of the nozzles a high speed is obtained by expansion assisted by the preheated nitrogen gas, as per the theory of gas dynamics [50]. The drag force exerted by the preheated nitrogen gas in a supersonic stream on the powders accelerates the powders to attain high speeds that

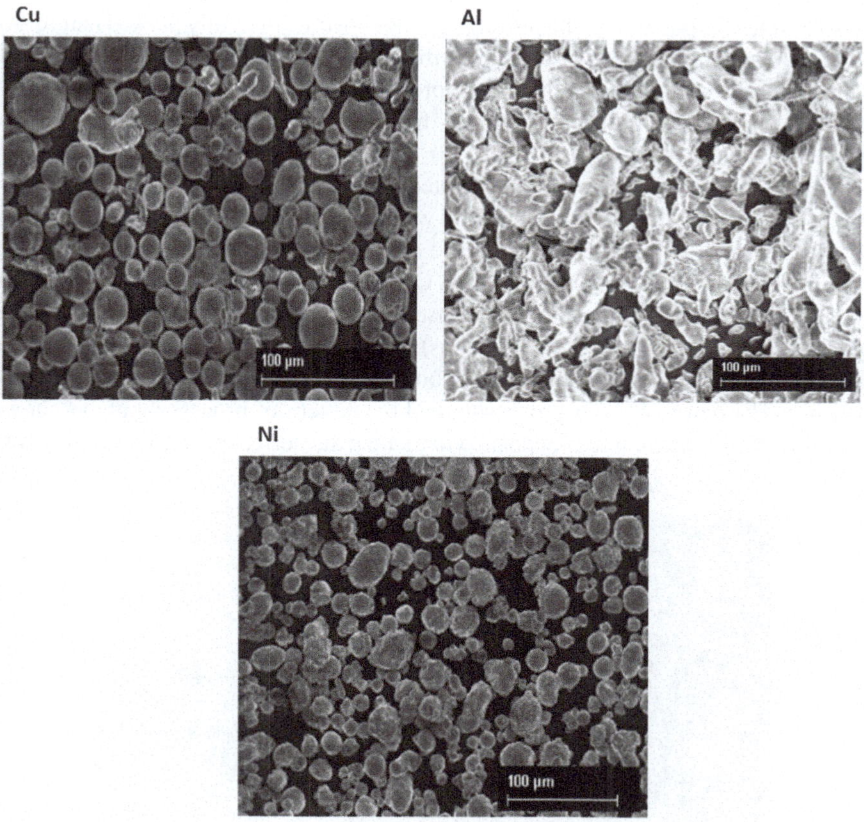

Fig. 4 Surface morphologies of the powders

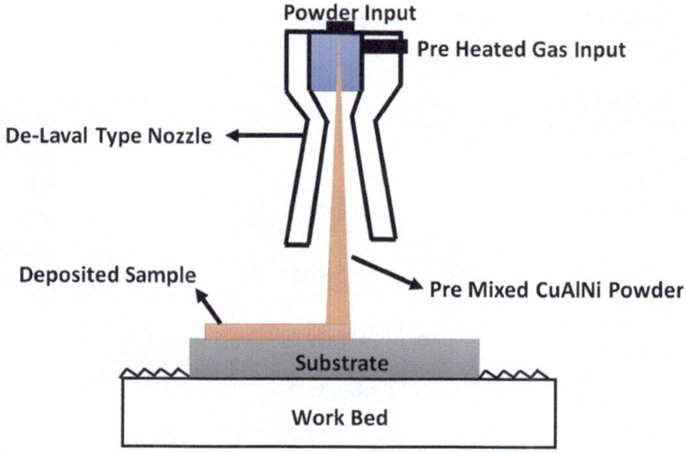

Fig. 5 Schematic diagram of cold spray setup

assist efficient deposition of the powders on the substrate in tightly packed nature. An efficient deposition can be assured once the powders travel above a high threshold value after absorbing enough drag force provided by the carrier gas (Nitrogen).

The deposition parameters deployed for the above-mentioned process are as quoted in Table 1.

The samples were successfully deposited on mild steel substrate, and the deposition is as shown in Fig. 6. The deposition was so closely bonded without any sought of porosity, and the bonding between the deposition and the substrate was visibly strong without any cracks in the interface. Once the deposition is done the deposition is done the samples were subjected to laser remelting carried out by two different types of lasers (i.e. pulsed and continuous) to form the perfect alloy formation. The laser power was maintained the same for both laser and the processing time was also uniform. The nature of laser processing and the pattern of processing are as shown in Fig. 7a. Once the sample is deposited to finalize the perfect power for melting sev-

Fig. 6 CuAlNi SMA fabricated by cold spray deposition

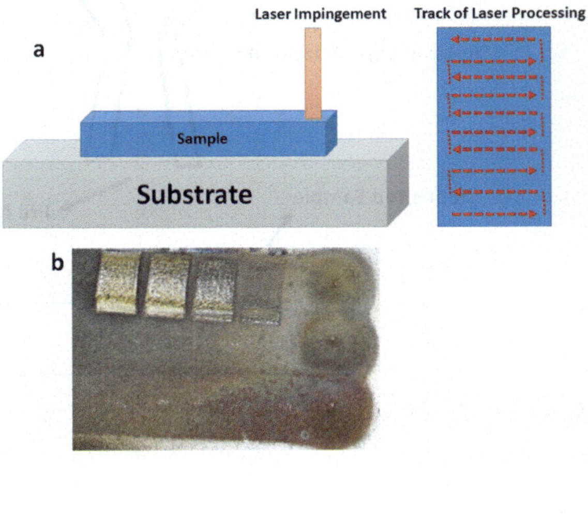

Fig. 7 The images of **a** the pattern of laser remelting of the cold spray deposited sample, **b** the laser process parameter optimization

eral trials were conducted with various laser powers as shown Fig. 7b. The analyses were carefully done to avoid ablation of the samples during the interaction with the laser. Also, the cross-section analyses were carried out to determine the efficiency of alloying by both the type of lasers. The nature of the samples was analyzed in detail post laser processing.

In a motive to analyze the nature of the samples alloyed by pulsed and continuous-wave laser, the surface morphological analyses, the mechanical properties and the phase transformation properties were analyzed. The surface morphological analyses are to study the nature of laser interaction with the sample and the alloying efficiency of the lasers. The mechanical properties of the sample are expected to analyze and to predict the sample's deployment in real-time application. Finally, the phase transformation ability of the sample is directly related to the shape memory effect in the sample. The details about the characterization results are discussed in detail in the sessions coming henceforth. The samples processed with pulsed laser is termed as CuAlNi PL and the sample processed with continuous-wave laser is termed as CuAlNi CW in the chapter ahead. The scanning electron microscopy (Make: Zeiss, Model: Supra55) attached with energy dispersive spectrograph (Make: Oxford Instruments, Model: X-mas), was used for the surface morphological analyses. The micro-hardness (Make: UHL, Model: VMH 002), was used to analyze the mechanical properties of the samples.

8 Results and Discussion

Once after the premixed powders were deposited the cross-section analyses were initiated for the deposits in order to confirm porosity free deposition. As shown in Fig. 8, the cross-section revealed a dense-packed deposition. The bonding at the intermediate layer is observed to be porous free, and the nature powder deposition is homogeneous by the assistance of powder deformation. Also, no pores were observed in the intermediate region, which indicates not much bowing effect has taken place as the standoff distance between the nozzle and substrate is good enough for high efficiency of deposition. Very less dark spots are seen in the image which may be attributed to the removal of powders during the polishing process. To alloy, the deposits two types of lasers opted pulsed and continuous-wave laser. The optimised range of power was opted for both pulsed and continuous-wave laser, to initiate the alloys of the powders. As the wave nature of both lasers is different, it is vivid that the alloying nature is not expected to be similar in nature. To investigate the nature of alloying a scanning electron microscopy (SEM) analyses were carried out. The results of the alloying process using pulsed laser are as shown in Fig. 9. The comparison of before and after laser processing is as shown in Fig. 9a. The impact of pulsed laser processing was visible to be highly efficient. Once when the alloyed surface was brought to a closer look, some sought of mild porosity of slanting pattern was observed on the surface as shown in Fig. 9b. The presence these patterns might

Fig. 8 The deposition of CuAlNi powders using cold spray process

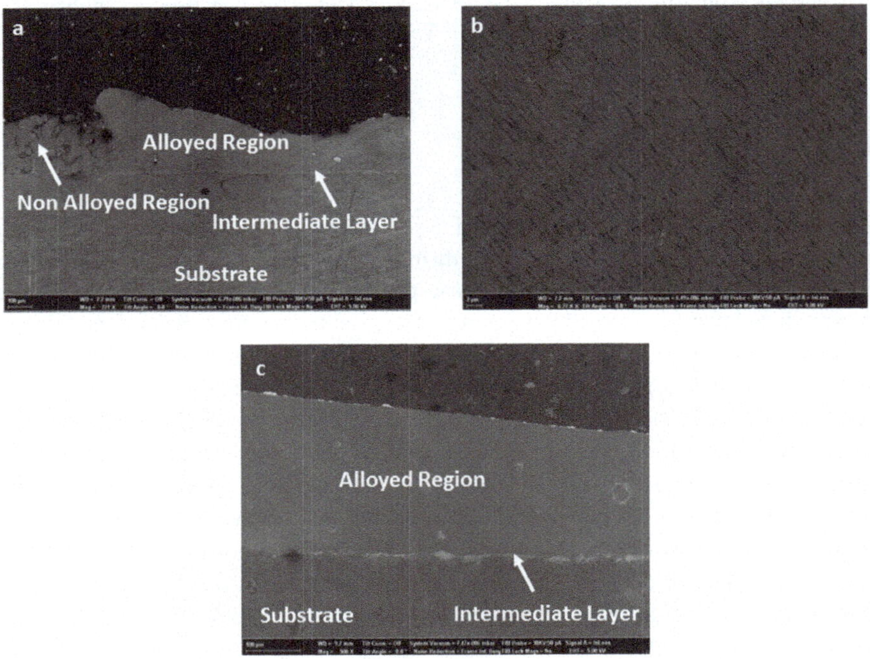

Fig. 9 SEM images of **a** pulsed laser processed sample, **b** the magnified image of pulsed laser processed samples, **c** continuous-wave laser processed sample

be due to the short life span of each pulse generated by the pulsed laser. The non-availability of enough dwell time to remelt and cover up the formed porosity for the alloys is observed.

But when it comes to continuous-wave laser melting the alloys seem to be having a homogeneous melting throughout the deposits as shown in Fig. 9c. The presence of any form of porosity is not observed, and this may be possible by the continuous wave pattern of the laser. Also, there are not many precipitates observed on the surface of the alloy. The alloying density is observed to be very high for continuous wave laser processed CuAlNi sample and the mechanical properties are expected to be useful as well.

To investigate the mechanical property of the alloys a preliminary micro-hardness test was conducted for the samples. The samples were bisected and prepared following the standard metallographic techniques. After the samples were finely polished, the samples were mounted in Vicker's hardness tester. The readings were noted at a distance of 25 mm at a load of 500 g. The impingement was carefully done ensuring to cover the entire surface area, without any overlap. The results presented in Fig. 10 reveal the micro-hardness test reveal pulsed laser processed samples have the highest value of 536 VHN comparing to continuous wave laser processed sample whose value to be 431 VHN. The nature of the pulsed laser processed sample indicates to be of much more of brittle nature. But the results of continuous-wave laser processed sample indicate the sample to have low micro-hardness values because of the proper alloying that had taken place during the processing.

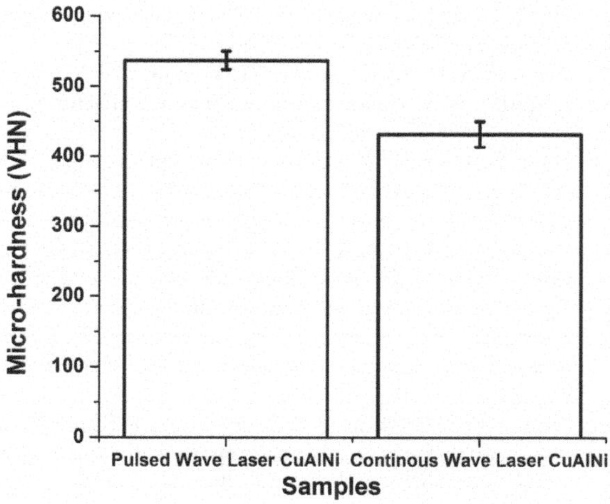

Fig. 10 The micro-hardness results of the samples

9 Conclusion

Cold spray deposition technique is an extremely versatile technique in the aspect of coatings. Deploying the same technique to build three-dimensional bulk structures. The novel approach of engineering the cold sprayed deposits using modern tools like lasers, where enough opportunities are availed to alloy the powders. The current chapter plays a crucial role in giving exposure to deploying two different types of lasers to obtain a common objective. Hence depending upon the application and requirement, the users have enough options. The CuAlNi shape memory alloy which is widely used for vibration damping applications. Hence the results reveal when continuous-wave lasers are deployed for the application they tend to have better alloying possibility which is much needed for the robust applications.

Acknowledgements The authors would like to thank Royal Academy of Engineering (RAE), London for assisting the cold spray deposition experimentations carried out at Institute For Manufacturing (IFM), University of Cambridge, UK, under the Newton Bhabha Project (Project No.: HEPI\1516\10). Thanks are also due to Sophisticated Instrument Centre (SIC) at Indian Institute of Technology Indore, India, for providing us with the required characterization facility.

References

1. Zavala-Arredondo, M., Boone, N., Willmott, J., Childs, D. T. D., Ivanov, P., Groom, K. M., et al. (2017). Laser diode area melting for high speed additive manufacturing of metallic components. *Materials and Design, 117,* 305–315.
2. Harun, W. S. W., Kamariah, M. S. I. N., Muhamad, N., Ghani, S. A. C., Ahmad, F., & Mohamed, Z. (2018). A review of powder additive manufacturing processes for metallic biomaterials. *Powder Technology, 327,* 128–151.
3. Zhou, Y. H., Zhang, Z. H., Wang, Y. P., Liu, G., Zhou, S. Y., Li, Y. L., et al. (2019). Selective laser melting of typical metallic materials: An effective process prediction model developed by energy absorption and consumption analysis. *Additive Manufacturing, 25,* 204–217.
4. Wei, X., Liu, Y0, Zhao, D., Mao, X., Jiang, W., & Ge, S. S. (2020). Net-shaped barium and strontium ferrites by 3D printing with enhanced magnetic performance from milled powders. *Journal of Magnetism and Magnetic Materials, 493,* 165664.
5. Kang, A. S., Grewal, J. S., & Cheema, G. S. (2017). Effect of thermal spray coatings on wear behavior of high tensile steel applicable for tiller blades. *Materials Today: Proceedings, 4,* 95–103.
6. López-Ortega, A., Arana, J. L, Rodríguez, E., & Bayón, R. (2018). Corrosion, wear and tribocorrosion performance of a thermally sprayed aluminum coating modified by plasma electrolytic oxidation technique for offshore submerged components protection. *Corrosion Science, 143,* 258–280.
7. Lashmi, P. G., Majithia, S., Shwetha, V., Balaji, N., & Aruna, S. T. (2019). Improved hot corrosion resistance of plasma sprayed YSZ/Gd2Zr2O7 thermal barrier coating over single layer YSZ. *Materials Characterization, 147,* 199–206.
8. Huang, C. J., Wu, H. J., Xie, Y. C., Li, W. Y., Verdy, C., Planche, M.-P., et al. (2019). Advanced brass-based composites via cold-spray additive-manufacturing and its potential in component repairing. *Surface & Coatings Technology, 371,* 211–223.

9. Tan, J.C., & Hashmi, M. S. J. (1995). High velocity oxygen fuel (HVOF) thermal spray: Prospect and limitation for engineering application. In *Current advances in mechanical design and production* (vol. VI, pp. 27–33).
10. Liu, M., Yu, Z., Zhang, Y., Wu, H., Liao, H., Deng, S. (2019). Prediction and analysis of high velocity oxy fuel (HVOF) sprayed coating using artificial neural network. In *Surface and coatings technology* (p. 124988).
11. Shukla, N., Trivedi, Harshit, Kumar, Hemant, & Yadav, Anant. (2017). Surface engineering analysis of D-Gun sprayed cermet coating in aggressive environment. *Materials Today: Proceedings, 4,* 10212–10215.
12. Davis, J. R. (2004). *Handbook of thermal spray technology*. Cleveland, OH: ASM International.
13. Ando, Y., Kindole, D., Noda, Y., Mbiu, R. N., Kosgey, B. K., Maranga, S. M., et al. (2017). Alumina and titania films deposition by APS/ASPPS dual mode thermal spray equipment using Ar added N_2 working gas. *Vacuum, 136,* 203–208.
14. Salhi, Z., Klein, D., Gougeon, P., & Coddet, C. (2005). Development of coating by thermal plasma spraying under very low-pressure condition. *Vacuum, 77,* 145–150.
15. Bitzer, M., Rauhut, N., Mauer, G., Bram, M., Vaßen, R., Buchkremer, H. P., et al. (2015). Cavitation-resistant NiTi coatings produced by low-pressure plasma spraying (LPPS). *Wear, 328,* 369–377.
16. Wang, Q., Zhang, S., Cheng, Y., Xiang, J., Zhao, X., & Yang, G. (2013). Wear and corrosion performance of WC-10Co4Cr coatings deposited by different HVOF and HVAF spraying processes. *Surface & Coatings Technology, 218,* 127–136.
17. Pawlowski, L. (2008). *The science and engineering of thermal spray coatings*. Chichester: Wiley.
18. Li, C.-J., & Yang, G.-J. (2013). Relationships between feedstock structure, particle parameter, coating deposition, microstructure and properties for thermally sprayed conventional and nanostructured WC–Co. *International Journal of Refractory Metals and Hard Materials, 39,* 2–17.
19. Molak, R. M., Araki, H., Watanabe, M., Katanoda, H., Ohno, N., & Kuroda, S. (2014). Warm spray forming of Ti-6Al-4 V. *Journal of Thermal Spray Technology, 23,* 197–212.
20. Villafuerte, J. (2015). *Modern cold spray: Materials, process, and applications*. Berlin: Springer.
21. Shayegan, G., Mahmoudi, H., Ghelichi, R., Villafuerte, J., Wang, J., Guagliano, M., et al. (2014). Residual stress induced by cold spray coating of magnesium AZ31B extrusion. *Materials and Design, 60,* 72–84.
22. Chavan, N. M., Kiran, B., Jyothirmayi, A., Phani, P. S., & Sundararajan, G. (2013). The corrosion behavior of cold sprayed zinc coatings on mild steel substrate. *Journal of Thermal Spray Technology, 22,* 463–470.
23. Champagne, V. K. (Ed.). (2007). *The CS materials deposition process*. Cambridge, England: Woodhead Publishing Ltd.
24. Schmidt, T., Assadi, H., Gartner, F., Richter, H., Stoltenhoff, T., Kreye, H., et al. (2009). From particle acceleration to impact and bonding in cold spraying. *Journal of Thermal Spray Technology, 18,* 794–809.
25. Pathak, S., & Saha, G. C. (2017). Sustainable development of cold spray coatings and 3D additive manufacturing components for repair/manufacturing applications: A critical review. *Coatings, 7*(8), 122–149.
26. Tang, J., Saha, G. C., Richter, P., Kondás, J., Colella, A., & Matteazzi, P. (2018). Effects of post-spray heat treatment on hardness and wear properties of Ti-WC high-pressure cold spray coatings. *Journal of Thermal Spray Technology, 27*(7), 1153–1164.
27. Cui, L., Gerber, A. G., & Saha, G. C. (2019). Cold gas dynamic spray technology: The simulation of aerodynamics of flow. *Key Engineering Materials, 813,* 7–12.
28. Barati, H., Wu, M., Kharicha, A., & Ludwig, A. (2018). A transient model for nozzle clogging. *Powder Technology, 329,* 181–198.
29. Smith, M. F. (2007). Comparing cold spray with thermal spray coating technologies. In V. K. Champagne (Ed.), *The cold spray materials deposition process: Fundamentals and applications* (1st ed.). Cambridge: Woodhead.

30. Karthikeyan, J. (2007). The advantages and disadvantages of cold spray coating process. In V. K. Champagne (Ed.), *The cold spray materials deposition process: Fundamentals and applications* (1st ed.). Cambridge: Woodhead.
31. Rokni, M. R., Widener, C. A., Champagne, V. K., & Crawford, G. A. (2015). Microstructure and mechanical properties of cold sprayed 7075 deposition during non-isothermal annealing. *Surface & Coatings Technology, 276,* 305–315.
32. Rokni, M. R., Widener, C. A., Crawford, G. A., & West, M. K. (2015). An investigation into microstructure and mechanical properties of cold sprayed 7075 Al deposition. *Materials Science and Engineering A, 625,* 19–27.
33. Vlcek, J., Gimeno, L., Huber, H., & Lugscheider, E. (2005). A systematic approach to material eligibility for the cold-spray process. *Journal of Thermal Spray Technology, 14,* 125–133.
34. Kim, K., Watanabe, M., Kawakita, J., & Kuroda, S. (2008). Grain refinement in a single titanium powder particle impacted at high velocity. *Scripta Materialia, 59,* 768–771.
35. Coddet, P., Verdy, C., Coddet, C., & Debray, F. (2015). Effect of cold work, second phase precipitation and heat treatments on the mechanical properties of copper–silver alloys manufactured by cold spray. *Materials Science and Engineering: A, 637,* 40–47.
36. Luo, X. T., Yang, G. J., Li, C.-J., & Kondoh, K. (2011). High strain rate induced localized amorphization in cubic BN/NiCrAl nanocomposite through high velocity impact. *Scripta Materialia, 65,* 581–584.
37. Richer, P., Zúñiga, A., Yandouzi, M., & Jodoin, B. (2008). CoNiCrAlY microstructural changes induced during cold gas dynamic spraying. *Surface & Coatings Technology, 203,* 364–371.
38. Bae, G., Kumar, S., Yoon, S., Kang, K., Na, H., Kim, H. J., et al. (2009). Bonding features and associated mechanisms in kinetic sprayed titanium coatings. *Acta Materialia, 57,* 5654–5666.
39. Choi, W. B., Li, L., Luzin, V., Neiser, R., Gnaupel-Herold, T., Prask, H. J., et al. (2007). Integrated characterization of cold sprayed aluminum coatings. *Acta Materialia, 55,* 857–866.
40. Lange, F. F. (1989). Powder processing science and technology for increased reliability. *Journal of the American Ceramic Society, 72,* 3–15.
41. Gartner, F., Stoltenhoff, T., Schmidt, T., & Kreye, H. (2006). The cold spray process and its potential for industrial applications. *Journal of Thermal Spray Technology, 15,* 223–232.
42. Wong, W., Rezaeian, A., Irissou, E., Legoux, J. G., & Yue, S. (2010). Cold spray characteristics of commercially pure Ti and Ti-6Al-4V. *Advanced Materials Research, 89,* 639–644.
43. Zhang, Q., Li, C. J., Li, C. X., Yang, G. J., & Lui, S. C. (2008). Study of oxidation behavior of nanostructured NiCrAlY bond coatings deposited by cold spraying. *Surface & Coatings Technology, 202,* 3378–3384.
44. Kang, K., Yoon, S., Ji, Y., & Lee, C. H. (2008). Oxidation dependency of critical velocity for aluminum feedstock deposition in kinetic spraying process. *Materials Science and Engineering A, 486,* 300–307.
45. Yin, S., Wang, X., Li, W., Liao, H., & Jie, H. (2012). Deformation behavior of the oxide film on the surface of cold sprayed powder particle. *Applied Surface Science, 259,* 294–300.
46. Binder, K., Gottschalk, J., Kollenda, M., Gartner, F., & Klassen, T. (2011). Influence of impact angle and gas temperature on mechanical properties of titanium cold spray deposits. *Journal of Thermal Spray Technology, 20,* 234–242.
47. Pattison, J., Celotto, S., Khan, A., & O'Neill, W. (2008). Standoff distance and bow shock phenomena in the Cold Spray process. *Surface & Coatings Technology, 202,* 1443–1454.
48. Bae, G., Xiong, Y., Kumar, S., Kang, K., & Lee, C. (2008). General aspects of interface bonding in kinetic sprayed coatings. *Acta Materialia, 56,* 4858–4868.
49. Pauly, S., PeiWang, U. K., & Kosiba, K. (2018). Experimental determination of cooling rates in selectively laser-melted eutectic Al-33Cu. *Additive Manufacturing, 22,* 753–757.
50. Yandouzi, M., Richer, P., & Jodoin, B. (2009). SiC particulate reinforced Al–12Si alloy composite coatings produced by the pulsed gas dynamic spray process: Microstructure and properties. *Surface & Coatings Technology, 203,* 3260–3270.

Dimensional Analysis and Laser-Ultrasonic Inspection of Cold Spray Additive Manufacturing Components

C. V. Cojocaru, P. Vo, D. Levesque, C. Bescond, M. Rivard, J. Boisvert, G. Lamouche, M. Martin, and E. Irissou

Abstract The use of cold spray (CS) in metal additive manufacturing (AM) offers well recognized advantages with typical commercial drivers being a rapid build rate, low process temperature and wide range of usable alloys. For cold spray, technology-specific considerations must be factored into each of the processing steps and in particular, an effective build strategy and toolpath are critical to moving towards near-net shape parts. Inspection and quality control of such complex parts is a challenge and new strategies have to be developed. For this purpose, this study looks to combine optical techniques for dimensional analysis with laser ultrasonics for volume probing.

Keywords Cold spray · Metal additive manufacturing · Non-destructive testing · Dimensional analysis · Laser ultrasonics

1 Introduction

The use of the cold spray in metal additive manufacturing (AM) offers several attractive competitive advantages in comparison with other additive techniques. Typical commercial drivers are rapid build rates, which can be on the order of kg/h, and low process temperatures that are well below material melting points [1]. Also, a wide range of alloys can be employed. Several reviews of specific benefits and considerations for AM by cold spray are available [2–8]. Fundamentally, cold spray additive manufacturing (CSAM) can be considered as a direct energy deposition (DED) process, as the material is deposited on a surface in a layer-by-layer approach to building a 3D structure. Due in part to its inherent spray characteristics, however, CSAM to date has often employed a methodology of rapid material deposition, with or without masking, into relatively simple shapes and wide tolerances that can lead

C. V. Cojocaru (✉) · P. Vo · D. Levesque · C. Bescond · M. Rivard · G. Lamouche · M. Martin · E. Irissou
National Research Council of Canada, 75 de Mortagne Blvd., Boucherville, QC J4B 6Y4, Canada
e-mail: cristian.cojocaru@cnrc-nrc.gc.ca

J. Boisvert
National Research Council of Canada Ottawa, 1200 Montreal Rd, Ottawa, ON K1A 0R6, Canada

S. Pathak and G. C. Saha (eds.), *Cold Spray in the Realm of Additive Manufacturing*, Materials Forming, Machining and Tribology,
https://doi.org/10.1007/978-3-030-42756-6_8

to constraints in part geometries and/or significant post-spray machining. Thus for cold spray, technology-specific considerations must be factored into each of the processing steps, and in particular, an effective build strategy and toolpath are critical to moving towards near-net-shape parts. These aspects are briefly addressed in Sect. 2 of this chapter. Inspection and quality control of such complex parts is a challenge, and new strategies have to be developed not only for off-line but also for on-line inspection. For this purpose, this study looks further to combine optical techniques for dimensional analysis with laser ultrasonics for volume probing of cold sprayed 3D structures. Dimensional measurements can be performed at various intermediate stages during the buildup of an intricate part, showing the value of in-process monitoring. In Sect. 3, various optical techniques for dimensional analysis are presented and compared: triangulation, structured-light and Optical Coherence Tomography (OCT). All techniques provide good results; nevertheless, OCT presents the advantage of requiring a single point-of-view compared to the two other techniques. This allows for the design of a compact scanner well suited for online process monitoring. Laser ultrasonics technique (LUT) is very attractive due to its non-contact nature and is well adapted to online implementation. In this context, we discuss in Sect. 4 non-destructive inspection performed off-line on metallic parts produced by the CSAM process. Laser ultrasonics is used to detect flaws using the synthetic aperture focusing technique (SAFT), and through-thickness distributed porosity is investigated using the backscattered signal. Also, laser shockwave is used to characterize bond strength at the interface between the cold spray deposits and the substrate. Inspection results from either the top layer or the underside of the substrate are presented. Laser-ultrasonic measurement can also be performed during post-heat treatment of cold spray AM metallic parts to capture microstructural and phase changes.

Work presented throughout the chapter stems from the activities carried out by the *National Research Council of Canada* within its *Cold Spray Additive Manufacturing Industrial R&D group* (*CSAM*). Among the efforts undertaken to develop 3D build-up capabilities and process windows to deposit new materials for cold spray, online and off-line testing of the deposits were identified as essential R&D avenues to deliver a robust AM process.

2 Development of Build Strategies and Advanced Toolpath Generation

In this section, we present a layer-by-layer build strategy and toolpath planning for producing more complex geometries and improving shape fidelity. The toolpath planning and programming required to build 3D structures with complex geometry and low (overspray) tolerances intrinsically need precise manipulation of the spray gun or part as applicable. This can often introduce an added layer of complexity versus that required for typical cold spray applications, which can often be programmed manually by a skilled operator using the robot teach pendant. Coating a surface, for

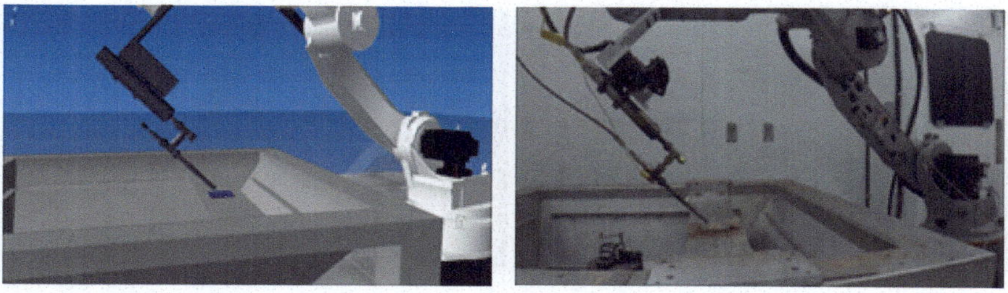

(a) (b)

Fig. 1 **a** Simulated and **b** spray setup

example, commonly employs a straightforward x-y index for stationary substrates or a linear displacement along the axis of rotating substrates.

Very similarly to other techniques such as laser-cladding, wire-arc additive manufacturing, etc., cold spray requires a digital thread to successfully build a 3D part starting from its 3D digital model. This thread includes the following steps: (i) generation of a part 3D model through CAD software and/or 3D scanning; (ii) slicing in a suitable orientation; (iii) path planning for each layer; and (iv) generation of machine codes. This needs different software to be achieved, and additional steps can be introduced, such as design optimization, manufacturing process simulation, etc. The use of software-based robot code programming provides flexibility and efficiency in toolpath planning as changes in path type, path sequencing, point density, etc. are more easily generated for evaluation. In addition, the simulation capabilities provided by commercial codes are often useful for verification and troubleshooting of build plans, e.g., collision avoidance, robot workspace, etc., as depicted in Fig. 1.

However, some specifics have to be considered in the development of build strategy and path planning in order to successfully manufacture 3D structures with cold spray additive manufacturing, e.g.: (i) the profile of the deposited materials that leads to significant overspray if not corrected; and (ii) cold spray is a high deposition rate, continuous process that cannot be stopped or started instantaneously. We will briefly explain these two points and see later how they can interfere with the overall manufacturing process.

2.1 *Profile Control*

With a basic cold spray setup, variations in the gas/particle flow across the diameter of a nozzle often manifest in a well-known tapering of the sides of a deposit. A 'basic' setup here refers to a standard commercial nozzle with axisymmetric geometry, no masking, and typically spray to a substrate surface. A typical deposit formed after repeated gun traverses along a single line of travel showed this tapering effect, Fig. 2a. Beyond a threshold number of traverses, deposition no longer occurred as

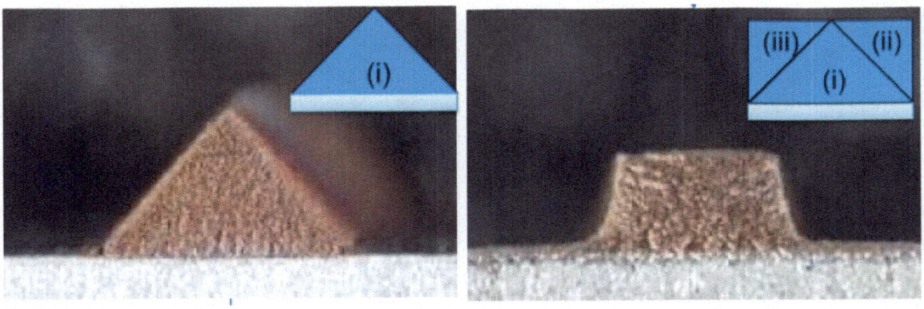

Fig. 2 Typical and altered deposit profiles: Cu sprayed at multiple angles to the substrate surface

the impact angles became increasingly off-normal, e.g., impacting the angled sides of the triangular deposit. In practice, this deposition profile effectively limits edge precision, and forces spray of a significantly larger footprint than required by designs as the tapering propagates with each additional layer, i.e., tapered edge grows as deposit thickness increases. This results in drawbacks such as significant amounts of excess material sprayed for large parts and a relatively large minimum size for fine, as-sprayed features. In order to limit these drawbacks, spray angles can be changed to avoid off-normal impacts, as in the case of producing vertical walls [2].

Profile alterations for line deposits of Cu were produced using a succession of spray angles, Fig. 2b. A rectangular profile was obtained for the Cu line deposit by spraying repeated passes normal to the substrate followed by repeated, alternating passes at $\pm 45^\circ$. The starting deposit profile was dependent on materials, nozzle (equipment), process conditions, etc. However, spray angles and/or positions were adjusted accordingly to perform profile alterations.

The development of unique toolpaths for every type of material deposit can be resource-intensive, requiring extensive experimentation and/or modelling. One approach to simplify toolpath planning is to employ a standard set of toolpaths for profile alteration, e.g., successive spray angles normal and $\pm 45^\circ$ to the substrate. This method can be implemented in a straightforward manner and designs adjusted accordingly. While efficient in practice, there are potential tradeoffs in terms of build accuracy and deposit properties if starting profiles deviate significantly. Consequently, a more practical approach was taken, which balanced these two extremes; the development of different sets of toolpath combinations that produced specific, predictable profile alterations from typical starting deposit profiles.

In a general layer-by-layer approach, it is important to note that the toolpaths employed for an alteration of the layer profile maintain similar spray conditions to that used to build the body of the layer. With spray angles maintained normal to a target surface (e.g., spray at 45° to the substrate is equivalent to normal direction from the triangle edge in Fig. 2), deposit properties for both the layer fill and alteration should be similar. Full characterization is an area of further investigation to validate that the profile alterations do not degrade performance.

2.2 Spray Path Planning

A build within each layer is deconstructed into two sub-layer structures, which are labelled as layer filling and layer alteration in the schematic, Fig. 3. We explained above how critical the layer alteration step is in order to improve the dimensional accuracy of the build and limit the needs for further machining. The path planning for this is most often reduced to a single trajectory, or a set of trajectories, that follows the profile of all open contours, with an adapted angle versus the base substrate surface. The layer fill refers to depositing material for the body of part, which is analogous to producing cold spray AM parts by spraying oversized deposits. Similar to what is seen in pocket milling with CAM software, different paths can be considered to cover the very same surface, i.e. zig-zag toolpath; zig toolpath; contour parallel toolpath; spiral-type toolpath. The decision on the type of toolpath can be related to the shape of the surface to fill but is also influenced by the second point mentioned in the specifics of cold spray, i.e. the fact that cold spray is a high build rate, continuous process. Bearing in mind that, for a constant powder feed rate, the deposited thickness is directly related to the traverse speed of the cold spray equipment over the substrate surface (or the substrate traverse speed under the cold spray gun, depending on what the robot is manipulating), the toolpath has to be generated with a proper strategy to minimize robot acceleration/deceleration and sharp direction change. For this reason, spiral-type toolpaths are often preferred versus zig-zag-type. Small radius turns, and intersections also represent challenges as they can lead to significant layer thickness variations that interfere with the continuation of the build sequence.

It is important to emphasize that although specific layering strategies are dependent on the design, the reverse also applies. A good design can greatly facilitate a

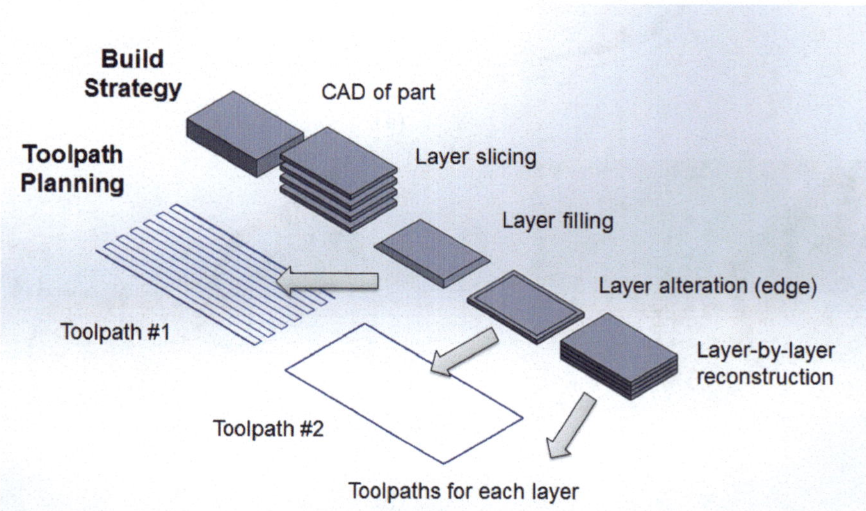

Fig. 3 Schematic of build strategy

cold spray build by simplifying the types of transitions and support structures that are required. The fabrication is also specific to materials, hardware, and process conditions and integration of these various elements is critical to a well-conceived, design-specific strategy.

Each sub-layer structure employs a different toolpath, with the complexity in toolpath planning heavily dependent on the geometries being produced. The toolpath accounts for the line-of-sight and proper access to the spray location. As build complexity increases and larger cold spray systems are used, this can become a significant challenge and designs become more constrained. In some cases, equipment changes, e.g., nozzles designed to produce uniform spray profiles, can be used to facilitate toolpath planning. The precision of a toolpath is also evaluated according to the design requirements as well as the capabilities of the manipulator (robot, gantry, etc.) and cold spray equipment (spray accuracy, process repeatability, etc.).

A layer-by-layer build strategy with profile alterations was employed to build several thin-walled features of copper on aluminum alloy substrates, Fig. 4. Figure 4a depicts the CAD drawings of such freeform stiffening structures and Fig. 4b the resulting cold sprayed structures.

The precision of the toolpath and process stability are two key factors in the uniformity of these as-sprayed features and comes in addition to layer thickness variations that could originate from variation in the traverse speed or from imperfect contour

(a)

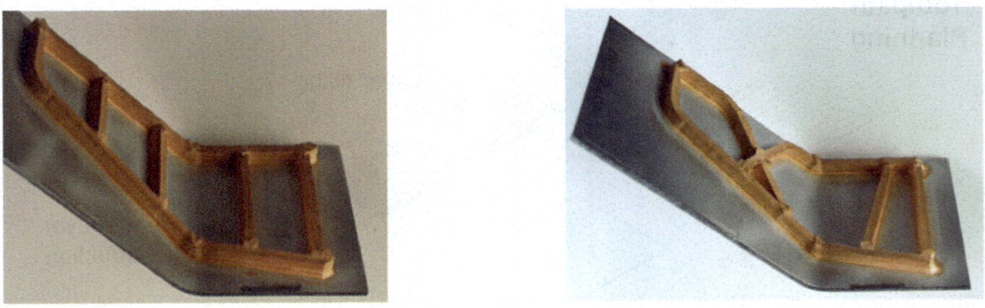

(b)

Fig. 4 **a** CAD drawings and **b** resulting in cold sprayed structures correction. In the case of low build heights, those variations in the layer thickness might not be so detrimental that they could cause the failure of the build. However, when more complex builds are intended, with tens of layers, these variations can become critical.

While improving the general stability of the process and carefully planning the path for the entire build would resolve many of these issues, having a solution to monitor online the geometry of added features and furthermore their quality (through representative properties assessment) is paramount. Such solutions could be used to tailor the toolpath as a correction of past variations, and even repair during build defaults. The next sections will present some advancements in the development of such diagnostic tools.

3 Swept-Source Optical Coherence Tomography (SS-OCT)

In this section, we evaluate the capability of Swept-Source Optical Coherence Tomography (SS-OCT) technology to provide real-time monitoring of the dimensional characteristics of a part fabricated by the cold spray process. We first describe the SS-OCT technology and present results from quasi-online monitoring of the process described in the previous section, highlighting the value that could be added by online inspection. SS-OCT is then compared to other techniques that could allow dimensional monitoring and its advantages are highlighted. We end this section by demonstrating the ability of SS-OCT to characterize large parts produced by the cold spray process.

Optical Coherence Tomography (OCT) is an interferometric imaging technique which has been developed initially for biomedical imagery applications [9] but has great potential for the monitoring of industrial processes [10–12]. Modern implementations of OCT, Spectral-Domain OCT (SD-OCT) and Swept-Source OCT (SS-OCT) allow the performance of real-time tomography of non-opaque materials, enabling the dimensional characterization of internal structures as well as the detection of internal defects. These technologies can also be used to perform surface profilometry, providing the external dimensions of a part and the identification of surface defects. Industrial applications include monitoring of laser machining [13], laser welding [14], polymer-based [15] and metal-based [16] additive manufacturing.

Figure 5a presents a basic SS-OCT system in a Michelson configuration. The light emitted by a swept-laser is sent to a coupler using single-mode fibre. Light is split between a sample arm and a reference arm. For illustrative purposes, the sample arm contains three partial reflectors illuminated by a collimated beam. The collimated beam is a simplified representation; in a real SS-OCT system, the light is usually focused in the region of the sample. The reference arm contains a fixed mirror. The light beams reflected in both arms return to the coupler where they are recombined to interfere before being sent to a photodetector. The SS-OCT signal is composed of the interference pattern recorded by the detection system as the wavelength of the laser is swept. Figure 5b illustrates a typical interference pattern referred to as

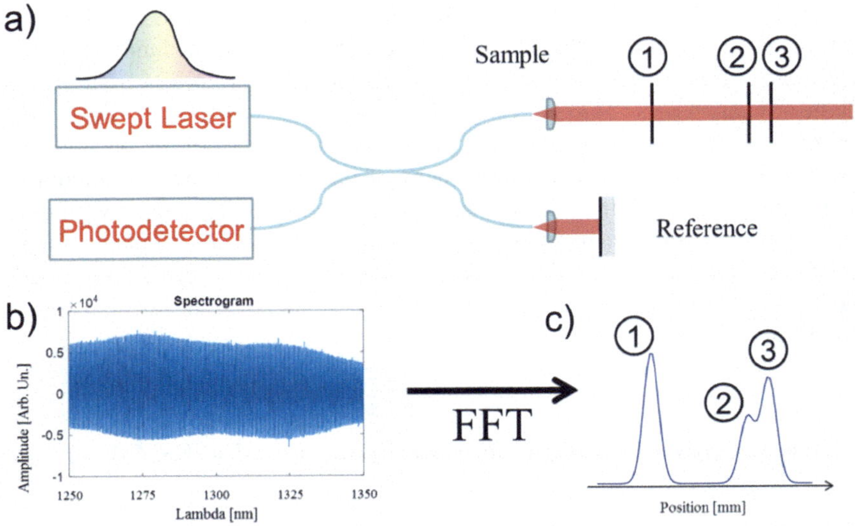

Fig. 5 **a** SS-OCT system. **b** Recorded interferogram signal. **c** Recovered sample structure after fast Fourier transform of the interferogram

a spectral interferogram. The location of any structure reflecting light in the sample arm is recovered from a fast Fourier transform (FFT) of the interferogram. This is illustrated in Fig. 5c where the three peaks provide the locations of the three partial reflectors in the sample arm.

When performing industrial monitoring, an optical scanning system is used in the sample arm to scan the surface of the part. A significant advantage of SS-OCT is that it requires a single point-of-view; a compact scanner can be designed to access any location, even hard-to-reach locations like corners and holes. Additionally, the scanner in the sample arm is connected to the other system components through an optical fibre and a few electrical wires. The scanner can thus be located close to a part while the other system components can be located away, at a more convenient location. This is a crucial feature when developing an online monitoring system for a challenging environment like the one found in the cold spray process.

Swept-lasers offered on the market today allow the design of SS-OCT systems that can achieve several hundreds of thousands of measurements per second. For 3D surface profilometry applications, that means several hundreds of thousands of points per second. This measurement rate is in the right order of magnitude for online monitoring of industrial processes.

3.1 Quasi-online Measurements

Quasi-online measurements were performed to demonstrate the added value of performing dimensional characterization while a part is built with the cold spray process. The experimental setup is illustrated in Fig. 6. Figure 6a shows the SS-OCT main unit, which was located just outside the spray booth. A one-dimensional galvanometer optical scanner was used to scan the surface of the part. The scanner was located within the spray booth and connected to the main OCT unit by an optical fibre and electrical wires passing through the wall of the spray booth. The part under fabrication was manipulated by a robot that moved the part between the fabrication location and the measurement location. At the fabrication location, the part was moved by the robot under the fixed nozzle, at a speed and orientation required by the spray pattern. At the measurement location indicated in Fig. 6b, the sample was translated by the robot in one direction while being scanned in a perpendicular direction by the optical scanner. These directions are indicated by the blue and red arrows in Fig. 6b. SS-OCT measurements were performed at a rate of 30,000 measurements per second. The optical scanner allowed 50 line profiles per second over a width of 15 mm, each line profile containing 600 points. The spacing between the line profiles was 0.05 mm, as dictated by the speed of the robot translation of 2.5 mm/s.

A rectangular ridge structure (freeform stiffening structure) was fabricated following the procedure described in the cold spray building strategy section with spray parameters that were far from optimal so as to provide challenging fabrication conditions, leading to a ridge with irregular corners. The fabrication conditions were intentionally challenging to provide clear features in the surface profilometry. An SS-OCT surface profile was performed after every five layers. In order to simulate results which would have been obtained from online monitoring during the fabrication, all surface profile measurements should have been performed with the same

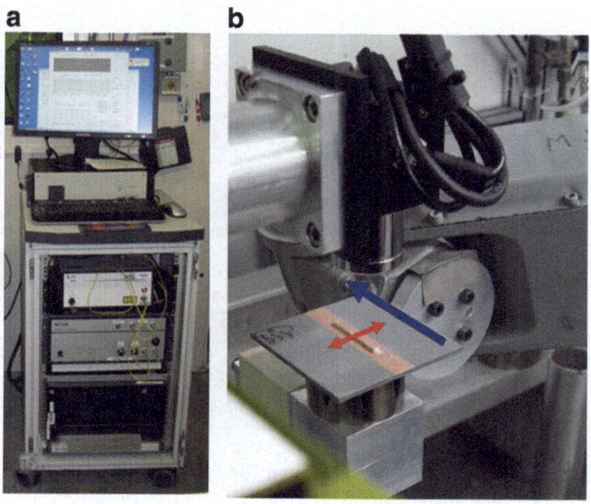

Fig. 6 Quasi-online setup. **a** OCT system. **b** OCT scanning head. The scan pattern is depicted by the red arrow. Robot movements are depicted by the blue arrow

angle as the one used for deposition. The measurement range of the SS-OCT system allowed measurements over a limited depth of 6 mm. This was not enough to capture the whole ridge structure for orientations of +45° and −45°, so surface profile measurements were performed at +20° and −20°. This is not a limitation of the technology, but a limitation of the specific SS-OCT system used for these measurements.

The surface profiles measured after each group of five layers are presented in Fig. 7. For each case, the bottom part shows the 2D rendering of the point cloud obtained with the PolyWorks software suite. The top left insert shows a line profile measured near the centre of the ridge while the top right insert shows a schematic representation of the structure after each five-layer step. Each surface profile is depicted with a specific colour and superimposed over the previous surface profiles. This gives a clear visualization of the contribution of each fabrication step towards the final structure and represents key information to optimize the fabrication process. When performed online, such information could allow process control through adjustments of the deposition parameters between each deposited layers. As an example, in the case shown in Fig. 7, it would have informed the operator to perform less than five layers for the two last steps which would have allowed a stop of the additive manufacturing process when the part reached a rectangle profile rather than overshooting to a trapezoid profile.

The added value of performing online monitoring to optimize the fabrication process is also illustrated in Figs. 7c, d. In Fig. 7c, the deposition at +45° created one sidewall of the ridge and overspray that results in a profile discontinuity to the

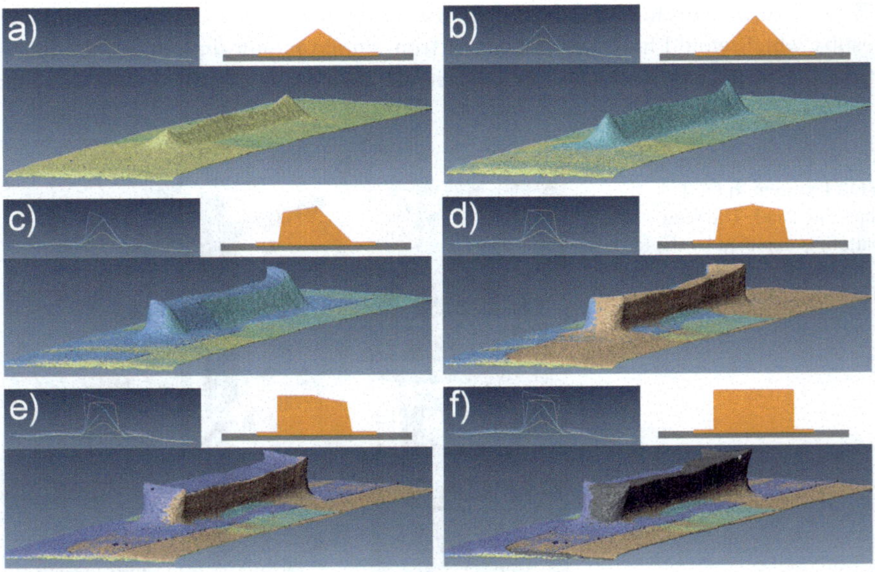

Fig. 7 Quasi-online measurement for fast deposition of a sample with a rectangular profile. **a–f** Point cloud surface, transverse profile and theoretical profile after successive fabrication steps

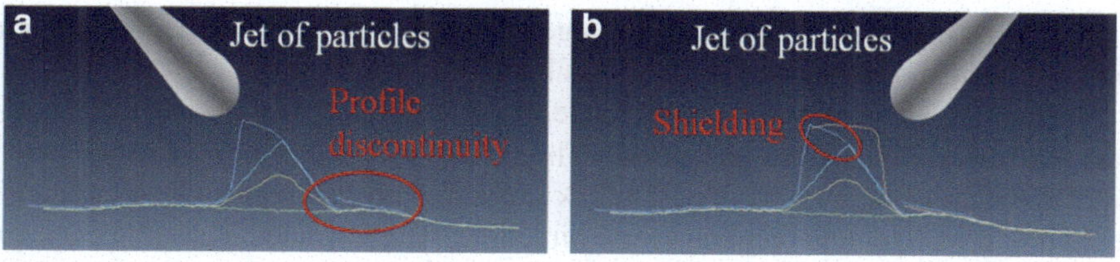

Fig. 8 **a** Sharp bump formed during fabrication, as seen in Fig. 7c. **b** The other side was shielded by the prior layer of material as seen in Fig. 7d

right of the ridge. As illustrated in Fig. 8a, this is caused by material being sprayed over the structure onto the base substrate. In Fig. 7d, the deposition at −45° created the other sidewall of the ridge. In that case, as shown in Fig. 8b, the previous layer insured shielding, so no material was sprayed over the structure to create a profile discontinuity on the left. Had online monitoring been performed between each layer, process parameters could have been adjusted as soon as the profile discontinuity creation of Fig. 7c was detected in order to minimize this unwanted feature.

3.2 Comparison with Other Techniques

Due to the nature of the fabrication process, only non-contact techniques should be considered to develop an online dimensional monitoring system with optical techniques being prime candidates. The surface roughness of parts produced by cold spray is large enough to ensure that any optical technique that relies on diffuse reflection should be capable of performing a surface profilometry. Dimensions should be measured with an accuracy of the order of few microns. This is not a stringent requirement since such accuracy can be achieved with most of the optical techniques. The main criteria to select one technology over another is the ease of integration of the online monitoring system. The technology must also allow measurements over a range large enough to ensure that the whole structure is monitored, especially when measured from an angle.

To illustrate a variety of optical techniques that could be employed, we performed measurements on the same part using SS-OCT, triangulation and structured light. The test sample contained 6 stripes representing various steps in the fabrication of a ridge structure. The dimensions of the sample were 56 mm × 77 mm and the largest stripe was 3.6 mm in width and 2.5 mm in height.

SS-OCT measurements were performed at a rate of 142,000 measurements per second with an illumination spot size of 100 μm (wavelength around 1.55 μm) over a scan width of 100 mm. The spot size provided a depth of field large enough which, combined with the large scan width, allowed the measurement of the whole sample at once. Scanning was performed with a galvanometer in one direction while the sample was translated by a motorized stage in the other direction. Spacing between measurements points was about 25 μm in both directions.

Triangulation measurements were made with a laser scan triangulation (LST) system. The measurement rate was 60 kHz. The surface was scanned by a galvanometer over a surface of approximately 30 mm × 20 mm, with spacing between measurement points of 10 μm in both transverse directions. The restricted measurement area is not a limitation of the technology, but a feature of the specific system used. The illumination spot size was 20 μm (wavelength of 405 nm). The system was equipped with an adaptive focusing system.

Structured light imaging does not require a surface scan; instead, it measures all the points in the field of view in parallel. Measurements were performed with a system capable of providing 24 million 3D points per second. Sections of approximately 21 mm × 28 mm were measured with a spacing of 7 μm in both directions between measurement points. The system used a LED source centered at 525 nm and was calibrated over 3 mm around the focal point.

We do not provide here the detailed analysis of the results as all techniques did perform well for these offline measurements. Figure 9a shows the SS-OCT surface profile measured in one shot over the whole surface sample from a normal (0° incidence) point of view. All points of the surface are clearly visible, except for the sharp edges of the large ridge structures on the right; the light was coming from above, travelling parallel to the edges, making ridge wall surface measurement very challenging. In Fig. 9b, structured-light measurements are overlaid over the SS-OCT point cloud. The whole structure was not imaged with the structured light system at once. Three measurements are shown in different colours. A cross section of the surface profiles

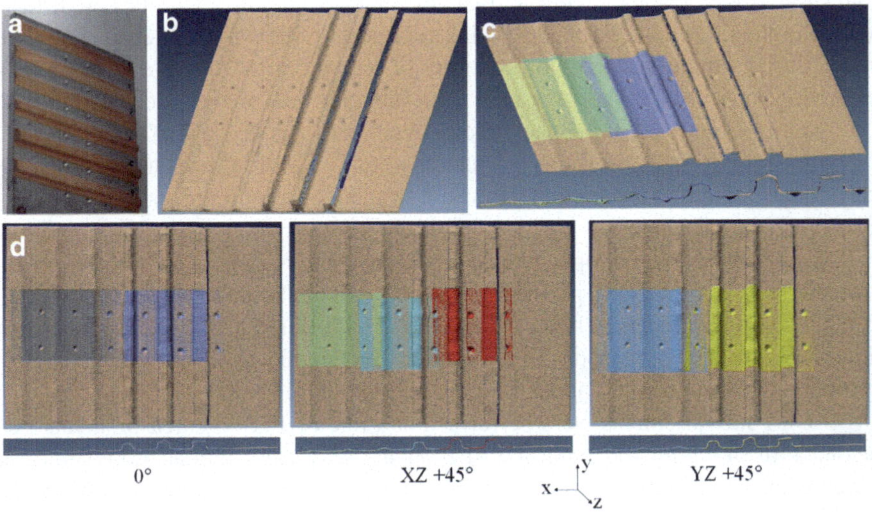

Fig. 9 CSAM copper ridges sample scanned with three different profilometry imaging techniques. **a** Picture of round-robin CSAM copper ridges sample. **b** SS-OCT 0°. **c** Structured Light compared to SS-OCT. **d** Laser scan triangulation compared to SS-OCT

Table 1 Comparison of online imaging techniques attributes

	Laser line triangulation	Laser scan triangulation	OCT profilometry	Structured light (phase shift)
Measurement quality	+++	+++	+++	+++
Cost	$	$$	$$$	$$
Adapted to online measurement	+++	+++	+++	*
Speed	+++	+	++	++++
Parasite reflection handling		+	++	
Compactness	+	+	+++	*

is also shown below the point clouds. There is good agreement between both imaging modalities. Figure 9c compares the laser scan triangulation measurements with SS-OCT. Again, measurements performed over different areas are shown with different colours. Triangulation measurements are provided for a scanning performed from the normal of the surface (Z-axis), as well as for orientations tilted by +45° in the XZ plane and tilted by +45° in the YZ plane. Cross-sectional views of the profiles are provided under the point clouds. Again, there is good agreement between results obtained with SS-OCT and triangulation. One side of the ridge structures are not visible for triangulation measurements performed with orientations of +45°, a result which is expected and common to all-optical inspection techniques since the structures create shadowing when inspected from an angle.

Results of Fig. 9 confirm that the various optical measurement techniques can provide good results when performing an offline inspection of cold sprayed parts. The challenge resides in how efficiently these can be integrated into an online environment. This is addressed in Table 1, where the strengths and weaknesses of online monitoring are summarized for each technique. We have added Line Laser Triangulation (LLT) to the techniques considered above.

In terms of cost, SS-OCT is the most expensive technology, while LLT is the least. The cost of LST can vary, depending upon the complexity of the system since features like adaptive focusing can be used for enhanced capability, but these come at a cost. In terms of measurement speed, structured light is the most efficient since it provides the full 2d profile at once. LLT can easily reach a rate of 1000 line profiles per second. LST and SS-OCT both rely on scanning which can be a limitation, although they can both provide measurements rate large enough for cold spray online monitoring. All techniques have been implemented in various online monitoring applications, so nothing intrinsic to these technologies could prevent them from being used online. In the case of cold spray monitoring, the compactness of the sensor is an important aspect. Structured-light requires that the whole sample be visible at once and can only be used to inspect a layer after it is entirely deposited. LST and LLT both require two points of views which requires space. One significant advantage of SS-OCT is that it requires a single point of view; the same optical path is used for both illumination and detection. This allows the design of very compact scanning heads which can be easily integrated with a fabrication system. The ability to handle parasite reflection is also important when fabricating complex parts. Multiple reflections can provide unwanted artefacts. LLT and structured light are more sensitive to these parasitic reflections. LST systems can be customized to handle parasitic reflections at the cost of a more complex system. SS-OCT provides a full measurement through depth. Parasitic reflections will lead to additional peaks in this depth profile. These peaks can easily be identified and disregarded, making SS-OCT robust regarding parasitic reflections.

3.3 *Capacity to Measure Large Objects*

OCT imaging was initially developed to take high-resolution images of small areas in biological samples. But, like other laser profilometry techniques, it can be suited to scan larger parts such as those shown in Fig. 10. These reinforcement-like structures measure 150 mm × 70 mm × 40 mm. The second sample was produced using the same cold spray process as the as-sprayed sample but underwent a machining fabrication step after cold spray deposition to smooth its surfaces. By stitching together two separate scans of a sample, the topography of the whole part can be reconstructed.

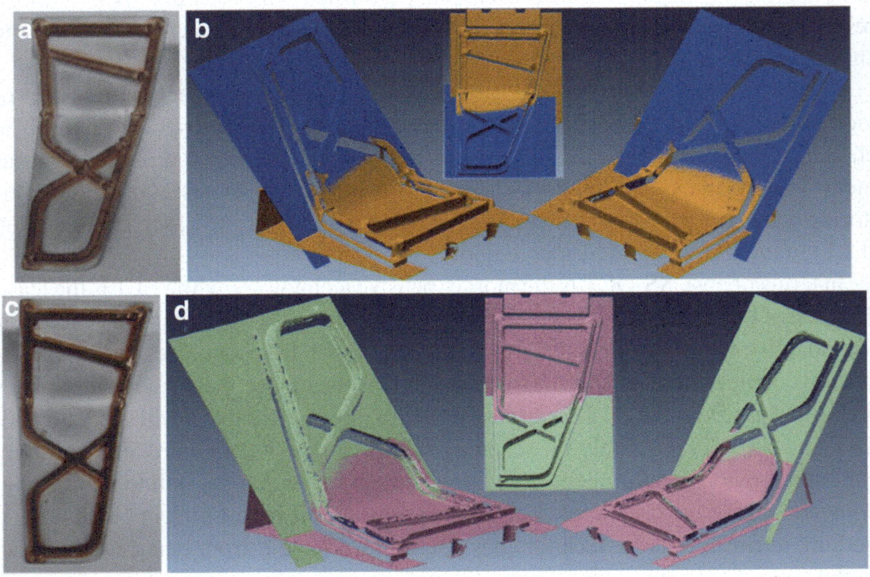

Fig. 10 OCT measurements of CSAM reinforcement-like structures. **a** Picture of the as-sprayed sample. **b** Stitched OCTscan. **c** Picture of the machined sample. **d** Stitched OCT scan

Additional scans from other points of view would be needed to capture the vertical surfaces on the part as well. Even larger parts could also be measured online using optical imaging techniques by tracking the position of the sensor relative to the part and through image stitching.

4 Laser-Ultrasonic Inspection of Cold Spray Additive Manufacturing Components

In this section, we present results of laser ultrasonic testing (LUT) performed off-line, on metallic structures deposited by the cold spray AM process. LUT was employed to detect flaws using the synthetic aperture focusing technique (SAFT) [17], and through-thickness distributed porosity was investigated using the backscattered signal [18]. The latter approach was applied in recent years to study porosity in composite materials [19], as well as grain shape and size distribution in steel [20–22]. Also, a laser shockwave technique was applied to characterize bond strength [23, 24] at the interface between the cold spray deposition and the substrate. Finally, for the post-heat treatment of cold spray AM metallic parts, we show an example of how LUT can be performed to capture the microstructural and phase changes. Inspection results from either the top layer or the underside of the substrate are discussed in the following.

4.1 Laser-Ultrasonic Inspection for Flaw Detection

Laser ultrasonics combined with SAFT was employed for the flaw, and porosity detection in Al cold spray coated samples. A set of three aluminum (Al) cold spray coatings deposited with conditions to intentionally produce different porosity levels was prepared. Figure 11a shows one such sample. Sample #1 (8.8 mm thick coating) has low porosity, sample #2 (5.5 mm thick coating) has an intermediate porosity and sample #3 (4.1 mm thick coating) has a high porosity with possible flaws present. Each coating is deposited on an Al6061 4.8 mm thick substrate.

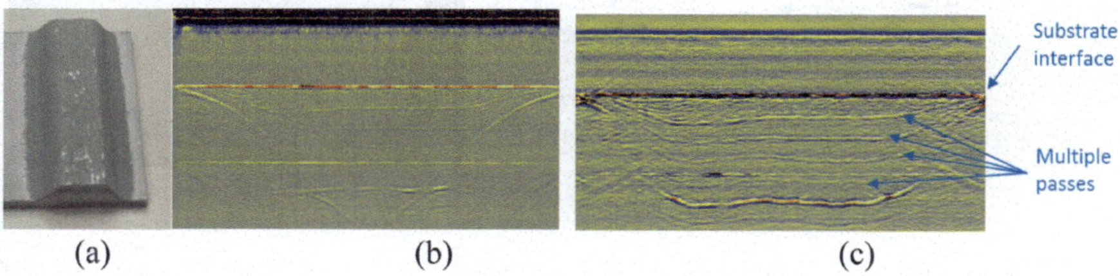

Fig. 11 **a** Photo of one Al/Al6061 cold spray sample used for testing. B-scans on Al/Al6061 sample #2 from the substrate, **b** raw data and **c** after SAFT reconstruction

For use with SAFT, the generation and detection zones overlapped at the surface for 1D or 2D scanning from either the top deposited layer or the underside of the substrate. Ultrasound generation was performed in the slight ablation regime (less than 1 μm) with a short pulse Nd:YAG laser in its 2nd harmonic (532 nm wavelength) to achieve high frequencies. For detection, a long pulse Nd:YAG laser (1064 nm wavelength) and a small spot size were used. The phase demodulator was a 1-m long confocal Fabry-Perot interferometer in reflection mode. Frequency content up to 80 MHz was successfully generated and detected in the above test samples. Mechanical scanning along single lines up to 30 mm length was performed for data acquisition of the waveforms with a step size of 0.1 mm. SAFT reconstruction was performed with an aperture angle of 30° and a frequency bandwidth from 1.5 to 80 MHz using the longitudinal mode.

Using the above test conditions, line scans were performed on the underside of the substrate of the three samples over a length of 30 mm (Fig. 11). B-scan images from the raw data and after SAFT reconstruction of a single scan line are shown respectively in Fig. 11b, c for sample #2 of intermediate porosity, starting with the substrate on top. Weak indications are observed from the raw data image, some of them as hyperbola shapes, and are more clearly resolved after SAFT reconstruction. The more or less continuous indications observed in Fig. 11c are attributed to imperfect contact between layers associated with the multiple passes of the cold spray AM at suboptimal spray conditions. Also, a strong indication of the substrate interface with the coating is apparent in both B-scan images. A similar indication of lower amplitude from the substrate is observed close to the last cold spray pass (bottom of B-scan), related to an echo propagating back and forth in the substrate. The longitudinal velocity difference between the Al alloy substrate and the Al coating is small and such reflections are more related to the bond integrity of the coating with the substrate. This is also investigated further later.

Figure 12a shows a B-scan image after SAFT reconstruction of a particular scan line from the coating side over a length of 15 mm. As expected, the continuous indications attributed to the multiple passes are observed while a bit less resolved.

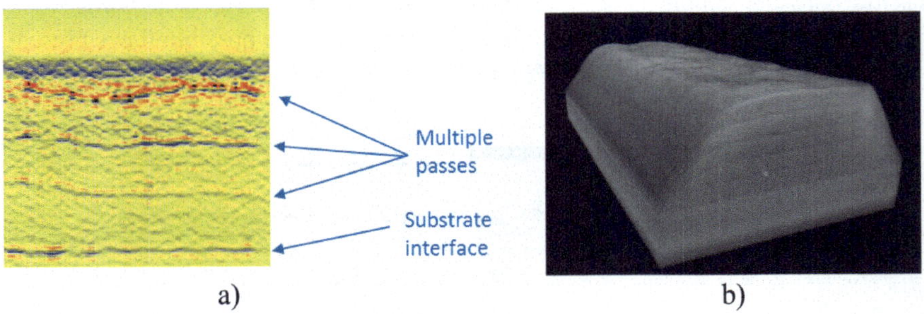

Fig. 12 **a** B-scan from coating after SAFT reconstruction for Al/Al6061 sample #2 and **b** corresponding X-ray μCT image

The indication between last cold spray passes on top of the coating is missing due to the blind zone related to the large surface displacement in laser-ultrasonic signals. A further investigation was made on these samples using X-ray μCT as well as a series of micrographs. Figure 12b shows an X-ray μCT image after processing the radiographic data. The agreement between the observations in this image and the laser-ultrasonic results is fairly good. Not shown here, the micrograph image shows porosity everywhere throughout the thickness, with a slightly larger pore concentration for the interface between each pass. This point is investigated in the next section.

4.2 Laser-Ultrasonic Backscattering for Porosity Evaluation

To further study porosity, the spectral analysis of the laser-ultrasonic backscattered signals obtained from the top surface of the coating was considered. Figure 13 presents the principle of measurement with the backscattered amplitude (BSA) spectrum after averaging 100 spectra from a line scan, and the decay slope in the spectrum estimated between the two cursors. As shown in Fig. 13, the BSA method can be applied over the full-time window between the large surface displacement and the first interface echo arrival, as an indication of average porosity found through the sample thickness. In a further study, an analytical model for backscattering [19, 20] could be developed to relate such decay slope with the porosity level.

Additionally, a sliding small time window of typically 0.6 μs can be applied in the same time interval to estimate the decay slope in successive BSA spectra. Using the single scattering hypothesis, the time window location can be associated with an

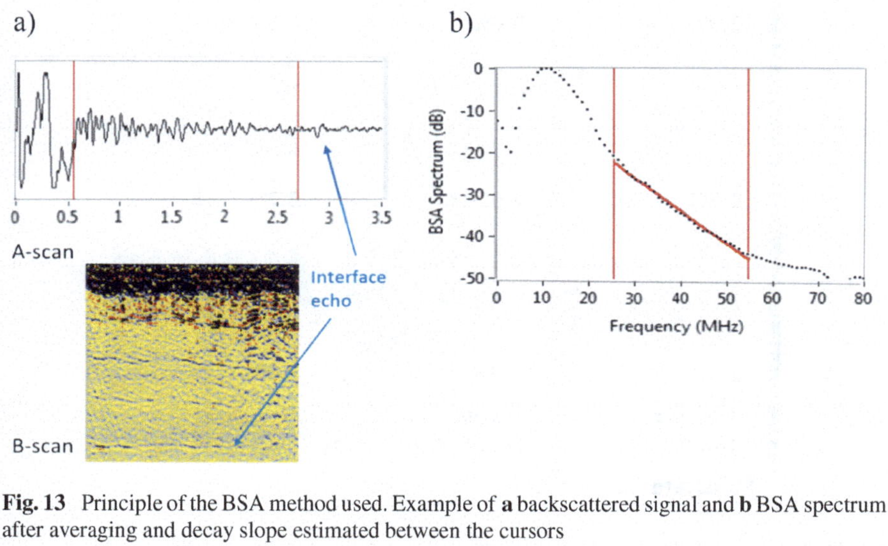

Fig. 13 Principle of the BSA method used. Example of **a** backscattered signal and **b** BSA spectrum after averaging and decay slope estimated between the cursors

equivalent depth, where the BSA decay slope indicative of local porosity is estimated. Figure 14 shows the BSA decay slope as a function of depth and Fig. 15 shows the series of micrographs analyzed through the thickness of the three Al/Al6061 cold spray samples. Compared to micrographs, consistent results are found between samples as well as through the thickness of each sample, with a larger negative slope corresponding to higher porosity near the top surface and a decreasing slope toward

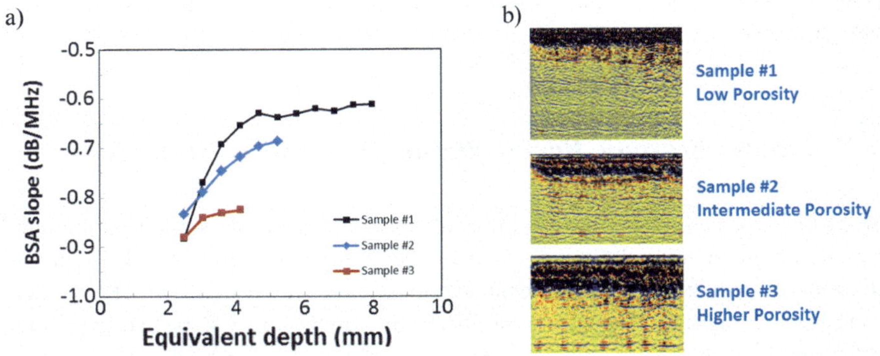

Fig. 14 **a** BSA decay slope as a function of depth through the thickness of the Al/Al6061 samples, and **b** corresponding B-scan images

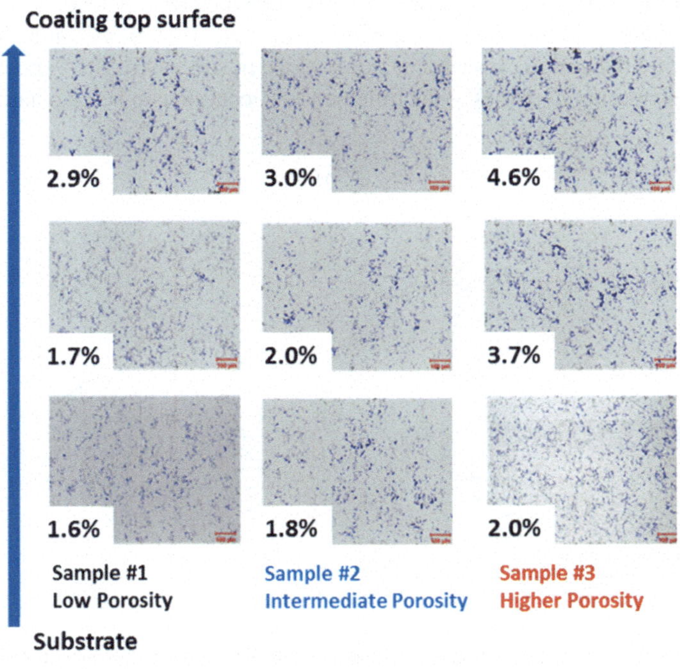

Fig. 15 Series of micrographs through the thickness of the Al/Al6061 samples

the substrate interface. Not shown here, an opposite behaviour (smaller to larger negative slope) is found scanning from the substrate side and analyzing backscattered signals after the interface echo.

While testing on a limited set of samples, the approach appears promising and will be further investigated. The method should work better when porosity near the surface is not too large to avoid a substantial reduction of the BSA signal deeper in the material. Also, the effect of multiple scattering should make the method not work well at typically more than 4–5 mm deep.

4.3 Laser-Ultrasonic/Shockwave for Bond Integrity Assessment

Another aspect to be considered with the cold spray AM process is the bond integrity between the coating and the substrate. This is particularly important for AM applications such as part reinforcement, dimensional restoration and structural repair. For these tests, a set of Al/Al6061 cold spray samples were prepared with a pulsed laser surface pre-treatment to improve the adhesion between the coating and the substrate [25].

Line scans were performed on the underside of the substrate of the cold spray samples over a length of 42 mm. B-scan images after SAFT reconstruction of a single scan line of 4 samples are shown in Fig. 16, starting with the substrate on top. A clear indication of the substrate interface with the coating is apparent in all cases but of variable amplitude. As in Fig. 11c, such reflections should be related to bond integrity and may vary depending on the surface treatment during the AM process. Also, there is a need to quantitatively determine the bond strength at the interface in each case.

For this purpose, a laser shockwave proof test method is considered with the principle shown in Fig. 17. With this technique, a high-energy pulsed laser produces a compressive large-amplitude wave at the top surface of the sample. The compressive shock wave travels through the sample and after reaching the lower surface, the compressive wave is changed by the free surface into a tensile wave that travels back through the sample. A shock wave amplitude that reaches over a certain threshold will create tensile stresses large enough to debond interfaces in the component. By

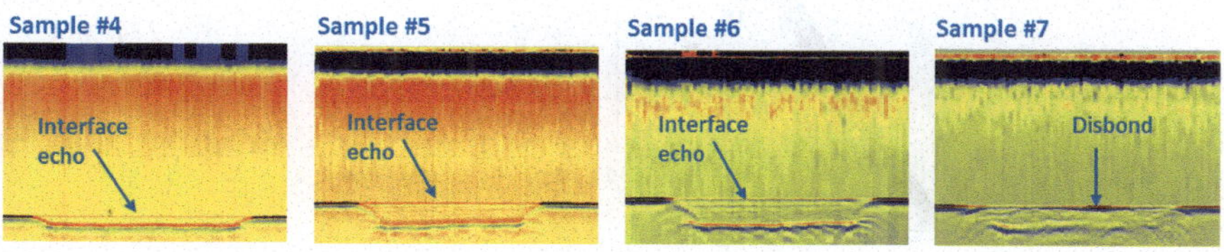

Fig. 16 B-scans from the substrate on the Al/Al6061 samples with surface treatment after SAFT reconstruction

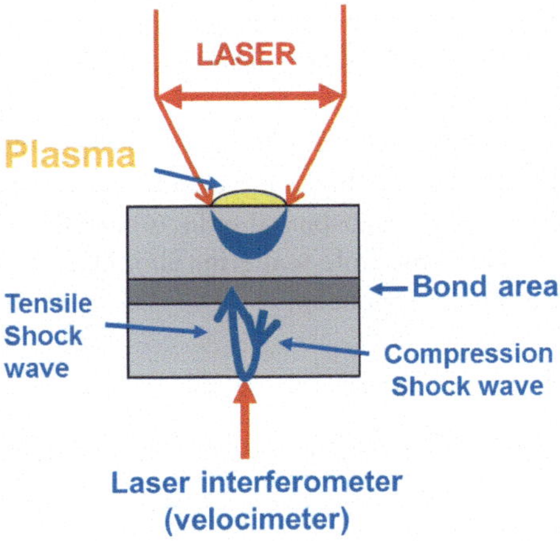

Fig. 17 Principle of the laser shockwave proof test

measuring the back surface velocity response with a Fabry-Perot etalon (velocimeter) [26] for different laser energy and using a model for shock wave propagation [27], the stress required to delaminate the coating from the substrate can be assessed. Figure 18 shows an example of such a calculation for one of the cold spray samples. Calculated stress values are generally larger than in quasi-static mechanical tests since laser shock wave measurement involves high strain rate, typically of 10^5–10^6 s^{-1}. To confirm that the system debonded the interface a post-shock laser-ultrasonic inspection is used to perform C-scan imaging of the sample. The debonded interface will keep the ultrasound from propagating to the backside of the sample, and this will be captured in the resulting image.

To determine bond strength a laser shockwave technique was applied on the cold spray AM samples #4–6 that do not present a disbond in Fig. 16. Laser energies from 0.2 to 2.2 J were tested at different locations from the substrate side of each sample. Figure 19 shows the back surface velocity from the velocimeter (on the coating) at

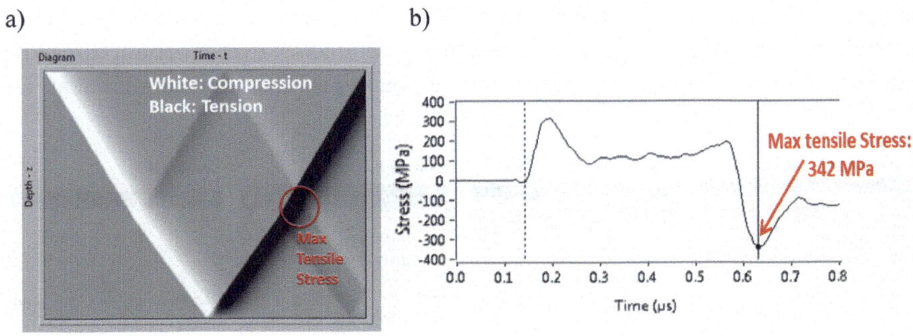

Fig. 18 Example of **a** ray tracing shock calculation with time-depth representation and **b** stress evaluated at the interface from measured back surface velocity

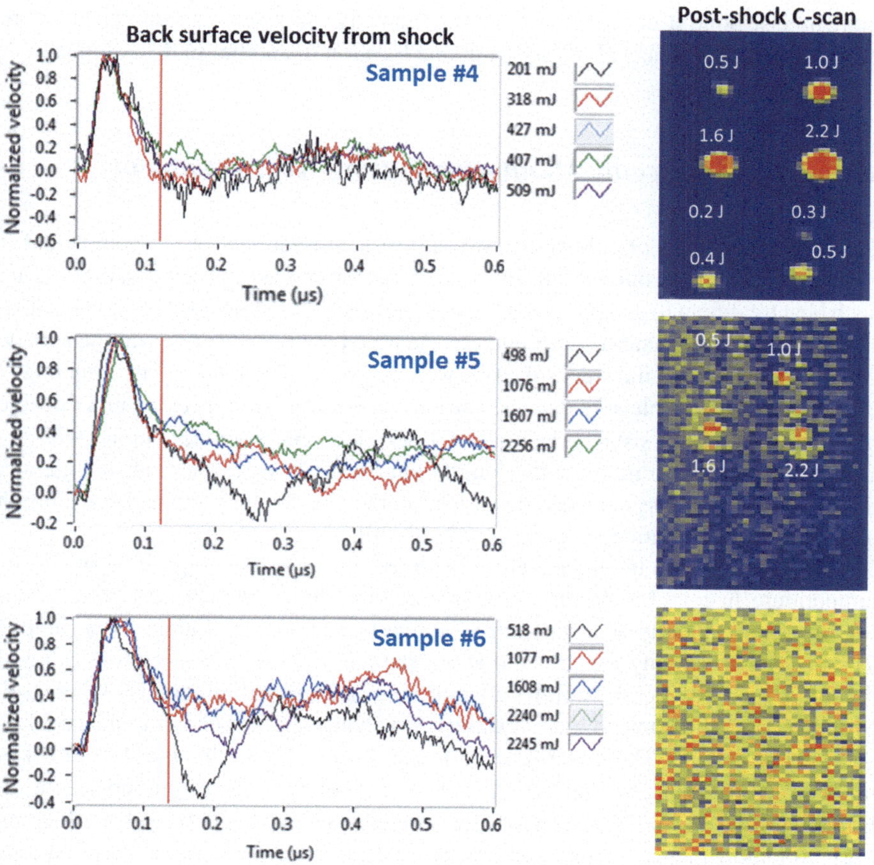

Fig. 19 Back surface velocity at different laser energies and post-shock laser-ultrasonic C-scans of the Al/Al samples

the different laser energies (indicated on the right of each graph) and the post-shock laser-ultrasonic C-scans on these samples. For sample #6, the bond did not open for the laser energies tested. The back surface velocity response at the threshold laser energy producing a disbond is then used to calculate stress at the interface. Stress values calculated for corresponding laser debonding energy are given in Table 2. A

Table 2 Calculated stress near debonding at the interface of the Al/Al alloy samples

Al/Al sample ID	Coating thickness (mm)	Debonding energy (J)	Min stress (MPa)	Max stress (MPa)
4	0.7	0.3	104	242
5	0.9	0.5	271	393
6	1.2	2.2+	358	N/A

more precise bond strength value could be obtained by testing each sample with more energy levels, but it may also vary from one location to another in the component.

4.4 Laser-Ultrasonic Monitoring During Heat Treatment

In some applications of cold spray AM, a post-heat treatment of the component is required. Laser-ultrasonic monitoring can be performed at high temperature during such heat treatment to understand better and optimize the occurrence of recrystallization, phase transformation, sintering, etc. As a first study, cold spray samples made of H13 tool steel on mild steel substrate were prepared. Then, the 3.3 mm thick H13 coating of each sample was detached from the substrate and inserted in a Gleeble machine for direct resistance heating. A laser-ultrasonic system, coupled with the Gleeble, was used to monitor the longitudinal velocity as a function of time and temperature. Such a system was developed in the past to study grain size evolution and phase transformation in steel [21, 22, 28–30].

Two different heat treatments are considered for these tests. Figure 20 shows the temperature history for the first heat treatment of long duration and corresponding ultrasonic velocity as a function of temperature on a stand-alone H13 sample. The ultrasonic velocity for the cold spray sample starts with a quite low value of 4.5 mm/μs and reaches 6.0 mm/μs after cooling, as expected for bulk H13 tool steel. Moreover, the ultrasonic velocity exhibits two changes in its behaviour near 550 °C and later at 800 °C. These changes are found attributable to the apparent effect of microstructural changes. The 550 °C is near a standard tempering temperature for H13 [31] while the 800 °C is near a phase change according to the H13 phase diagram [32]. Not shown here, the ultrasonic velocity during heating of short duration may show a quite different dynamic behaviour, and this has to be further investigated.

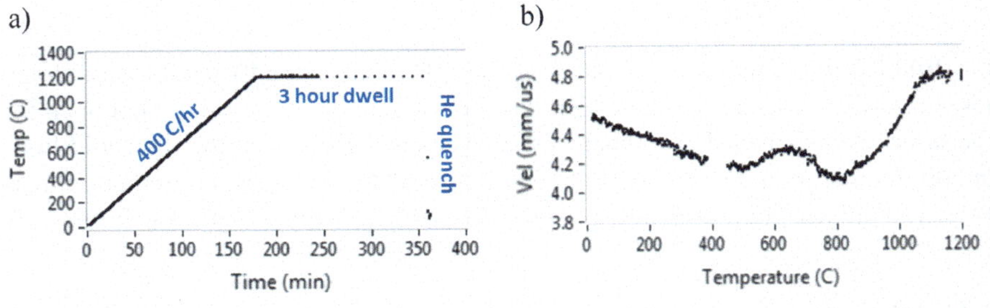

Fig. 20 **a** Temperature history of first heat treatment and **b** ultrasonic velocity as a function of temperature

5 Conclusions

To approach near-net shape features of cold spray some specific considerations with respect to deposit profile control and spray path planning must be considered. This was demonstrated by building freeform stiffening structures using a digital thread similar to other AM processes. The precision of the toolpath and process stability are two key elements in the uniformity of these as-sprayed features. While improving these factors for the entire build would resolve many issues, having a solution to monitor online, the geometry of added features and furthermore their quality, through representative properties assessment, is critical. The build requires both online and offline inspection.

A comparison of various dimensional inspection techniques, which all showed good results, prompted OCT as being the technique of choice for online inspection due to the compact scanning head design which can ultimately be easily integrated with a fabrication system. Moreover, since OCT provides full through-depth measurement, it is more robust regarding parasitic reflections than the other techniques presented. A quasi-online OCT measurement demonstrated the feasibility of implementation of the technique for real-time inspection, which allows adjustment of the spray parameters during the build.

It is equally essential to be able to perform remote inspection and to confirm the integrity of the sprayed part, e.g., porosity, adhesion, etc. Consequently, laser-ultrasonic testing was employed in a parallel investigation of a variety of cold sprayed samples. LUT, combined with SAFT, was used to detect flaws in Al/Al alloy samples from both the substrate and deposit sides (surfaces). Indications of multi-passes and variations in bond integrity with the substrate were observed. This was further validated by X-ray μCT showing larger pore concentration between each pass. Another aspect considered was capturing information regarding through-thickness distributed porosity using a spectral analysis of the laser-ultrasonic backscattered signal. Using a sliding time window a BSA decay slope was found promising to provide an indication of porosity variations through the sample thickness as confirmed by a series of micrographs. Also, the laser shockwave technique was used to characterize bond strength at the interface between the cold spray deposition and the substrate. Finally, laser-ultrasonic testing was shown to be useful to monitor microstructural changes such as phase changes, recrystallization and sintering phenomena during post-heat treatment of cold spray AM.

Acknowledgements This work has been conducted within the NRC's Cold Spray Additive Manufacturing Industrial R&D Group CSAM. The authors are grateful to Mr. M. Zeman and Mr. D. de Lagrave for their contribution in the sample preparation and performing the series of micrographs, as well as to Mr. C. Brosseau and Mr. M. Lord for their participation in the laser-ultrasonic measurements of the cold spray samples tested in this work.

References

1. Schmidt, T., Assadi, H., Gartner, F., Richter, H., Stoltenhoff, T., Kreye, H., & Klassen, T. (2009). From particle acceleration to impact and bonding in cold spraying. *Journal of Thermal Spray Technology, 18,* 794–808.
2. Pattison, J., Celotto, S., Morgan, R., Bray, M., & O'Neill, W. (2007). Cold gas dynamic manufacturing: A non-thermal approach to freeform fabrication. *International Journal of Machine Tools and Manufacture, 47,* 627–634.
3. Sova, A., Grigoriev, S., Okunkova, A., & Smurov, I. (2013). Potential of cold gas dynamic spray as additive manufacturing technology. *International Journal of Advanced Manufacturing Technology, 69,* 2269–2278.
4. Villafuerte, J. (2014). Considering cold spray for additive manufacturing. *Advanced Materials and Processes, 172,* 50–52.
5. Yin, S., Cavaliere, P., Aldwell, B., Jenkins, R., Liao, H., Li, W., & Lupoi, R. (2018). Cold spray additive manufacturing and repair: Fundamentals and applications. *Additive Manufacturing, 21,* 628–650.
6. Li, W., Yang, K., Yin, S., Yang, X., Xu, Y., & Lupoi, R. (2018). Solid-state additive manufacturing and repairing by cold spraying: A review. *Journal of Materials Science and Technology, 34,* 440–457.
7. Raoelison, R. N., Verdy, C., & Liao, H. (2017). Cold gas dynamic spray additive manufacturing today: Deposit possibilities, technological solutions and viable applications. *Materials and Design, 133,* 266–287.
8. Li, W., Cao, C., Wang, G., Wang, F., Xu, Y., & Yang, X. (2019, April). 'Cold spray +' as a new hybrid additive manufacturing technology: A literature review. *Science and Technology of Welding and Joining, 24*(5), 420–445. https://doi.org/10.1080/13621718.2019.1603851.
9. Huang, D., et al. (1991). Optical coherence tomography. *Science, 254*(5035), 1178–1181.
10. Dufour, M. L., Lamouche, G., Vergnole, S., Gauthier, B., Padioleau, C., Hewko, M., et al. (2006, September 8). Surface inspection of hard to reach industrial parts using low-coherence interferometry. In *Proceedings of SPIE, Photonics North 2006* (Vol. 6343, p. 63431Z).
11. Dufour, M. L., Lamouche, G., Detalle, V., Gauthier, B., & Sammut, P. (2005). Low-coherence interferometry—An advanced technique for optical metrology in industry. *Insight, 47,* 216.
12. Stifter, D. (2015). Nondestructive material testing using OCT. In W. Drexler & J. Fujimoto (Eds.), *Optical coherence tomography*. Cham: Springer.
13. Ji, Y., Grindal, A. W., Webster, P. J., & Fraser, J. M. (2015). Real-time depth monitoring and control of laser machining through scanning beam delivery system. *Journal of Physics D: Applied Physics, 48,* 155301.
14. Dupriez, N. D., & Truckenbrodt, C. (2016). OCT for efficient high quality laser welding. *Laser Technik Journal, 13,* 37–41.
15. Gardner, M. R., et al. (2018). In situ process monitoring in selective laser sintering using optical coherence tomography. *Optical Engineering, 57*(4), 041407.
16. Kanko, J. A., Sibley, A. P., & Fraser, J. M. (2016). In situ morphology-based defect detection of selective laser melting through inline coherent imaging. *Journal of Materials Processing Technology, 231,* 488–500.
17. Lévesque, D., Blouin, A., Néron, C., & Monchalin, J.-P. (2002). Performance of laser-ultrasonic F-SAFT imaging. *Ultrasonics, 40,* 1057–1063.
18. Kruger, S. E., Moreau, A., Lévesque, D., & Lord, M. (2001). Laser ultrasonic measurements of scattered waves in steel. In D. O. Thompson & D. E. Chimenti (Eds.), *Proceedings, Review of Progress in Quantitative Nondestructive Evaluation* (Vol. 20, pp. 1298–1305).
19. Karabutov, A. A., & Podymova, N. B. (2013). Nondestructive porosity assessment of CFRP composites with spectral analysis of backscattered laser-induced ultrasonic pulses. *Journal of Nondestructive Evaluation, 32,* 315–324.
20. Lobkis, O. I., Yang, L., Li, J., & Rokhlin, S. I. (2012). Ultrasonic backscattering in polycrystals with elongated single phase and duplex microstructures. *Ultrasonics, 52,* 694–705.

21. Legrand, N., et al. (2015). Laser-ultrasonic sensor to monitor steel microstructure at elevated temperature: Applications to hot rolling. In *Proceedings, 4th International Symposium on Laser Ultrasonics and Advanced Sensing (LU2015)*, Evanston, IL, paper #53.
22. Lévesque, D. (2017). Laser-ultrasonic methods to characterize steel microstructure: Overview and recent developments. In *3rd International Workshop on Laser-Ultrasound for Metals*, Stockholm, Sweden.
23. Vossen, J. L. (1978). Measurements of film-substrate bond strength by laser spallation. *American Society for Testing and Material Special Technical Publications, 640,* 122–133.
24. Gupta, V., et al. (1990). Measurement of interface strength by laser-pulse-induced spallation. *Materials Science and Engineering: A, 126,* 105–117.
25. Christoulis, D. K., et al. (2010). Cold-spraying coupled to nano-pulsed Nd-YaG laser surface pre-treatment. *Journal of Thermal Spray Technology, 19,* 1062–1073.
26. Arrigoni, M., et al. (2009). Laser Doppler interferometer based on a solid Fabry-Perot etalon for measurement of surface velocity in shock experiments. *Measurement Science & Technology, 20,* 015302.
27. Perton, M., Lévesque, D., Monchalin, J.-P., Lord, M., Smith, J. A., & Rabin, B. H. (2013). Laser shockwave technique for characterization of nuclear fuel plate interfaces. In D. O. Thompson & D. E. Chimenti (Eds.), *AIP Conference Proceedings. Proceedings, 39th Annual Review of Progress in Quantitative Nondestructive Evaluation*, Denver, CO (Vol. 1511, pp. 345–352).
28. Irissou, E., et al. (2008). Review on cold spray process and technology: Part I—Intellectual property. *Journal of Thermal Spray Technology, 17,* 495–516.
29. Christoulis, D. K., Jeandin, M., Irissou, E., Legoux, J.-G., & Knapp, W. (2012). Laser-assisted cold spray (LACS), Chapter 5. In: D. C. Dumitras (Ed.), *Nd:YAG laser* (pp. 59–96). Intech. ISBN: 978-953-51-0105-5.
30. Kruger, S. E., & Damm, E. B. (2006). Monitoring austenite decomposition by ultrasonic velocity. *Materials Science and Engineering: A, 425,* 238–243.
31. ASTM 681-08 Standard Specification for Tool Steels Alloy, ASTM International (Warrendale, PA), 2015.
32. Besler, R., Bauer, M., Furlan, K. P., Klein, A. N., & Janssen, R. (2017). Effect of processing route on the microstructure and mechanical properties of hot work tool steel. *Materials Research, 20*(6), 1518–1524. https://doi.org/10.1590/1980-5373-MR-2016-0726.

Index

S. Pathak and G. C. Saha (eds.), *Cold Spray in the Realm of Additive Manufacturing*, Materials Forming, Machining and Tribology,
https://doi.org/10.1007/978-3-030-42756-6

The manufacturer's authorised representative in the EU is Springer Nature Customer Service Centre GmbH, Europaplatz 3, 69115 Heidelberg, Germany. If you have any concerns regarding our products, please contact ProductSafety@springernature.com

Printed and bound by CPI Group (UK) Ltd, Croydon, CR0 4YY
25/09/2026
02222904-0001